R00219 79604

CHICAGO PUBLIC LIBRARY
HAROLD WASHINGTON

R0021979604

REF
QE
720
.S77
Cop. 1

Structure and classification of paleocommunities

DATE DUE

REF
QE
720
.S77
cop. 1

SCIENCE DIVISION
BUSINESS/SCIENCE/TECHNOLOGY DIVISION

The Chicago Public Library

Received_____ AUG 9 1977

Structure and Classification of Paleocommunities

STRUCTURE AND CLASSIFICATION OF PALEOCOMMUNITIES

Edited by
ROBERT W. SCOTT
Amoco Production Company

RONALD R. WEST
Kansas State University

Dowden, Hutchinson & Ross, Inc.
STROUDSBURG, PENNSYLVANIA

Copyright © 1976 by **Dowden, Hutchinson & Ross, Inc.**
Library of Congress Catalog Card Number: 76-4587
ISBN: 0-87933-225-5

All rights reserved. No part of this book may be reproduced or transmitted in any form or by any means—graphic, electronic, or mechanical, including photocopying, recording, taping, or information storage and retrieval systems—without written permission of the publisher.

78 77 76   1 2 3 4 5
Manufactured in the United States of America.

```
REF
QE
720
.S77
cop. 1
```

LIBRARY OF CONGRESS CATALOGUING IN PUBLICATION DATA

Main entry under title:
Structure and classification of paleocommunities
  "The papers in this book were part of a symposium on 'structure and classification of ancient communities,' sponsored by the Paleontological Society at the 1974 annual meetings of the Geological Society of America in Miami Beach."
  1. Paleoecology—Congresses. 2. Biotic communities—Congresses. I. Scott, Robert William, 1936- II. West, Ronald R. III. Paleontological Society. IV. Geological Society of America.
QE720.S77   560   76-4587
ISBN 0-87933-225-5

Exclusive distributor: **Halsted Press**
A Division of John Wiley & Sons, Inc.
ISBN: 0-470-15073-4

Preface

Paleoecology has become a separate and distinct discipline of paleobiology since at least 1957 when the Geological Society of America published its *Memoir 67* on marine ecology and paleoecology. This was the first attempt to synthesize the vast amount of ecological data available to the paleontologist. Prior to that time, some paleontologists made reference to the living relatives of their particular group of fossils; some marine biologists designed their studies to aid the paleontologist in comparing suites of species and environments. Although brilliant individual work was done prior to 1957, there was no coherent effort in the profession to study paleoecology as such. The year 1957 was also the date of the first textbook in paleoecology by R. F. Hecker in the U.S.S.R. The first text in Western Europe was that by D. V. Ager in 1963; both books are still important reading. A significant symposium, "Approaches to Paleoecology," was published in 1964 (Imbrie and Newell, 1964). In 1967, the American Geological Institute sponsored a short course in paleoecology (Bandy et al., 1967) that summarized much of what was known about foraminiferal paleoecology and distribution patterns, as well as techniques of using forams, radiolarians, and functional morphology to reconstruct ancient environments. The term "community" was indelibly imprinted upon the collective paleontological mind in 1969 by the symposium on the evolution of communities at the North American Paleontological Convention. Several significant papers showed that suites of fossils can be interpreted as representative of the ancient community of which they once were members (e.g., Bretsky and Lorenz, 1970). Indeed, community dynamics were interpretable using new concepts developing in marine ecology. The 1971 American Geological Institute short course, "Recent Advances in Paleoecology and Ichnology," (Howard et al., 1971) presented community concepts and taphonomy to the broad spectrum of paleontologists and sedimentologists; by 1973, the new theoretical community approach to paleoecology was fully developed by James W. Valentine in *Evolutionary Paleoecology of the Marine Biosphere*. This book was the first attempt to integrate ecological theory and paleoecological data. Most recently, a 1974 short course, "Principles of Benthic Community Analysis" (Ziegler et al., 1974), was offered by the Comparative Sedimentology Laboratory at the University of Miami. This short course signaled the widespread application of community analysis to a broad spectrum of geological problems by both industry and academia.

Two informal seminars also served to further blend ecological and paleoecological approaches. At the 1970 Penrose Conference of the Geological Society of America, 60 ecologists and paleoecologists exchanged views on community ecology (Bretsky, 1971). Ecologists discussed how their theories could be tested in the fossil record, and paleoecologists attempted to use the wealth of theory available. Both groups were impressed by the abundance of descriptive data and the paucity of theoretical frameworks. In 1973, ecologists and paleoecologists again met at the first International Congress of Systematic and Evolutionary Biology to discuss the community concept in both fields (Scott and West, 1974). By this time, paleoecologists had begun to think in terms of energy flow, community equilibrium, and community evolution.

Within the framework of the growth of paleoecology, it seemed timely to ask "Whither goest paleoecology?" Objectives of this "state of the art" symposium have been (1) to review and evaluate the techniques of grouping taxa into paleocommunities, (2) to summarize structural properties of paleocommunities that are potentially useful in their recognition, and (3) to consider methods of comparing and classifying paleocommunities and living communities.

A *paleocommunity* is the suite of preservable taxa comprising a community, as opposed to a living community, which consists of all taxa in the community.

Major trends of the symposium are revealed by a brief scan of the papers. Kauffman and Scott define fundamental terms and analyze the logic of paleoecology. Scott reviews classification schemes and proposes a trophic classification. A geomorphic classification of environments and their communities is outlined by Parker. Macdonald compares properties of living and paleocommunities in different environments. Stanton compares structural properties based on total and preservable taxa of some Holocene communities, while Warme et al. compare living-shelled assemblages with the dead-shelled assemblages in the same environments. Using some of these properties, West compares seven communities and demonstrates some properties that can be used in comparing communities spatially and temporally. Quantified diversity measures of ostracode associations are analyzed by Brondos and Kaesler for the late Paleozoic and by Lister for the Pleistocene. Finally, in two examples, Hanley and Broadhead illustrate the application of these techniques and properties to Tertiary nonmarine mollusk associations and Carboniferous marine associations, respectively.

The papers in this symposium show that community structure, such as species composition, relative abundance, rank dominance patterns, spatial distribution patterns, diversity, trophic structure, and substrate niche preferences, are recognizable in the fossil record, and that these properties can be interpreted in light of ecological theories of diversity, stability, spatial and environmental variability, and resource levels. These approaches are applicable to Paleozoic as well as to Holocene communities. The species composition of a community can be determined by quantitative methods and communities can be classified by habitat and trophic structure. The community concept in paleoecology is analogous to the concepts of time in stratigraphy and environments in sedimentology. Neither time, nor environment, nor community are preserved in the geologic column; they are not to be collected in the outcrop; but communities can be interpreted and predicted from preserved paleobiological and paleoecological data.

If paleocommunity studies are to be used to investigate the evolution and biogeography of species and ecosystems, paleoecologists must concentrate on careful study design to gather meaningful data. Sound data are necessary for the determination of density, diversity, relative abundances, trophic relations, and substrate niches, among other properties. For example, trophic relations and substrate niche assignments of species are possible only with sound interpretations of functional morphology. For numerous taxa, we lack detailed studies of functional morphology. Furthermore, it is impossible to make meaningful community comparisons spatially and/or temporally until we know that we are comparing communities from the same or similar depositional environments. Environmental reconstruction requires the use of all the tools of the geologist's trade: stratigraphy, petrology, sedimentology, geochemistry, and "good old-fashioned" paleontology. Basically, meaningful paleoecology is a field study; removal of fossils from the outcrop greatly limits their paleoecological usefulness. This is not meant to preclude well-planned and meaningful laboratory studies, but, basically, community paleoecologists must get back to the outcrops and observe fossils in place. We are reminded of a very apt statement by L. Agassiz, "Study nature, not books."

The papers in this book were part of a symposium, "Structure and Classification of Ancient Communities," sponsored by the Paleontological Society at the 1974 annual meetings of the Geological Society of America in Miami Beach. We are grateful to Curt Teichert, Warren Addicott, Porter Kier, William Oliver, and Charles Rowett, officers of the society, for their aid and advice in planning the symposium. Many persons served as outside reviewers of these papers: D. V. Ager, J. F. Baesemann, R. J. Cuffey, J. R. Derby, E. G. Driscoll, J. A. Fagerstrom, R. G. Johnson, N. G. Lane, A. La Rocque, B. A. Masters, A. R. Ormiston, A. J. Rowell, J. T. Sprinkle, S. M. Stanley, and J. W. Vincent. Amoco Production Company and the Department of Geology, Kansas State University, provided support for many of the editorial tasks.

R. W. Scott
R. R. West

REFERENCES

Ager, D. V. 1963. Principles of paleoecology. McGraw-Hill Book Company, New York, 371p.

Bandy, O. L., J. C. Ingle, Jr., R. R. Lankford, and H. A. Lowenstam. 1967. Paleoecology. American Geological Institute, Short Course Notes, Washington, D.C., 138p.

Bretsky, Peter. 1971. Benthic marine communities. Geotimes 16:20-21.

Bretsky, P. W., and D. M. Lorenz. 1970. Adaptive response to environmental stability: a unifying concept in paleoecology. North Amer. Paleont. Conv., Proc., E:522-550.

Hecker, R. F. 1965. Introduction to paleoecology. American Elsevier Publishing Company, New York, 166p. (First published as Wedeniye V. Paleoekologiuy, Moscow, 1957).

Hedgpeth, J. W. (ed.). 1957. Treatise on marine ecology and paleoecology. Geol. Soc. Amer. Mem. 67, 2 vol., 2373p.

Howard, J. D., J. W. Valentine, and J. E. Warme. 1971. Recent advances in paleoecology and ichnology. American Geological Institute, Short Course Notes, Washington, D.C., 268p.

Imbrie, J., and N. D. Newell. 1964. Approaches to paleoecology. John Wiley & Sons, Inc., New York, 432p.

Scott, R. W., and R. R. West. 1974. Community concept debated in Boulder. Geotimes 19:25-26.

Valentine, J. W. 1973. Evolutionary paleoecology of the marine biosphere. Prentice-Hall, Inc., Englewood Cliffs, N.J., 511p.

Ziegler, A. M., et al. 1974. Principles of benthic community analysis. Sedimenta IV, Comp. Sed. Lab. Univ. Miami, 148p.

Contents

Preface   *v*

BASIC CONCEPTS OF COMMUNITY ECOLOGY AND
PALEOECOLOGY   *1*
    E. G. Kauffman and R. W. Scott

TROPHIC CLASSIFICATION OF BENTHIC COMMUNITIES   *29*
    R. W. Scott

CLASSIFICATION OF COMMUNITIES BASED ON
GEOMORPHOLOGY AND ENERGY LEVELS IN THE
ECOSYSTEM   *67*
    R. H. Parker

PALEOCOMMUNITIES: TOWARD SOME CONFIDENCE LIMITS   *87*
    K. B. Macdonald

RELATIONSHIP OF FOSSIL COMMUNITIES TO ORIGINAL
COMMUNITIES OF LIVING ORGANISMS   *107*
    R. J. Stanton, Jr.

RAW MATERIAL OF THE FOSSIL RECORD 143
  J. E. Warme, A. A. Ekdale, S. F. Ekdale, and C. H. Peterson

COMPARISON OF SEVEN LINGULID COMMUNITIES 171
  R. R. West

TEMPORAL CHANGES IN A PLEISTOCENE LACUSTRINE OSTRACODE ASSOCIATION, SALT LAKE BASIN, UTAH 193
  K. H. Lister

DIVERSITY OF ASSEMBLAGES OF LATE PALEOZOIC OSTRACODA 213
  M. D. Brondos and R. L. Kaesler

PALEOSYNECOLOGY OF NONMARINE MOLLUSCA FROM THE GREEN RIVER AND WASATCH FORMATIONS (EOCENE), SOUTHWESTERN WYOMING AND NORTHWESTERN COLORADO 235
  J. H. Hanley

DEPOSITIONAL SYSTEMS AND MARINE BENTHIC COMMUNITIES IN THE FLOYD SHALE, UPPER MISSISSIPPIAN, NORTHWEST GEORGIA 263
  T. W. Broadhead

Taxonomic Index 279

Subject Index 287

Structure and Classification of Paleocommunities

Basic Concepts of Community Ecology and Paleoecology

Erle G. Kauffman
Robert W. Scott

U.S. National Museum
Amoco Production Company
Research Center

*The term "animal community" is really a very elastic one, since we can use it to describe on the one hand the fauna of the equatorial forest and on the other hand the fauna of the mouse's caecum (*Elton, 1927, p. 17*).*

ABSTRACT

Approaches to the study and classification of ecological units are diverse, and emphasize different aspects of the whole unit. Among these concepts only the holistic concept incorporates sufficient data to define the complex structure and ecological interactions of each unit. A holistic community is defined by its total biotic composition, structure of species interactions and energy flow, relations with the surrounding environment, and the biotic and/or physicochemical character of its boundaries. In practice many ecologic and paleoecologic studies fail to describe one or more of these elements because of incomplete sampling, research methods, time limitations, or incomplete organism preservation. The holistic concept is more complete than community concepts based solely on associated physicochemical parameters, or statistically recurring taxa sets, or species of one life habit. Despite loss of data in paleontology, the holistic concept may be applied to the definition of paleocommunities through a methodology herein described.

Organisms, their traces, and their mode of preservation are the basic data for inferring sets of recurring taxa. Numerous analytical steps lead to the definition of ecological units such as communities and paleocommunities. Both are interpretive units derived by a logical process that separates observation from inference and interpretation. A hierarchial classification of these and other community-type units is integrated with biogeographic units and consistently serves ecology and paleoecology.

INTRODUCTION

Biology and paleobiology are in the midst of an era of ecological research. Field and laboratory studies in organism interrelations, trophic structure and energy flow in ecological units, and the evolutionary development of ecological units have led to the reexamination of ecological concepts, the construction of dynamic models, and the application of ecological data in understanding natural and man-imposed environmental systems through time. Ecology and paleoecology today are dynamic scientific disciplines of equal importance to systematics, biochemistry, and morphological analyses.

Widespread activity in ecological and paleoecological research requires a clear and consistently applied set of ecological concepts at basic levels — the determination, classification, and definition of ecological units. Our initial quote from Elton (1927) is fitting, and characteristic of the loose manner in which ecological units such as the community are regarded (e.g., see Fig. 1); but even Elton's broad range of possibilities would not satisfy the holistic ecologist who recognizes that plants must be included in any concept of an ecological unit.

We lack a common language of ecology that will allow theoretical, experimental, and observational ecologists and paleoecologists to interface and to apply consistently the same conceptual meaning to common terms. We also seem to lack an understanding of the fundamental concepts of logic and methodology behind much of ecological and paleoecological research. Despite the obvious desirability of not rigidly restricting ecological concepts and methodology, some consistent form of communication is important for the advancement and application of ecological science and for meaningful exchange of ideas among its practitioners.

The purposes of this paper are (1) to define the logical steps and general methodology underlying the study of ecological units — from species pairs to whole ecosystems — and (2) to define and classify ecological units in a way that will consistently fit the dominant concepts of the science, as well as be equally applicable to the varied problems and approaches of both ecologists and paleoecologists. We present this as a backdrop to the diverse papers that follow, in hopes that it will provide some unifying concepts. Throughout, it has been necessary to contrast the application of these concepts to neontological and paleontological situations because of the very different scope of data, and problems peculiar to each discipline.

Ecological units that are fully defined by their composition, physicochemical parameters, spatial distribution, biological structure, energy flow, and history are powerful tools in biological and geological interpretations. For example, the interpretation of the population variation, behavior, functional morphology, and adaptation of a given species is strengthened by comprehension of the overall structure and complexity of its biotic interactions in an ecological unit. Furthermore, the ecological unit and the adaptations of its component organisms are in part confined by physicochemical factors; consequently, the structure of the ecological unit alone permits prediction of these environmental factors. The definition and evolution of ancient environments and of natural selective forces can, therefore, be interpreted from the structure and composition of "paleocommunities," even where direct evidence of some or most physicochemical factors has been lost by diagenesis. Finally, like individual species populations, ecological units are definable biological entities that show dynamic spatial (biogeographic) and temporal (successional and evolutionary) changes. Although the patterns and causes of these changes are only now under serious study, they promise to shed light on major conceptual issues in evolution and biogeography. This "interpretive" phase of ecological research is ultimately dependent on the stability and consistent use of basic concepts by ecologists and paleoecologists.

HOLISTIC COMMUNITY CONCEPT

Ecological units are categories erected by the scientist to recognize different levels or types of ecological relations (e.g., the community). Many ecological units are defined in widely varying ways and thus are of minimal value in communicating ideas among ecologists. The inconsistencies in definition relate both to the biological content of a unit and to the inferred environmental controls on organisms comprising an ecological unit. For instance, some define community by a set of distinctive and widespread taxa, or by a set of numerically recurring taxa, or by taxa of only one suprageneric group. For others a community consists of all organisms within sharply defined biotic and/or physicochemical boundaries, regardless of their internal patterns of distribution. Still others define a community by its peculiar internal structure and energy system. Other types of ecological units, such as province, share some of the above problems and are equally defined in an inconsistent manner. A return to original concepts sheds light on the problem and provides important guidelines to the consistent definition of ecological units.

Ecology is the study of the totality of relations between organisms and their surrounding environment, the distribution of organisms within their environment, and the nature of their interactions. It is concerned mainly with natural groups of organisms, their structure, and their function within the confines of the natural environment (e.g., see Ricklefs, 1973, p. 11; Odum, 1971, p. 3). The word "ecology" is derived from the Greek *oikos*, meaning house, or place to live, and the suffix *-logy*, meaning knowledge of, study of, or science (Odum, 1971, p. 3). Whereas Charles Darwin may be regarded as the first great ecologist, Karl Möbius (1871, 1877) is generally credited with developing the concept of an ecological unit—the biocoenosis, or living community (Schäfer, 1972, p. 453). Möbius emphasized that the unit is an integrated "living congregation" of diverse organisms with both biotic and abiotic characteristics. Schäfer (1972, p. 453) has reviewed the history of the biocoenosis concept and concludes: "Since a *biocoenosis* (or ecosystem) encompasses a biotope and a community of all organisms living in it, biofacies is a closely related concept. It is therefore necessary to study biocoenosis to understand biofacies." Schäfer's concept stands as a largely unfulfilled clue to paleoecology.

Four major elements distilled from these basic ecological concepts serve as important components in the definition and study of ecological units: (1) biotic composition, (2) species interaction, (3) environmental parameters, and (4) unit boundaries. A major part of the early definitions of ecological units is *biotic composition*. Early workers considered an ecological unit to be a unique grouping of species recurrent in time and space. Consequently, all, or at least a great diversity of, its component organisms, both animal and plants, small and large, soft- and hard-bodied, comprised an ecological unit.

Second is the concept of *interaction* among organisms, and between organisms and their environment. Biological interaction refers to intra- and interspecies dynamics — the reaction of one organism to another in a complicated food chain and/or behavioral web. Interaction imparts structure to the ecological unit that controls the energy flow within the ecosystem. Consistent identification of an ecological unit depends upon maintenance of its structural integrity. The organisms interact in a certain way under a certain energy flow system and environmental parameters (food chain, nitrogen cycle, behavioral patterns, etc.). *The study and definition of an ecological unit should, therefore, be deeply concerned with its structure.* This is not achieved in many "community" analyses that do not ask *why* these taxa co-occur.

Equally important in the concept of ecological units is the definition of *interactions between component organisms and the surrounding environment*. The environmental limits and the environmental optima for maintenance of an acceptable level of energy flow are part of the concept of the ecological unit. Ecological units have relatively consistent demands upon the environmental resources, and predictable constraints are placed on them by the environment. In paleobiology the sedimentary matrix and the mode of preservation provide clues to the physicochemical conditions associated with an ecological unit. It is, therefore, recognized from the start that both external physicochemical and biological factors of the environment, as well as the internal biological structure and energy flow, are critical to the concept of an ecological unit. Both Forbes (1844) and Möbius (1871) considered this when they described biotic and abiotic characteristics of marine biocoenoses.

A fourth point inherent in original concepts of ecological units is that of *unit boundaries*. The physical, chemical, and biological conditions limit the distribution of component organisms in an ecological unit. Boundaries may also be partially controlled by the internal structure of the community (i.e., the competitive boundaries of adjacent communities acting as "superorganisms," a point of considerable debate). Whatever the cause, sharp to gradational interfaces exist with adjacent ecological units or biotopes. Ecological unit boundaries are measured both by taxonomic discontinuity and by the character of the environmental gradient across the interface.

The *holistic concept of ecological units* consists of four elements: (1) total biotic composition (as complete an inventory as possible), (2) structure defined by species interaction and energy flow, (3) the nature of constraining and interacting environmental parameters, and (4) the character of unit boundaries. In practice, however, both biologists (e.g., Thorson, 1957) and paleobiologists (e.g., Ziegler, 1965) have strayed far from this concept, commonly defining ecological units only by the most obvious and/or most abundant organisms, or even on a single group of organisms. Commonly, the numerically dominant microbiota and rarer taxa, such as large predators which may be ecologically dominant, are ignored. In paleontology, where sediment diagenesis and lack of preservable hard parts commonly deplete the biota of a once-living ecological unit, many fossils (especially trace fossils and microorganisms) are still excluded from the normal definition of "paleocommunities." If we are to hold that an ecological unit is characterized by its structure and energy flow, then the sum total of its biotic constituents (or the best possible approximation) is necessary for the complete definition of any single unit. Anything less sacrifices the "ecology" of the unit for the ease and efficiency of defining consistently statistically recurring groups of organisms in the living and fossil record.

Often, this approach is prescribed by logistics, preservation, or time restrictions. We do not argue here that more simplistic analytical methods and resultant definitions of "ecological units," as published by numerous workers, are not valid studies of ecological phenomena. The fact that such units are recurrent in space and time and consistently identifiable suggests that they are some kind of biological unit with ecological significance. They have proved to yield significant clues to the presence of more complex, more truly ecological units in a study area. Furthermore, as we shall demonstrate in a later section, such units can be easily accommodated into the conceptual hierarchy of ecological relationships without compromising their importance. We merely point out that, to the degree that these analyses ignore portions of the biota as well as interrelations among community members and with the surrounding environment, such biological aggregates are, to varying degrees, only partially ecological in concept.

We do argue that it is important for the ecologist and paleoecologist to recognize the *holistic and dynamic aspects of ecological units*. The study of ecological structure, whether of a highly interdependent nature or of a simple food chain, and of energy-flow characteristics is needed in the definition of *ecological* units. This should be the ultimate aim of all neontological and paleontological studies of ecological units. Definition of these factors is not impossible, even with fossils, although the paleoecologist must carefully separate fact from inference. Modern texts display analytic methodology for the testing and interpretation of population dynamics, interspecies relations, reconstruction of food chains, and energy-flow pathways. Many of these tests can be used directly or modified, to paleontological situations. The "holistic biota" approach is most easily developed by simply studying all component organisms of modern or ancient ecological units. However, not all fossil deposits retain enough data to enable the reconstruction of a holistic paleocommunity. In some situations the paleoecologist must settle for an incomplete interpretation.

In addition to these basic concepts of biologic structure, environmental constraints, boundary zones, and uniqueness of composition, ecological units have other inherent characteristics that aid in their study and recognition. A compilation of the voluminous literature on the subject reveals the following:

1. Ecological units must be recurrent in space and time to be easily and consistently recognized and to be useful in environmental analysis. Presumably, the complex evolution of biological characteristics, organism interrelations, and, thus, structure of an ecological unit takes some time to develop and occurs in response to a recurrent and/or long-term set of environmental parameters. It is possible, but unlikely, that any true ecological unit (especially communities and larger-scale units) would have evolved only once, in one place, at one time. Recurrence can be subjectively defined from full taxonomic lists coupled with relative abundance data, or through mathematical tests of many samples using similarity coefficients and other statistical tests. By the same logic stated above, it can be assumed that at least major ecological units will have significant longevity in space and time.

2. Ecological units are dynamic in space and time. Communities, for example, go through successions or seres from early colonizing to "climax" phases during which taxonomic composition and relative abundance change significantly, and often in a patterned manner. At any stage of development, environmental or biological perturbations may, and commonly do, disrupt the temporary balance of the community, and the entire structure must adjust, in some cases by returning to an earlier phase of the succession. Taxa and interspecies relations that form part of the community structure evolve; organisms migrate in and out of communities through time; and in the long term, environments are never uniform, so that the idealistic concept of an absolute climax community and of community stability is not practical.

3. Ecological units cannot be easily defined or classified by size or biogeographic spread. They must be allowed a broad size range at all levels of the ecological hierarchy. Thus, our initial quote from Elton (1927) has this message: an ecological unit, defined by its internal structure, biological composition, and its interaction with and restriction by the surrounding environment (producing a boundary), can theoretically be of any size. A few small rocks scattered widely over a sea bottom may each contain all elements of a hard surface community within a few centimeters, whereas thousands of square miles of deep ocean bottom may be covered by a single community of equal biological distinctiveness and structural uniqueness.

4. Ecological units, despite their characteristic structure and suite of organisms, are spatially variable. Variations in structure and composition are a response to small-scale environmental differences and to stochastic distribution processes. Near boundaries with adjacent ecological units, ecotone populations result from competi-

tion and migration. Variation is accentuated toward the distributional margins of an ecological unit; on the one hand, ecological units may merge gradually if the environmental gradient between them is gradual and competition is not severe; on the other hand, these boundaries may be extremely sharp, as between a reef surface and surrounding mobile bioclastic sand.

5. Most ecological units have certain broad dominance patterns that characterize their structure. Ecological dominance refers to low-density, high-dominance-level predators at the top of the food chain, that is based upon high-density, low-dominance-level crop organisms. Dominance may also be expressed in absolute numbers, biomass, or relative abundance. Whether or not dominance of this type should be used to characterize and identify an ecological unit is still a matter of debate; many ecologists prefer only to use presence–absence data in analysis. Obviously, the numerically dominant organisms in an ecological unit are those organisms (normally microorganisms) at the base of the food chain.

Dominant species are not to be confused with common and obvious (i.e., large) species of an ecological unit. Petersen (1924) called these the first-order characteristic species of his level-bottom marine community; others have called them identifier species or dominant species. Many ecologists and paleoecologists have considered characteristic species to be the main basis for the definition of ecological units. Petersen's classification continues with "second order characteristic species," which are uncommon but widespread and occur at most localities (i.e., ecologically dominant predators), "third-order characteristic species," which are commonly but not always present, and "fourth-order characteristic species," which includes all other more common forms.

In summary, the concept of ecological unit which we find most viable, and which best fits both the original intent of ecologists and the bulk of modern theory, is that of a holistic, dynamic entity. A community, or any other ecological unit (ecosystem, assemblage, association, etc.) is *ecologically* defined on the basis of all component organisms: their diversity characteristics; the uniqueness of their composition (including normal variations in time and space); the sum total of their interactions with each other (trophic, food chain, and behavioral relations) and with the containing environment (acceptable environmental parameters and relation to material cycles)—that is, their structure and energy-flow characteristics; the nature of the boundary that separates one ecological unit from the next, measured by restrictive environmental and biological parameters; the nature of their spatial and temporal distribution and the changes that take place within these parameters. This concept may be applied to something as simple as a two-organism symbiosis (a low-level *association*), to the basic unit of ecology—the *community*, and to something as large and complex as an oceanic or continental biota. It is important to note that, even though the original data and materials of an ecological unit are greatly depleted in most fossil deposits, the holistic concept of ecological units is applicable to paleoecology. Although rarely practiced at present, it is strongly recommended as an improved method of paleobiological research, which more clearly interfaces with ecological studies of living organisms.

DIVERSE ECOLOGIC AND PALEOECOLOGIC COMMUNITY CONCEPTS

An extensive survey of ecologic and paleoecologic literature demonstrates broadly variable and inconsistent concepts and usage of such basic ecological terms as "association," "assemblage," and especially "community." Most usage is inconsistent with the holistic concept of ecological units previously described, and is, to varying degrees, less purely ecological in nature. It is pertinent to recognize the scope of this variation and patterns of usage before proposing a consistent set of definitions and a classification of ecological terms. To do this we have selected examples of divergent usage of the most basic ecological unit, that of the *community*. These uses are compared in Fig. 1.

Whittaker (1970) provides one of the best short definitions of the holistic community concept: "A system of organisms living together and linked together by their effects on one another and their responses to the environment they share." An alternative view of community is promoted by Gleason (1926), who believed that a community results from the random migration of reproductive units in response to environmental conditions. In Gleason's view, species distributions form a gradational continua resulting from stochastic processes rather than species interaction. Distinct groups of species are not recognizable and communities are not real. We believe, however, that much evidence in this symposium and elsewhere demonstrates that definite, recurring groups of co-occurring species are real. In the following discussion, the three major concepts of "community" currently in use will be compared with the holistic concept to demonstrate differences, weaknesses, and strengths in these alternative systems. Examples are selected from various published ecological and paleoecological analyses; to centralize the theme, each is related to a single complex ecosystem—the coral reef.

Physically Defined Communities

Petersen (1911-1918, 1924), the great Danish pioneer in the study of level-bottom community structure, and before him Forbes (1844), stated that the main controls on community distribution were physical factors related to depth, substrate, salinity, water currents, and other environmental parameters. This view has been held, even more strongly, by a number of workers since Petersen's time. For example, in numerous works the term "coral reef community" implies that all organisms living within the confines of a coral reef constitute a single community defined mainly by the physical existence of the reef structure and the obvious physical boundary between the reef surface and adjacent sedimentary surfaces.

In common ecological and paleoecological practice, this translates simply to the determination of physicochemical environmental divisions of a large area *first*. Then, all organisms that live within these parameters are included in one community on the assumption that they respond to habitat conditions the same way. This is often done without ever testing independently for the existence of more than one biologically-ecologically defined community within the single set of physico-

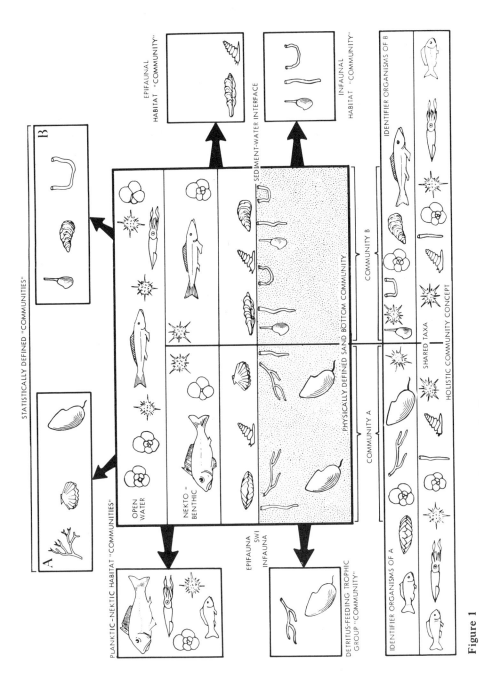

Figure 1
Variations in the use of the term "community."

chemical parameters. In other cases, such tests are always secondary to determination of physicochemical environmental divisions.

Modern ecological studies of the marine realm, such as those of Murina et al. (1973), seem to follow this concept and produce such terms as "sand biotope" and "mud biotope," or, in other publications, terms like "lagoonal," "rocky shore," "mud flat," "shallow shelf," "sea grass," and "deep basin" communities. In paleoecology Walker (1974, p. 9.5) and Anderson (1974a, p. 3.1-3.4) among others have been proponents of this methodology. Walker states (1974, p. 9.5), "Physical environmental subdivisions were worked out first, and communities were placed in their context," although ultimately he has recognized a low degree of patchiness in single Ordovician marine paleoenvironments of Tennessee. Close correlations between single "communities" and single major paleoenvironments (lithofacies) in the Paleozoic studies of Sutton et al. (1970, Figs. 7-9), Anderson (1974b, p. 11.2-11.4), and others seem to reflect similar concepts and relationships.

We would not argue that these are not units of ecological significance, or that physicochemical parameters do not exercise a strong control on the distribution of communities and, independently, their component organisms. But it is unusual for only a single community to inhabit a single broad lithotope except in highly stressed or environmentally poor situations (deep ocean muds, intertidal environments, hypersaline lagoons, etc.), or during the earliest phases of colonization of a new barren substrate. Thus, a typical coral reef "community," physically defined, is a complex of ecologically defined communities of the lagoon (several), back-reef areas, reef crest, fore-reef area (two or more), and reef caves (e.g., see Ginsburg and James, 1974), each with characteristics that fit the holistic concept of a community. Similarly, both recent and ancient shallow shelf deposits (Kauffman, 1974, Fig. 12.2) are a mosaic of communities on similar or identical substrates and certainly within single major lithofacies. Nonrandom, but patchy, distribution of species populations of interacting species clusters and of communities is the *normal* pattern of biological dispersion in a broadly uniform ecosystem.

Failure to recognize these normal marine mosaics may stem from too limited a sampling program, or from analysis of only one or a few major types of "first-order" organisms (e.g., brachiopods; Ziegler, 1965), which may have broad facies ranges and actually be shared between benthic communities characterized by combinations of other taxa. It may also reflect postmortem mixing of elements from several communities through current and wave action, especially in paleontological samples and modern analyses utilizing dead-shell data. Finally, one cannot exclude the basic human prejudice that might arise from the initial assumption that a single major environmental division *should* contain a single community. Thus, one fails to search for recurrent species-clustering patterns within the biota of a single lithotope.

Further, if ecological units such as communities are defined, and restricted, by *biological* factors (e.g., structure, competition) as well as by physicochemical environmental factors, it is unlikely that a one-to-one relation will exist between major lithotopes and single communities in normal environmental situations. This is especially true in light of the many microniches that exist in a given lithotope today and presumably in the past. The mosaic of communities observed in many major lithotopes probably reflects the effect of microniches, or even of colonization of a single, environmentally uniform substrate by different communities (as in the early phases of reef colonization), which then biologically compete for space and resources along biologically controlled community boundaries.

In few studies of "physically defined ecological units" is the biological aspect of the community properly emphasized in analysis, nor are the structure and energy flow of the community considered. "Communities" so defined are commonly blends of selected organisms (usually large identifier taxa and/or those that preserve well) from two or more ecologically defined communities within the same broad lithotope. In the classification subsequently presented, most of these broad units would be considered "assemblages," as properly used by Parker (1959, 1964) for broad, environmentally similar areas of the Gulf Coast and the Gulf of California. Such assemblages have ecological meaning in that they prescribe the general environmental and ecological conditions of a large-scale lithotope (e.g., a reef or a shallow shelf sand), and their study is an important first step in defining more purely ecological units in a study area.

Quantitatively Defined Communities

The dominant school of thought in the study of marine communities and paleocommunities has been that of Petersen (1911-1918, 1924) and Thorson (1957, 1966, and references therein). Their "level-bottom community" concept refers to a recurrent statistical unit comprised of a few identifier or first-order characteristic macroorganisms, mainly mollusks, echinoderms, arthropods, and annelids, common to nearly all samples. The data upon which it was originally based were from North Sea meter-square grab samples initially designed to analyze food resources for bottom-feeding fish. Emphasis in the study was on large, obvious organisms; the microbiota and uncommon taxa were excluded. These identifier taxa were not selected from holistic community studies; just the reverse was true. The fact that they consistently occurred together in the North Atlantic, and that similar taxa sets were widely found in the world's oceans, was taken to indicate that they naturally represented single communities. The biological structure, species interrelationships, and energy-flow characteristics of these communities have not yet been determined. The Petersen-Thorson "community" is therefore clearly a statistically reproducible unit rather than an ecological unit because it is based upon a small portion of the total community biota. To their credit, Petersen (1924) and Thorson (1957) recognized that these "macrobenthic communities" or "macrofaunal communities" were a first step prior to determination of holistic communities.

Brilliant for its time, this work was the first real indication that there might be globally distributed ecological units in the marine realm comprised of similar or identical suites of species at widely scattered localities. It is, nevertheless, unfortunate that this concept has served as the model for so much subsequent work by community ecologists and paleoecologists (e.g., Ziegler, 1965; Bretsky, 1969;

Sutton et al., 1970; Kauffman, 1967, 1969). The approach is not holistic, only broadly ecological, and is largely based on the untested assumption that these statistical units are abstractions of ecologically defined, complexly structured communities.

In our modern coral reef example, this would be equivalent to defining a community on the basis of the common and recurrent occurrence of the visually obvious *Acropora palmata* and *Millepora* sp. In fact, these taxa are rather ubiquitous with a wide distribution on the reef complex and are found in association with more than one ecologically defined community. The community schemes of Ginsburg and James (1974) for Florida and Caribbean reefs employ the Petersen-Thorson principle of recurrent statistical units comprised of a few obvious and common taxa, without extensive knowledge of their interrelationships or less obvious but diverse biotic associates. These units (i.e., the shallow marginal reef community characterized by *A. palmata* and *Millepora*, the intermediate marginal reef community characterized by *Montastrea*, *Diploria*, *Porites*, and *Agaricia*, etc.) may, in part, actually correspond to benthic community units of the reef comprised of complexly linked suites of organisms. In other examples, they may be characterized by common species that are shared among two or more benthic communities.

In the classification and definition of ecological units subsequently proposed, most of these "communities" are either *associations* of taxa that are obviously abstracted from a single community or *assemblages* of taxa shared by two or more communities. But because of the failure to determine the entire biota or to interpret interspecies relationships and structure, these are not specifically communities, as previously defined.

Life-Habit Communities

Some communities have been defined on only one portion of the biota that is related in habitat, trophic behavior and food source, or morphological adaptations and behavior. Thus, we have many published references to "animal communities" and "plant communities" in the same forest, and to "infaunal communities" and "epifaunal communities" occupying the two sides of the same sediment-water interface, which commonly feed and compete at the same interface. Rhoads et al. (1972) and Levinton (1972), for example, have differentiated deposit-feeding assemblages or communities from suspension-feeding assemblages as end products of an environmental and community succession in organic-rich muds, where both are present in the same sediment during most of the history of the succession. In our modern reef example, this would be equivalent to considering reef borers (bivalves, echinoids, worms, etc.) as being a different community than the living reef surfaces from which they bore and at which they feed, or to consider predators as distinct from prey in separate communities. To divide biotas into separate plant and animal communities is to break up the community food chain into producers and consumers—an unrealistic ecological division. As Allee and Schmidt (1937) note, "To split assemblages into plant and animal communities involves disregarding the essential ecological unity."

In all the above cases, the "communities" or "assemblages" are merely related functional, trophic, or habitat groupings of interacting taxa primarily *within* single community structures; rarely do they stand alone as an entire community. They do have ecological significance in that they represent one major division of the structure and energy-flow system of a community. In the classification and definition of ecological units subsequently presented, each of these "communities" would probably comprise an *association*—or natural division of a community.

A holistic, ecologically defined marine-bottom community contains diverse organisms with diverse adaptive strategies as a means of dividing the available ecospace and resources most efficiently. It consists of infaunal organisms that are both detritus feeders and suspension feeders, epifaunal organisms with diverse feeding habits, epibenthic swimmers, floaters, and saltaters, predators and prey, and so on. The sum total of interaction between these diverse organisms is necessary for definition of the community.

Whereas there are additional and less commonly used community concepts in the literature, the preceding four—holistic, environmentally defined, quantitatively defined, and life-habit "communities," "associations," or "assemblages"—comprise the main range of concepts applied by ecologists and paleoecologists in recent years. The holistic community concept is preferred because it presents the maximum amount of biological and ecological information, follows closely the original concepts of ecology and ecological units, and it alone incorporates data that allow the description of structure and an energy-flow system in the community. The holistic concept is the only totally ecological statement among the various concepts cited. It is therefore taken here as the base unit for ecological and paleoecological research, and for construction of a broadly applicable system of classification.

DEFINITION AND CLASSIFICATION OF ECOLOGICAL UNITS

Ecological units range in size from the global biota to simple two-species symbiosis. All levels should be accommodated in a system of classification. Whereas the holistic community concept is preferred, other concepts of ecological units exist in practice which describe some degree of ecological relation that should be accommodated in a classification of ecological units. Furthermore, the raw materials of ecology differ significantly between paleobiology and neontology; the incompleteness of the fossil record definitely hinders ecological analysis utilizing the holistic concept of ecological units. Yet it is necessary and desirable to share common concepts and terminology between these disciplines; a system of concepts must, therefore, accommodate the realm of paleobiology. We propose the following classification and definitions to attempt to deal with all these problems; these are schematically compared in Fig. 2, and ranked on the right side of Fig. 3. Conscious adherence to a broadly applicable scheme of ecological units will go a long way toward providing the common language that we now lack in the science. The terms used below are all well known in the ecological literature, but like the term "community," they have

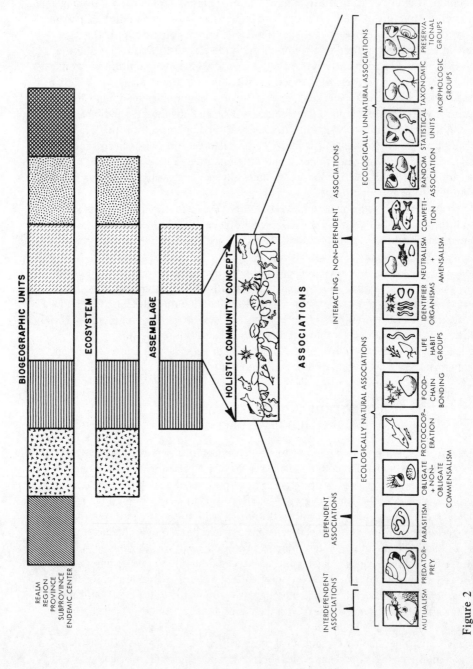

Figure 2
Outline of classification of ecological units (each pattern = one community).

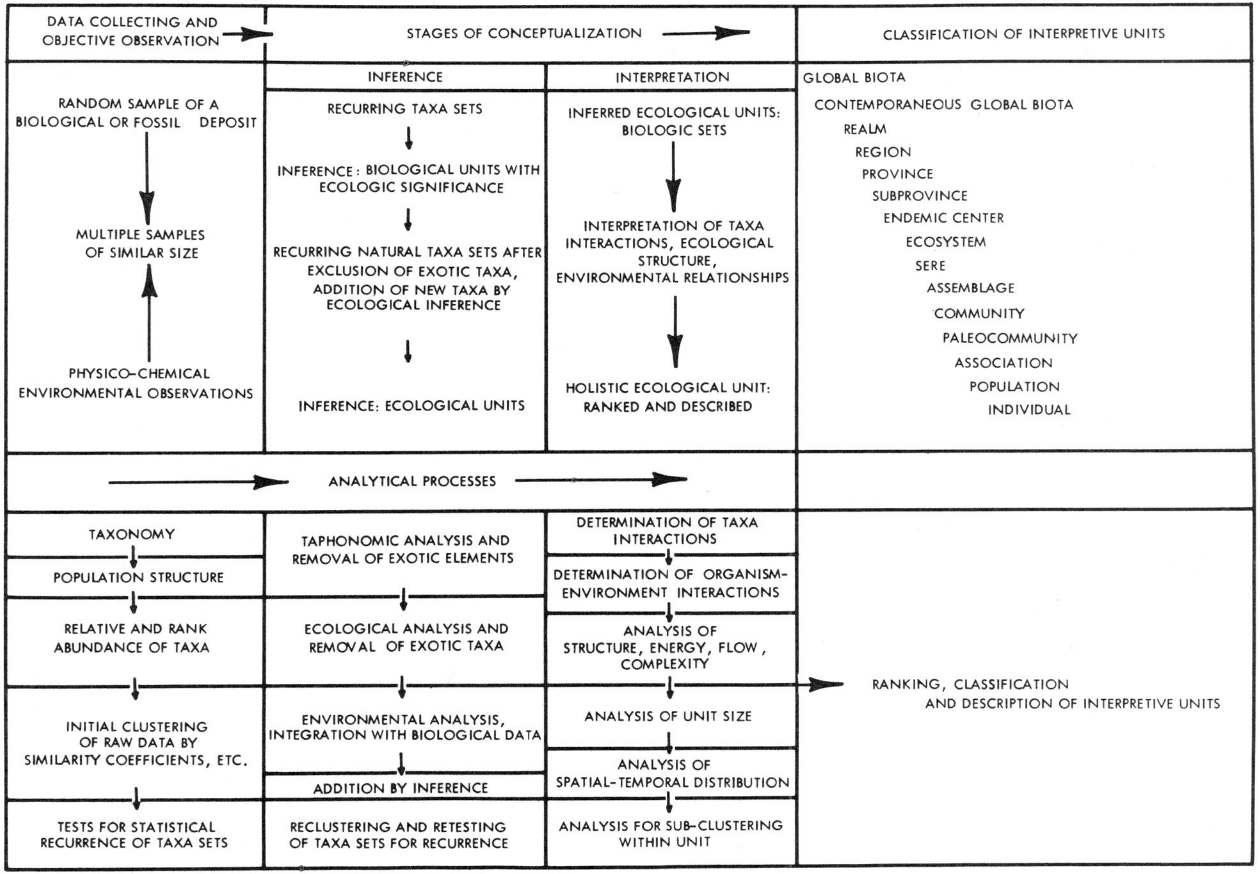

Figure 3
Outline of logical sequence of thought processes and analytical processes in the recognition, description, interpretation, and classification of ecological units.

been variously used and even substituted for some concepts of "community." The definitions given are based on a combination of original intent and common usage of each term in the literature. The classification extends from biogeographic units, which Valentine (1973) has pointed to as extensions of smaller-scale ecological units, to populations, the basic biological unit.

Global Biota

All organisms on the face of the earth, living and fossil, extant and extinct, constitute the global biota. A case can be made that this unit and all biogeographic units listed below are also ecological units in the holistic sense, with organisms interacting throughout the unit, at least at low levels.

Contemporaneous Global Biota

All organisms existing on the face of the earth at any one time.

Realm

The largest biogeographic unit, commonly circumpolar or encompassing an entire physiographic feature such as a continent, ocean basin, or sea. The limits are generally a major climatic zone boundary and/or a major physiographic boundary. Internal endemism is high, commonly exceeding 75 percent of the component biota (Kauffman, 1973). Examples are the Tropical or Tethyan Realm, or the Atlantic versus the Pacific Realm. Marine biotas are commonly differentiated from nonmarine biotas in describing realms (e.g., the Tropical marine realm).

Region

The largest subdivision of a realm, normally with 50 to 75 percent internal endemism (Kauffman, 1973), and with boundaries coincident to major climatic zone boundaries. Examples would be the North Atlantic Region of the North Temperate (or "Boreal") Realm, as opposed to the North Pacific Region (physiographically separated regions) or the South Atlantic Region (climatically separated).

Province

The main working unit of biogeography and paleobiogeography has in theory between 25 and 50 percent endemic biota, but in practice commonly between 20 and 25 percent endemism or nonoverlap of taxa at the Province boundary. Integrated biological studies using the holistic, structural approach to ecological units generally occur at the province and lower levels in the hierarchy of ecological units. The province is a major, climatically and biologically defined division of a region, for example, the Floridian (Tropical) and Carolinian (Subtropical) provinces of the North Atlantic marine region.

Subprovince

A local division of a province within a single geographic system is characterized by around 10 percent endemism, rarely higher, and/or between 10 and 25 percent nonoverlap of component taxa at the boundaries with adjacent biogeographic units (Kauffman, 1973).

Endemic Center

A level of 5 to 10 percent endemism characterizes genetically isolated, unique local biotas within subprovinces. The Red Sea is such a center; in some centers endemism is considerably higher (e.g., Lake Baikal, Lake Tanganyika, the Salton Sea). Highly distinct ecological systems develop around these taxa as they interact with one another through time to achieve an ecological balance. Several communities may develop within even small endemic centers and may structurally parallel more widespread communities of normal oceanic and continental areas where severe isolating mechanisms are lacking.

Ecosystem

The ecosystem comprises the collective interaction among a number of environmentally related communities that interface in a particular area of broadly similar environmental conditions (e.g., a lake, a desert, an atoll). The collective interaction (energy-flow and material-cycle characteristics) among these communities and those of the surrounding environment is much less than within the ecosystem. The term "ecosystem" has been commonly used at all levels of the ecological hierarchy (from the global ecosystem to the ecosystem of an intestinal tract). As originally intended, it had a more restricted connotation, however, and we use it here in that sense.

Sere

The sere is the entire group of communities that comprise a succession from its colonization phase to its "climax" or most mature phase of development. This is a special kind of ecological unit that can be determined only by monitoring natural successions of communities through time, either in rapidly changing modern situations or in the fossil record. Environmental changes of the sere are at least in part created by the community structure at each step in the succession. The sere differs from the assemblage, which is normally based upon less data, and consists of selected taxa from two or more communities. The sere also differs from the ecosystem and various biogeographic–ecological units in which contemporaneous communities are numerous.

Seres of *short-term successions* involving a few hundred years at the most probably differ from *long-term successions* involving thousands to millions of years and mainly evidenced in the fossil record. This problem is currently under study. Of interest to modern ecologists is the concept that seres, and other ecological units, should follow Walther's Law of Correlation of Facies, as Ziegler (1974, p. 1.8) has pointed out in regard to individual communities. The succession of communities forming a sere through time should also be evident along any single time plane by observation of the sere and its environmental sequence in various stages of development.

Assemblage

The assemblage consists of organisms derived from more than one community. The assemblage may contain all organisms from two or more communities, or selected interacting organisms, or randomly selected representatives from each community, such as a beach deposit of shells after a storm. Subdivisions of assemblages are thus possible, each with a more restricted connotation. The term "assemblage" is one of the most loosely applied ecological terms. It has been substituted for community, association, and ecosystem, for parts of individual ecological units, and even for random associations of living and fossil taxa that have little or no ecological interrelations. We define it here in its most commonly applied sense.

It is apparent that many fossil deposits and modern deposits of dead organisms represent assemblages. But "communities" prescribed primarily by physicochemical parameters may actually represent assemblages of several communities occupying a widespread lithotope. Likewise, "communities" derived by statistical analysis of first-order characteristic, identifier taxa may actually be based on common taxa shared by two or more communities, which are best defined on the basis of the other, less obvious organisms.

Community

The community, the basic unit of ecology and paleoecology, is a unique congregation of diverse organisms having a unique structure based on organism interactions, and in some cases interdependence, as well as on energy flow; the community is adapted to and restricted by a particular suite of environmental parameters (physical, chemical, and biological). Communities can be of any size and are recurrent in space and time. They are bounded by sharp to gradually developed interfaces with other communities, from which they are in part biologically distinct. Community boundaries may represent increased physicochemical and/or biological gradients, or in some cases may be formed by competition with other communities. The community so defined is a holistic structure. Its identity basically involves analysis of all component organisms, their structure and energy flow pathways, and the nature of the boundaries with adjacent communities. Some ecologists believe that a community has biological characteristics exceeding the sum total of its component organisms, measured individually.

Paleocommunity

Paleocommunity is a community that existed in geologic history and is represented only by fossils or other evidences of past organisms. The taxonomic composition, habitat relations, and species interaction of the preserved members of the community can be studied and analyzed. Our understanding of the ancient community should be as close an approximation to its holistic Holocene counterpart as is allowed by the fossil evidence.

A holistic paleocommunity differs from a living community by the absence of certain of its members and by the absence of active energy flow. Many factors act to deplete the potential data available for the definition of paleocommunities (Kauffman, 1974, p. 12.3). First, many organisms do not possess preservable structures and are preserved only in very unusual circumstances. Some of these organisms leave indirect evidence of their existence so they can be included in the known biota by the technique of addition of inference (to be discussed). Second, the biologic activity of burrowing, scavenging, and sediment ingestion, among other processes, destroys another portion of the biologic record. Third, selective removal and mixing of potential fossils by physical and biological processes may produce a deposit of mainly exotic species. Finally, diagenetic processes take a heavy toll on organic remains prior to and during fossilization. These processes have been discussed by Johnson (1960), Fagerstrom (1964), and Lawrence (1968).

Furthermore, the methodology of many paleoecologic studies has diluted the ecological significance of many "paleocommunities." Quite obviously many workers have used samples with many distinct living levels, which they have combined as a single set of data. Many have made no attempt to sort out the exotic from the native elements of a sample through taphonomic or ecologic compatibility analyses. Most have not analyzed the whole biota that is preserved in the fossil deposit, especially evidence for soft-bodied organisms, and the microbiota. Few have attempted to analyze the ecological relations and structure of organisms comprising their paleocommunities, but rather have emphasized statistical recurrence. To the degree that it is possible, paleocommunities should also be defined on the basis of population dynamics and the interrelations of component species, determination of relations with the paleoenvironment (monitored with the same care as the biological data), and, thus, their structure and energy-flow patterns. In short, the paleocommunity is reconstructed by the most detailed possible analysis utilizing the holistic concept of the living community, and is determined through the same conceptual and analytical processes that go into the analysis of living communities (see subsequent discussion).

The image is discouraging, and it is valid to ask what hope there is for applying modern ecological concepts to fossil material. The answer lies in the method utilized and in a more rigorous definition of "paleocommunity," which parallels that used for "community" in this paper.

Association

An association is a group of organisms derived from a single community. If the community is viewed as a holistic concept, involving all taxa and their interrelations with each other and with the environment, recognition of the association becomes extremely important because it encompasses many uses of the word "community." The term "association" seems to stem initially from, and have dominant use in relation to, small-scale species interactions and groupings, especially with regard to various symbioses. By defining association to incorporate any combination of species derived from a single community, the term can accommodate a great variety of biotic co-occurrences. The concept of *association* would include most statistically defined, Petersen-Thorson type "communities" comprised mainly of first-order characteristic taxa, most described "paleocommunities" that are not redeposited assemblages of shells from two or more communities, and virtually all life-habit "communities" or "assemblages" (infaunal and epifaunal "communities," detritus feeding and suspension feeding "assemblages," etc.). Many "paleocommunities" of co-occurring taxa that persist together through long spans of geologic time are associations. The species interaction and energy pathways may never be known because the taxa are extinct. However, these distinctive suites of taxa clearly were a part of a paleocommunity.

There is considerable precedent for incorporation of such a broad range of concepts in our definition of association. Modern ecologists have defined diverse types of interspecies relationships, that is, associations, within communities (e.g., Odum, 1971, p. 211, Tables 7 and 8). All these relations, based on biological and paleobiological observations, can be accommodated under the concept of association as follows:

Interdependent Associations. Both parties benefit and relation is obligate, called *mutualism* (e.g., the relation between hermatypic corals and zooxanthellae).

Dependent Associations. One is dependent upon other(s) for survival, while the other(s) is nondependent. All are natural ecological associations.

1. *Parasitism.*
2. *Predator-prey associations* (it can be argued that this is an interdependent relationship because the prey benefits from cropping, as a species population).
3. *Food-chain-dependent associations* (grazing gastropod and algae). The argument under item 2 might also apply here.
4. Some cases of commensalism where host specificity has evolved.

Nondependent Associations ("Interacting" Associations). These include two or more interacting or "neutral" organisms from a community that are not obligate in any way in their relationship. The relationship may consequently be less highly evolved, and interaction takes place at a relatively low level.

1. Natural ecologic association:
    a. *Life-habit associations.* Portions of the community with similar life habits, trophic level, or behavioral characteristics.
    b. *Identifier-organism associations.* Associations comprised of the abundant and most obvious biotic elements of the community regardless of their size, taxonomic affiliation, habitat, or morphology.
    c. *Neutral associations.* Commonly co-occurring taxa with no apparent effect on one another, or very low levels of interrelationship.
    d. *Competitive associations.* Co-occurring taxa or taxa sets whose usually sharp population boundaries are set by competition for space or resources.

e. *Amensal associations.* One taxon or taxa set is inhibited, the other passive.
   f. *Nonobligate predator-prey associations.* The predator may be a single species while prey species are normally diverse.
   g. *Nonobligate commensal associations.* One organism or taxa set benefits; the other is passive.
   h. *Protocooperation associations.* All organisms benefit without obligate relationships. Many communities may be regarded as having widespread protocooperation.
2. Ecologically unnatural associations:
   i. *Random associations.* Drawn randomly from the entire community without any particular reference to its structure.
   j. *Statistical associations.* Associations of organisms, in many cases initially selected for some special trait (size, taxonomic group, dominance, habitat) and identified by recurrence of the more obvious elements (first-order characteristic species) in combination through various statistical tests.
   k. *Preservational association.* An association of organisms, usually in the fossil record, that is found together because of a common type of preservation, or morphological-structural characteristics, that saves it from destruction through direct environmental forces or diagenesis. Many paleocommunities fall into this category.
   l. *Taxonomic or morphologic associations.* All organisms within the community that belong to a particular taxon, or taxa set, or which have like morphology.

These associations allow incorporation of the many concepts of "community," "assemblage," and other ecological units commonly found in the literature, but which are divisions of single, holistically defined communities, within a system of classification that is consistently applicable to neontological and paleontological situations (Figs. 2 and 3). At the same time, they allow the concept of the holistic community to stand in its fullest sense as a dynamic ecological unit, so that all demands on accuracy in concept, yet breadth and consistency of ecological language, are satisfied if such a classification is consistently applied.

LOGICAL STAGES OF ECOLOGIC ANALYSIS

The identification and definition of ecological units must follow a definite logic and methodology if one is to arrive at the dynamic holistic concept of these units. This scheme is illustrated in Figure 3 and was initially outlined by Kauffman (1974) and Kauffman and Scott (1974).

The scheme distinguishes the stages of deduction of ecological units from the analytical procedures used to generate evidence for progressing from one stage to the next. The analytical processes are a variety of data manipulation techniques currently in vogue and described by other contributors to this symposium (Macdonald, Brondos and Kaesler, Lister, and Hanley). Four stages of logic proceed from concrete data to progressively more abstract concepts: (1) data collecting and observation, (2) inference, (3) interpretation, and (4) classification.

Data Collecting and Observation

Ecologic analysis begins with specimens in the outcrop; these are the only real and tangible materials available to the paleoecologist. The ecologist may measure energy flow through time, as well as numerous physicochemical parameters of the environment. But the paleoecologist has nothing more than a set of specimens in the rock at a given locality—a *fossil deposit* (fossil assemblage of Fagerstrom, 1964). He then proceeds to collect multiple bulk samples of a standard size at closely spaced intervals within the ranges of the ecologic unit or otherwise defined study area. At each fossil deposit, observations are recorded about the density, dispersion, orientation, preservation state, and relation of fossils to depositional surfaces. These observations are the raw data of taphonomic analysis and part of the data for environmental analysis. Census studies of the bulk samples are the basis of taxonomic composition, population structures, and relative and rank abundances. Observations on the physicochemical parameters of the enclosing sedimentary rock provide clues to the environment, as well as do autecologic studies of the associated species.

Ideally, sampling should be of a single life surface or synchronous surface and should include the microbiota and the macrobiota, soft-bodied organisms as well as those with hard parts, those organisms living in, on, and just above the substrate, and, where possible, those rare organisms that are ecologically dominant in the community. Practically, such sampling of Holocene communities is commonly impossible, owing to logistics and time limits; paleoecologic work is hampered by these factors, plus the expected depletion of the biota by fossilization and diagenesis. But it sets a standard for ecological sampling for which we should strive.

Second, multiple sampling at closely spaced intervals in space and time provides a basis for the demonstration of statistical recurrence that is inherent in the definition of ecological units. Inasmuch as boundaries of ecological units are critical in defining the limits and nature of each unit, sampling across these boundaries is important.

In paleoecology, these sampling criteria have to be adjusted to fit the nature of the problem and the fact of differential fossilization. Numerous bulk samples collected so as not to exclude any organism purposefully remains the important objective of sampling. Randomness of sampling is desirable but must be tempered by the character of the outcrop area or well samples. The great danger in sampling fossil deposits lies in the tendency to collect biologically or mechanically time-averaged samples, to collect three-dimensional strata samples rather than single life-habitat surfaces, and thus to incorporate organisms from several time planes in a single sample. Great care must be taken in sampling to compensate for the fact that a few millimeters of strata commonly represents hundreds to thousands of years. Species transported into the community must be recognized, as well as inferred taxa (see the next section).

Inference

Stages of conceptualization are mental processes of induction and deduction; they are stages of progressive abstraction that move successively from data to conclusions. Inference, the first stage of conceptualization, is the process of assembling mental constructs out of the data. These groupings are closely tied to the data, but are subjective conclusions that are open to revision as new data are gathered or as dictated by the experience of the scientist.

Inferential groups of recurring taxa are derived from clustered matching coefficients and are called *taxa sets*. The similarity coefficients are calculated from taxa lists of multiple samples. Much subjective analysis is used by the paleoecologist in allocating taxa to these clusters. For instance, the functional morphology and autecology of the species are used to decide which species may have possessed compatible biologic requirements. Taphonomic analysis aids in the recognition of exotic species possibly derived from other ecological units. Environmental analysis is based upon sedimentologic and stratigraphic studies that result in depositional models. The compatibility between such models and the taxa sets is subjectively evaluated.

One somewhat controversial operation can be performed at this time which is of greater significance to paleobiological studies than to those of Holocene communities, where theoretically the entire biota can be sampled. This has been called "addition by inference" (Kauffman, 1974) and involves the addition of taxa to the known biota of the ecological unit (as determined by actual sampling) because of indirect evidence that they are a normal part of the unit structure even though they did not show up in the sampling (Fig. 4). The indirect evidence may be such things as predation marks (e.g., Kauffman and Kesling, 1960) or simply the knowledge that the ecologically dominant predators normally associated in very small population densities with such ecological units were probably there but too rare to show up in the sampling employed (Scott, 1970). Other examples would be the presence of a living food source (e.g., algae) inferred from the presence in the sample of a grazing gastropod, or the presence of a common obligate symbiont species inferred from the occurrence of its symbiotic pair species.

Interpretation

This stage of conceptualization consists of deductive reasoning. The inferred taxa sets are concluded to represent and to have been a part of past ecological units. In studies of Holocene communities, this stage is normally achieved by direct observation over a period of time. Where such observations are not possible, as in studies of bathyal and abyssal communities, for instance, the logical steps are the same as deducing paleocommunities.

At this point in the analysis, clusters of organisms that maintain their compositional integrity and statistical recurrence, and which are prescribed by a clear and relatively consistent set of environmental parameters, may be considered to represent ecological units. Their *interpretation* depends upon one additional set of

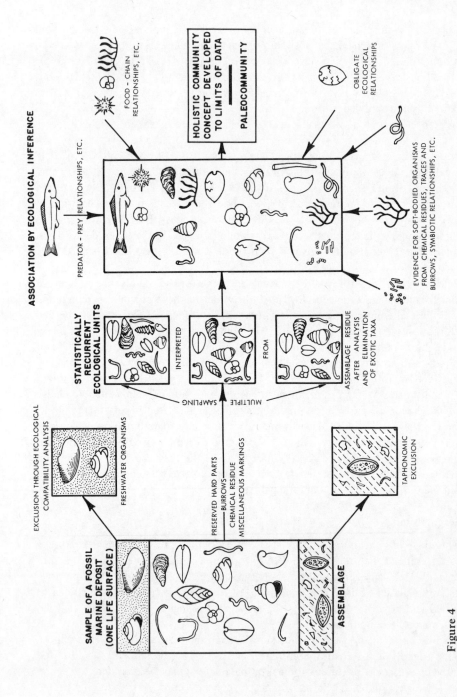

Figure 4
Outline of holistic community approach to paleoecology.

analyses, which, once completed, will allow the units to be identified, described, and classified in the hierarchy of ecological units.

The final tests and analyses to which ecological units are subjected before classification are those designed to determine the extent of interspecies relations among component taxa, the structure of the ecological unit (pattern of all interspecies relations), the energy-flow system (food-chain relations), and the three-dimensional distribution patterns, including spatial and temporal variations. These then permit an analysis of size, homogeneity of biological distribution (uniform or clustered), and ranking within the hierarchy of ecological units. Initial tests here are largely concerned with species associations to determine symbioses, food-chain relations, and ecological structure, and with population dynamics in time and space among component taxa to determine the extent of interaction among them. Some questions to be asked are what happens to the ecological unit when one population "explodes," for instance, or becomes severely restricted, or evolves a new adaptive character? Collectively, this is a measure of the size and complexity of the ecological unit, upon which classification is dependent.

The *interpretive phase* of conceptualization attempts to define interspecies relations among component organisms in the purified data clusters that result from testing at the inference level. From these relations, the structure and energy flow—the dynamics of the ecological unit—are constructed for each cluster, and relations with associated environmental parameters are determined. These characteristics are compared among clusters of data (ecological units) and the nature of unit boundary zones determined. A final test for subclustering within each unit and an analysis of size and complexity among clusters allows them to be classified in the ecological hierarchy, and identified and described as associations, communities, assemblages, or larger-scale ecological units.

Classification

Classification in ecology is concerned with ranking consistently defined ecological units into a hierarchy on the basis of the size and structural complexity of the units, diversity of the component organisms, and the nature of subclustering within the units. This is possible only after clear demonstration that ecological units exist (through the data-collecting and inferential phases of research) and a careful analysis of their structure (interpretive phase of research).

Successful classification depends upon a set of standards for comparison of each entry; for this purpose we have selected the dynamic, holistic concept of the *community* as the standard. Divisions of the community (low-level subclusters) are various kinds of *associations*, depending upon the nature of the interspecies relations; the simplest community division is the species population. Clusters of two or more communities are assemblages, seres (progression of communities in a long-term succession), ecosystems, and larger-scale units (see Fig. 2, and the right column of Fig. 3).

The classification proposed here (Figs. 2 and 3) employs a definition of terms and ordering of units that represent a combination of the principal usage among ecologists and paleoecologists (with the possible exception of "community" as used in practice) and also the original concept of each unit insofar as we can determine from the literature. Its greatest value lies in the fact that all types of ecological or partially ecological clusters of organisms recognized under a variety of names in the literature can be accommodated in this system, including paleoecological units with all their inherent problems involving preservation of the biological and ecological record. This classification presents for the first time a universally applicable language of standard terms and definitions for ecological units, and we urge its use for the sake of accuracy of communication within the science.

ACKNOWLEDGMENT

Kauffman's research was supported by a grant from the German government while he was Visiting Professor at the Institut fur Geologie und Paläontologie, University of Tübingen. This is Publication 41 of the research program, "Fossil-Vergesellschaftungen" within the Sondersforschung Bereich 53 (Palökologie) of the University of Tübingen. Financial support is gratefully acknowledged.

REFERENCES

Allee, W. C., and K. P. Schmidt. 1937 (reprinted in 1951). Ecological animal geography. John Wiley & Sons, Inc., New York, 715p.

Anderson, E. J. 1974a. Community patterns. Principles of benthic community analysis. Sedimenta IV, Comp. Sed. Lab. Univ. Miami, 3.1–3.11.

———. 1974b. Stratigraphic models: the Lower Devonian and Upper Silurian of the Central Appalachian Basin. Principles of benthic community analysis. Sedimenta IV, Comp. Sed. Lab. Univ. Miami, 11.1–11.10.

Bretsky, P. W. 1969. Central Appalachian Late Ordovician communities. Geol. Soc. Amer. Bull., 80:193–212.

Elton, C. 1927. Animal ecology. Sidgwick and Jackson Ltd., London, 209p.

Fagerstrom, J. A. 1964. Fossil communities in paleoecology: their recognition and significance. Geol. Soc. Amer. Bull., 75:1197–1216.

Forbes, Edward. 1844. Report on the Mollusca and Radiata of the Aegean Sea, and on their distribution considered as bearing on geology. Report of the British Association for the Advancement of Science for 1843, p. 130–193.

Ginsburg, R. N., and N. P. James. 1974. Spectrum of Holocene reef-building communities in the Western Atlantic. Principles of benthic community analysis. Sedimenta IV, Comp. Sed. Lab. Univ. Miami, 7.1–7.22.

Gleason, H. A. 1926. The individualistic concept of the plant association. Torrey Bot. Club Bull., 53:7–26.

Johnson, R. G. 1960. Models and methods for analysis of the mode of formation of fossil assemblages. Geol. Soc. Amer. Bull., 71:1075–1086.

Kauffman, E. G. 1967. Coloradoan macroinvertebrate assemblages, Central Western Interior, United States. *In* E. G. Kauffman and H. E. Kent, (eds.), Paleoenvironments of the Cretaceous Seaway in the Western Interior; a symposium. Colo. School Mines Publ., p. 67–143.

———. 1969. Cretaceous marine cycles of the Western Interior. The Mountain Geologist, 6(4):227-245.

———. 1973. Cretaceous Bivalvia. *In* A. Hallam (ed.), Atlas of paleobiogeography. Elsevier Publishing Company, Amsterdam, p. 353-383.

———. 1974. Cretaceous assemblages, communities, and associations: Western Interior United States and Caribbean Islands. Principles of benthic community analysis. Sedimenta IV, Comp. Sed. Lab. Univ. Miami, 12.1-12.27.

———, and R. V. Kesling. 1960. An Upper Cretaceous ammonite bitten by a mosasaur. Contrib. Mus. Paleont. Univ. Michigan, 15(9):192-248.

———, and R. W. Scott. 1974. Basic concepts of community paleoecology. Geol. Soc. Amer. Abst., 6:815-816.

Lawrence, D. R. 1968. Taphonomy and information losses in fossil communities. Geol. Soc. Amer. Bull., 79:1315-1330.

Levinton, J. 1972. Stability and trophic structure in deposit-feeding and suspension-feeding communities. Amer. Naturalist, 106:472-486.

Möbius, K. 1871. Das Thierleben am Boden der deutschen Ost- und Nordsee. Samllg. Gemeinverst. Wiss. Vorträge, 6:3-32, Hamburg.

———. 1877. Die Auster und die Austernwirtschaft. Berlin, Hempel and Parry, 126p.

Murina, V. V., V. D. Chukhchin, O. Gomez, and G. Saurez. 1973 (translation; original date, 1966). Quantitative distribution of bottom macrofauna in the upper sublittoral zone of northwestern part of Cuba. *In* A. O. Kovalevskogo, Investigations of the Central American seas, Akad. Nauk USSR, Inst. Biol. Yuznikh Movei Im. Akad.; "Naukova Dunka," Kiev, p. 242-259 (trans. by Indian Nat. Sci. Doc. Centre, New Dehli).

Odum, E. P. 1971. Fundamentals of ecology, 3rd ed. W. B. Saunders Company, Philadelphia, 574p.

Parker, R. H. 1959. Macro-invertebrate assemblages of central Texas coastal bays Laguna Madre. Assoc. Petrol. Geol. Bull., 43:2100-2166.

———. 1964. Zoogeography and ecology of some macro-invertebrates, particularly mollusks, in the Gulf of California and the continental slope off Mexico. Vidensk. Medd. Dansk Naturh. Foren., 126:78p.

Petersen, C. G. J. 1911. Valuation of the sea. I. Animal life of the sea bottom, its food and quantity. Rept. Danish Biol. Sta., 20:1-81.

———. 1913. Valuation of the sea. II. The animal communities of the sea bottom and their importance for marine zoogeography. Rept. Danish Biol. Sta., 21: 1-44.

———. 1914. Appendix to report 21. On the distribution of the animal communities of the sea bottom. Rept. Danish Biol. Sta., 22:1-7.

———. 1915. On the animal communities of the sea bottom in the Skagerrak, the Christiana Fjord, and the Danish waters. Rept. Danish Biol. Sta., 23:3-28.

———. 1918. The sea bottom and its production of fish food. A survey of the work done in connexion with the valuation of the Danish waters from 1883-1917. Rept. Danish Biol. Sta., 25:1-62.

———. 1924. A brief survey of the animal communities in Danish waters. Amer. Jour. Sci., ser. 5, 7:343-354.

Rhoads, D. C., I. G. Speden, and K. M. Waagé. 1972. Trophic group analysis of Upper Cretaceous (Maestrichtian) bivalve assemblages from South Dakota. Amer. Assoc. Petrol. Geol. Bull., 56:1100-1113.

Ricklefs, R. E. 1973. Ecology. Chiron Press, Inc., Newton, Mass., 861p.

Schäfer, W. 1972. Ecology and paleoecology of marine environments (translation). C. Y. Craig (ed.), Trans. I. Oertel, University of Chicago Press, 568p.

Scott, R. W. 1970. Paleoecology and paleontology of the Lower Cretaceous Kiowa Formation, Kansas. Univ. Kansas Paleont. Contrib. Art. 52 (Cretaceous 1), 94p.

Sutton, R. G., A. P. Bowen, and A. L. McAlester. 1970. Marine environments of the Upper Devonian Sonyea Group of New York. Geol. Soc. Amer. Bull., 81:2975–2992.

Thorson, G. 1957. Bottom communities. *In* J. W. Hedgpeth (ed.), Treatise on marine ecology and paleoecology, v. I, Ecology. Geol. Soc. Amer. Mem., 67:461–534.

——. 1966. Some factors influencing the recruitment and establishment of marine benthic communities. Neth. Jour. Sea Res., 3:267–293, Den Helder.

Valentine, J. W. 1973. Evolutionary paleoecology of the marine biosphere. Prentice-Hall, Inc., Englewood Cliffs, N.J., 511p.

Walker, K. R. 1974. Community patterns: Middle Ordovician of Tennessee. Principles of benthic community analysis. Sedimenta IV, Comp. Sed. Lab. Univ. Miami, 9.1–9.9.

Whittaker, R. H. 1970. Communities and ecosystems. Macmillan Co., London, 158p.

Ziegler, A. M. 1965. Silurian marine communities and their environmental significance. Nature, 207:270–272.

——. 1974. The community technique. Principles of benthic community analysis. Sedimenta IV, Comp. Sed. Lab. Univ. Miami, 1.1–1.9.

——, K. R. Walker, E. J. Anderson, E. G. Kauffman, R. N. Ginsburg, and N. P. James. 1974. Principles of benthic community analysis. Sedimenta IV, Comp. Sed. Lab., Univ. Miami, 172p.

Trophic Classification of Benthic Communities

Robert W. Scott   Amoco Production Company
                                   Research Center

ABSTRACT

Classification of species into communities has been based on character species, recurring sets of species, dominant species, the physiognomy of component species, and habitat features. The community-unit theory explains that species are co-occurring because they have similar physical and biological needs. The theory of individualistic dissent states that species occur independently of one another.

Comparison and classification of communities are based upon trophic structure. Suspension feeders represent the grazing food chain, and detritus feeders make up the detritus chain; predators terminate both. These three feeding types are combined to form the feeding habit ternary diagram on which the percentage of species in each is plotted. The companion substrate niche diagram indicates the proportions of epifaunal to infaunal suspension feeders and vagrant detritus feeders. Together these two diagrams characterize the trophic structure of a community.

Trophic structure of contemporaneous communities changes in response to stability, resource level, and spatial variability. Deeper, open-sea mud substrates and lower shoreface environments support vagrant-infaunal, detritus-suspension feeding communities. Middle and upper shoreface environments support infaunal suspension feeding communities.

INTRODUCTION

Classification of natural communities has concerned ecologists since Humboldt (1805) defined plant communities by their growth forms as well as by associations of characteristic dominant species. In 1844, Forbes classified marine faunas into depth-influenced groups. Since these pioneer studies, many biotic communities have been identified and named for their characteristic taxa. Paleoecologists have continued this tradition of naming communities after taxa so that a list of paleocommunities reads like a faunal list.

Recognition of discrete sets of recurring species is an important basic step in every study. However, one cannot assume outright that taxa occur in interrelated groups. Botanists have developed two theories to explain the distribution of species. The *association-unit theory* or *community-unit theory* predicts that species groups are cooccurring because they are interrelated (Whittaker, 1956). This theory states that sets of species are not randomly associated and occur together in nature because of similar physical and biological needs. The *theory of individualistic dissent* (Ramensky, 1924; Gleason, 1926) predicts the independent distribution of species and is based on the principles of vegetation continuity and independently ranging species. Some ecologists have presented evidence showing that in certain environments species seem to have independent, disjunct or overlapping ranges rather than closely matched ranges (Stephenson, 1973). Whittaker (1962, p. 78-83) gives many examples of continuous plant distributions and groups of species with gradational and arbitrary boundaries. Although few paleoecologists have completely adopted this view, many have noted the broadly gradational nature of boundaries between paleocommunities (Bretsky, 1969a; Johnson, 1962, 1973; Scott, 1974). Numerical techniques such as binary matching coefficients and the chi-square statistic have been successfully used to recognize recurring sets of taxa (Johnson, 1962; Kaesler, 1966; Valentine and Peddicord, 1967; Fox, 1968; and Mello and Buzas, 1968). Consequently, the major thrust of paleocommunity classification has been the recognition of distinctive, recurring sets of taxa (Ziegler, 1965; Speden, 1966; Scott, 1970a; and West, 1972; among others).

The purposes of this paper are to review classification schemes used in community studies and to show some possible theoretical relations among environmental stability, community stability, and trophic classification concepts. This review is not exhaustive, but will attempt to point out certain major trends and themes used in dealing with the more common types of communities preserved in the geologic record. Botanists and ecologists have faced many of the same problems as have paleoecologists, and hopefully their experiences will show us which are fruitful paths and which are dead ends.

CONCEPTS OF CLASSIFICATION

Attention to classification marks a period of maturation in scientific disciplines. Paleoecology has had to deal with taphonomy and information loss due to incomplete preservation. The science is now at the stage of describing and classifying paleocommunities. This step precedes the development and application of unifying theories. Neither has ecology yet reached this next stage. However, current trends in concepts of stability of populations and communities (Margalef, 1968; Lewontin, 1969) suggest that such theories may be forthcoming.

The philosophy and procedures of classifying biotic communities have been thoroughly discussed by Whittaker (1962, p. 102-123). Classification, a process of confrontation between the classifier and the classified, is an ordering process influenced by both the properties of the objects and by the predilections of the scientist. The scientist must select the properties and features to be observed and used for discrimination of groups. This choice depends not only on the sophistication of the observation technology but also upon the scientist's concept of which properties have significance in relation to his theories. This choice is a judgment because "As a totality the environmental complex is unknowable and inexpressible" (Whittaker, 1962, p. 103). Thus, objects are allocated to classes by their perceived properties, not by the totality of their properties.

Community classes are groupings of fossil deposits that share some characteristics, such as dominant or identifying species. Such classes are mutually exclusive; together the classes are exhaustive and encompass all such fossil deposits. Boundaries are arbitrarily but objectively drawn even though most are gradational rather than discontinuous. Characteristic properties are chosen that are repetitive in several fossil deposits. These properties also result in relatively distinct groups to which samples can be allocated with a minimum of confusion.

CLASSIFICATION OF ORGANISMS INTO COMMUNITIES

The major trend in the classification of communities has been the recognition and naming of distinct suites of taxa according to the community-unit theory. This involves the classification of species into communities. Other trends have been to classify communities by their physiognomy, stage of succession, climate, soil, total habitat, community stability, trophic structure, and diversity (Table 1).

Character Species

The recognition of groups of characteristic or differential plant species, called *associations*, has been the basic goal of the Braun-Blanquet school of botanists (Whittaker, 1962, p. 149). Floristic composition or plant species composition is the nuclear unit in the definition and classification of terrestrial communities according to this school. Braun-Blanquet (1913) classified Alpine floras into groups of species having narrow environmental tolerances. The character species was restricted to the association, although not necessarily constant in each stand of the association. *Fidelity* described the degree of restriction to an association.

Critics of this school pointed out that in many areas no characteristic and restricted species could be found; plant species were interdispersed. In some places

Table 1
Criteria for defining communities.

Character Species	Recurring Species	Dominant Species	Physiognomy	Habitat	Succession
Properties					
Association: Floristic composition of restricted, not constant taxa	Taxa set, biofacies, "association": Species groups found in many samples	Dominance-type, biofacies "association": Common taxa conspicuous by their abundance	Formation, biome, "parallel com.": Group of communities w/similar growth forms and related environ. conditions: Feeding types Substrate types Shell forms	Multifactorial or landscape features: Geography Depth Climate Soil	Dynamic groups of species influenced by dominants or by environmental changes
Communities					
Holocene (Newell et al., 1959) Devonian (Johnson, 1974) Silurian (brachiopod, Ziegler, 1965; Ziegler et al., 1968) Cambrian (trilobite, Jago, 1973)	Holocene (ostracodes, Kaesler, 1966); (forams, Mello and Buzas, 1968), (mollusks, Warme, 1969, 1971) Cretaceous (Scott, 1970a, 1971, 1974) Pennsylvanian (Johnson, 1962; West, 1972) Siluro-Devonian (Boucot, 1970) Ordovician (Fox, 1968)	Devonian (Sutton et al., 1970) Devonian (Johnson, 1974) Ordovician (Walker, 1972b; and Walker and Ferrigno, 1973; Bretsky et al., 1969; Bretsky, 1969a)	Holocene (Bandy and Arnal, 1960, Phleger, 1960) Cretaceous (Scott, 1972, 1974; Rhoads et al., 1972) Pennsylvanian (West, 1972) Ordovician (Walker, 1972b; Walker and Ferrigno, 1973)	Holocene (Parker, 1956, 1960, 1964) Cretaceous (Scott, 1970b; Rhoads et al., 1972) Pennsylvanian (West, 1972) Devonian (Sutton et al., 1970) Ordovician (Walker, 1972b; Bretsky, 1969b)	Holocene (Clements, 1916; Clements and Shelford, 1939); Cretaceous (Scott, 1970a; Scott et al., 1973) Silurian (Halleck, 1973) Ordovician (Walker and Alberstadt, 1974)

fidelity was very low. Other critics felt that the association approach ignored too much relevant information on species abundance and habitat.

Although the Braun-Blanquet approach has not been applied directly to paleocommunities, character species have been used to define ancient communities in a few studies. Ziegler (1965) and Ziegler et al., (1968) used characteristic species of brachiopods to define Silurian benthic communities. Other parameters, such as diversity, relative abundance, and substrate relations, supplemented differentiation and distinction among communities. In another example, Cambrian agnostid trilobite communities (Jago, 1973) have been defined by the presence of certain taxonomic groups, either by themselves or in association with other types of organisms. More often this approach has been used in defining biofacies. Hedgpeth (1957b, p. 47), among others, has used the term "association" for both characteristic plant and animal species occurring together in a community.

Recurring Suites of Taxa

The concept of recurring suites of abundant taxa has been used most frequently in studies of ancient communities. Some botanists have used recurring suites of dominant species to define generalized plant communities (Crocker and Wood, 1947). However, this technique has been used mostly by marine zoologists studying benthic communities where boundaries between communities are more gradational than between stands of trees. Petersen (1913) pioneered this technique in his studies of Danish benthic communities. Thorson (1957) discussed the relation between these statistical units and counterpart animal communities.

Recurring sets of species generally are based upon quantitative and uniformly collected samples, such as Petersen grab samples, bedding plane quadrats, or bulk samples of many specimens. Species groups found repeatedly and in significant abundances in many samples are then taken to represent the total community. These groups of recurring species can be recognized empirically, but the bulk, uniform sampling method lends itself to quantitative treatment, such as binary similarity coefficients (Kaesler, 1966; Mello and Buzas, 1968), or chi-square statistics (Johnson, 1962). This technique has been used by many paleoecologists to define modern and ancient benthic communities (Newell et al., 1959; Johnson, 1962; Valentine and Mallory, 1965; Kaesler, 1966; Valentine and Peddicord, 1967; Fox, 1968; Mello and Buzas, 1968; Warme, 1969, 1971; Scott 1970a, 1974; West, 1972).

Assumptions and techniques of defining recurring taxa sets, usually called biofacies or associations (Kauffman and Scott, this volume), have been carefully analyzed by Johnson (1962) and Kaesler (1966). Normally, samples containing many species and species found in more than three to four samples tend to form well-defined clusters in Q-mode and R-mode analyses. Low diversity samples and species of restricted occurrence form spurious or misleading groups. The technique of cluster analysis is powerful in that it defines recurring sets of taxa; however, the nature of the mathematics results in cluster formation even when the species or samples are part of the same group, biotopes, biofacies, association, or taxa set.

Consequently, the technique must be backed up by other relationships, such as stratigraphic, lithologic, and geographic.

Dominant Species

Plant ecologists have used dominant species as an alternative method of defining and classifying species into community groups. Whittaker (1962, p. 141) defines dominant species as "major species of the uppermost stratum of the community, sometimes of a lower stratum." Vegetation stands, according to this technique, can be characterized by one or two conspicuous and abundant species, many of which have wide ecological tolerances. An advantage of classification by dominance types is that useful and easily recognizable groups can be defined. Whittaker (1962) illustrates seven different patterns of dominance types and discusses various units of subdivision based upon patterns of relative abundance. However, several problems arise in the use of dominant species to define communities. Some dominant plants have different characteristic associates in different stands indicating different community structures, and some species are dominant in more than one environment (Whittaker, 1962, p. 142). Similar problems when using dominance types to define benthic marine communities are discussed by Bloom et al., (1972). Some organisms were dominant outside of their communities or were not consistently dominant within multiple samples of communities recognized by recurring suites of taxa.

Dominant species were used to define four of seven Devonian communities in New York (Sutton et al., 1970). No clearly dominant species were present in the other biofacies. The stratigraphic-geographic ranges of these communities were closely correlated with lithologic and, hence, substrate and other environmental conditions. Walker (1972a) proposes five techniques to measure *biovolume*, the relative abundance of species in statistical samples. Walker claims that biovolume is proportional to calcified biomass, which is only part of community biomass and unattainable in paleocommunities. Species with high relative and rank abundances characterize certain Ordovician shelf and reef communities (Walker, 1972b; Walker and Ferrigno, 1973; Walker and Bambach, 1974). Mean rank abundances of bryozoan genera distinguished six communities in flank and core facies of an Ordovician reef (Walker and Ferrigno, 1973). Rank abundances from sample to sample of the flank facies were fairly consistent; but in the core facies, rank abundances fluctuated widely between the basal and upper parts, which suggests some type of succession (Walker and Alberstadt, 1974). Conditions on the flank presumably were more constant so that successional changes were not as marked. In summary, species dominance as measured by mean rank abundances and relative abundances is a useful tool to classify species into community groupings; however, most paleoecologists combine this technique with others.

Physiognomy

The physiognomy of component species has been a major approach to plant communities. Physiognomy refers to the overall character of the plant community; its basic unit is the *formation*. Botanists define formations by the dominant life form or growth form species (Whittaker, 1962). These growth forms are particularly adapted to the specific environmental conditions. The *growth form* of plants refers to the characteristic plant morphology: leaf form, size, texture, and arrangement, ratio of evergreen to deciduous trees, and ratio of woody to herbaceous growth. In contrast, the *life form* of a flora refers to the character of the buds and seeds. Communities so defined are closely linked with the environment; wherever a similar environment is found, the same growth forms will adapt to it. Consequently, formations can recur from continent to continent on a worldwide scale (Whittaker, 1962, p. 7-8).

An analogous approach was taken by Thorson (1957) in defining "parallel" animal communities that inhabit the same benthic marine substrate in different parts of the world. He found that different species, usually of the same genera, characterize the parallel *isocommunities*. Isocommunities possess similar morphological structure and ecological characteristics, such as feeding types, growth rates, population dynamics, productivity, and substrate relations. Thorson's parallel communities were related to environmental factors of salinity, depth, substrate type, and variability of the environment. These parallel communities occupy the relatively uniform level sea floor; these level-bottom communities are discussed by Johnson (1964). Parallel communities exist also in the intertidal zone. Stephenson and Stephenson (1950) described four "zones" in the supratidal to intertidal environments of the Florida Keys. The zones are also recognized in the Bahamas, where some species were replaced because of geographic or local climatic differences (Newell et al., 1959).

The concept of growth form as used in paleoecology is much broader and covers many more facets. The grouping of foraminifers by test structure and habit, arenaceous-calcareous and planktic-benthic, has been extremely successful in delimiting depth zones (Bandy and Arnal, 1960; Phleger, 1960).

Most paleoecological studies, however, define communities by one of the preceding techniques and then discuss the growth-form characteristics of the important taxa (Johnson, 1962; Ziegler et al., 1968). Walker (1972b) and Walker and Ferrigno (1973) used substrate relations and trophic structure to characterize communities already defined by dominant species. West (1972) used four feeding types and three modes of life to characterize Pennsylvanian marine benthic communities. Scott (1972) has proposed a classification scheme of communities already defined by other techniques; this system is based upon three factors: feeding habit, substrate niche, and diversity. Scott's system distinguishes communities of different Early Cretaceous environments in the southern Western Interior (Scott, 1974). Rhoads et al. (1972) also used trophic structure and substrate relations to differentiate Late Cretaceous communities of the Western Interior. Trophic structure differentiates taxonomically defined Jurassic communities and enables their comparison with Late Cretaceous communities (Wright, 1974).

In summary, the physiognomic characters of ancient communities include substrate niches, feeding habits, and shell forms. These features can successfully discriminate among paleocommunities consisting of species groups defined by other techniques, such as recurring species, character species, or dominant species.

Habitats

Communities have also been classified by habitat conditions or multifactorial landscape features (Whittaker, 1962). This approach is based on the assumption that organisms and environment form a functional unit. Systems of classification are based upon features of both the organisms and habitat, not just vegetation properties. Whittaker summarizes the concept (1962, p. 59-60):

> The ideal of this multi-factoral [sic] or landscape approach to classification is an integrative conception that takes into account all properties of ecosystems, and evaluates them in relation to one another to recognize these groupings of ecosystems based on most critical factors and with the broadest possible significance.

In some classifications certain environmental properties are emphasized over others. Climate and soil type have been used effectively to classify terrestrial communities. Whittaker (1962, p. 56) gives some key references. An integrated classification of terrestrial and freshwater habitats was proposed by Elton and Miller (1954). Their classes were "broad scale" habitat units "in which the degree of integration and intensity of dynamic action are at a high level" (p. 479). Population relations, species interaction, and energy flow are concentrated in these habitat units. Their intent was to recognize these units by descriptive characters, and at the same time, acknowledge the levels of interaction among different communities in an ecosystem. Within these habitat units, interaction was more intense than between them. Elton and Miller defined classes by means of vegetation and life form, called *formation types*, which subdivided the major habitat systems into (1) terrestrial, (2) aquatic, (3) aquatic-terrestrial transition, (4) subterranean, (5) domestic, and (6) general. "Domestic" includes all habitats strongly influenced by man; "general" includes small, yet distinct communities such as rotting logs and dung. The authors do not apply this scheme to marine habitats, although they suggest that it could be done.

In the study of marine communities a distinction is made between habitat, organism community, and biocoenosis (Newell et al., 1959, p. 198). A *habitat community* consists of the set of organisms in a specific environment. Examples in the Bahamas are rocky shore, reef, rock pavement, unstable sand of outer platform, unstable oolite sand bar, stable sand of shelf lagoon, and muddy shelf lagoon (Newell et al., 1959). Boundaries of *organism* or *biotic communities* (i.e., a recurring set of species), partly coincide, and within most habitat communities several organism communities can be recognized. On the Gulf Coast, assemblages of organisms are clearly related to well-defined environments (Parker, 1956). A *biocoenosis* is a dynamic community defined by the mutual interaction and interdependency of species; the species occur together because of their strong symbiotic requirements. For a synoptic classification of communities by environments see Parker (this volume); for a classification of marine habitats, see Hedgpeth (1957a). In general, ancient habitats can be recognized primarily by sedimentological parameters, and paleocommunities need to be studied independently lest the inferred relation between substrate and organisms becomes a "ruling theory" rather than an hypothesis.

Miscellaneous Concepts

Several other approaches to the classification of communities need to be mentioned. The ordination technique separates species populations within a two-dimensional framework (Shaffer and Wilke, 1965). The position within the framework is based upon similarity coefficients such as the Jaccard coefficient. Population patterns relative to environmental or other gradients can be delineated within continuously integrating communities. It is a powerful technique that has been little used in paleoecology.

Community structural parameters, such as dominance, diversity, homogeneity, equitability, and environmental stability, can be used to define general categories supplemental to the taxonomic classification base. Species diversity is commonly measured by a coefficient such as the information theory coefficient, H, of Shannon (Beerbower and Jordan, 1969). Alternatively, Sanders (1968) expressed diversity by means of rarefaction curves constructed from plotting the cumulative number of species and cumulative number of specimens in successively less diverse samples. Effects of sample size on diversity tend to be negated by differing slopes, which reflect differing diversities. This approach has helped in characterizing Holocene communities off the Mississippi delta (Stanton and Evans, 1972), coral communities in the Red Sea (Loya, 1972), and Cretaceous bay to shoreface communities (Scott, 1974). Rather broadly defined categories like species-diverse and species-dominant communities (Scott, 1972) should be replaced by more precisely defined units. This approach would be useful as a supplement to others.

Equitability equations (Beerbower and Jordan, 1969) distinguish equally diverse communities in which species dominance differs. Homogeneity (Stanton and Evans, 1972) is the average of matching coefficients of all pairs of species within a community; it is the average bonding of the species. As diversity increases, the number of pairs increases and the average bonding decreases, so that homogeneity is roughly the inverse of diversity. Low-dominance and high-dominance communities (Margalef, 1960) are distinguished by the relative abundance of species and are related to community diversity. Relative abundance is an approximate measure of community order and niche size (Deevey, 1969). These parameters, however, do not lend themselves to the definition of absolute classes, but are useful in comparing communities in a study.

Two major types of communities, *physically controlled* and *biologically accommodated*, reflect the relative impact upon community structure of physical-habitat forces versus biological-adaptation forces (Sanders, 1968). Physically controlled communities develop within unstable environments where conditions fluctuate widely and unpredictably. In contrast, biologically accommodated communities characterize environments that are stable for long periods of time. Hedgepeth (1957b), Slobodkin and Sanders (1969), and Bretsky and Lorenz (1970) relate stability to species diversity and suggest that, in general, diversity is directly related to environmental predictability.

Biotic succession predicts that communities will replace each other in an orderly sequence controlled by the type of dominant species. Clements (1916) championed the view that succession results in the growth and development of a formation, and each growth stage was identified by a seral plant community; the final stage was the climax community. Critics have challenged both the analogy of the formation with ontogeny of an organism and the monoclimax theory. Some successional sequences may lead to one of several climax communities and some seral stages may, under certain circumstances, be relatively permanent. Biotic succession has been little studied in paleocommunities except in reef environments (Lowenstam, 1957; Newell et al., 1953; Nelson, 1973; Kauffman and Sohl, 1974).

TROPHIC CLASSIFICATION

Communities recognized and defined by recurring suites of taxa, characteristic species, or any other method can be compared using any of several structural properties. The act of comparison creates a need for the standardization of data and of the treatment of structural properties. Such a classification should be based upon observable characteristics that can be uniformly defined. These characteristics should be recognized independently of, but also be suitable for integration with, current theories. Feeding habits, substrate niches, and diversity are such structural properties. They are readily related to trophic and stability theories and serve to classify communities rather than species. Such classification will enable the comparison of structural components of communities consisting of different species groups.

Trophic Theory

Trophic relations deal with the nourishment of the community. The trophic structure of a community consists of the pattern of feeding habits of the species, which transmits energy through the community and results in the metabolism and growth of species populations.

$$food \longrightarrow feeding\ habits \longrightarrow energy\ flow$$
$$\ \longleftarrow metabolism\ and\ growth \longleftarrow$$

Feeding habits are separated into two food chains based upon fundamentally different food sources: the grazing chain is based on green plants and the detritus chain on dead organic matter (Odum, 1971) (Fig. 1). Within each food chain, species of various feeding habits can be arranged into trophic levels at each stage of energy transfer. The energy loss at each transfer is usually depicted by means of the Eltonian food pyramid (Odum, 1971, p. 79-83). But because many species can feed at more than one trophic level, the complex feeding relations among species comprise the food web. Thus, a basic distinction must be made between feeding habit and trophic level. Feeding habit is what a species does and trophic level is the position of an organism in the steps of energy transfer. Odum (1971, p. 63, 66) specifically states that trophic-level classification does not classify species, only their position. The trophic level of a species may vary during ontogeny because of secular environmental changes, or because of other factors.

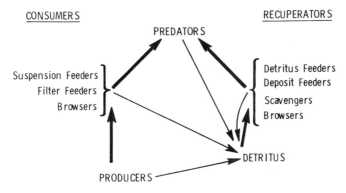

Figure 1
Energy-flow pathways (heavy arrows) and feeding habits within the grazing food chain of consumers and the detritus food chain of recuperators. (Thin arrows indicate donors of detrital organic matter.)

Feeding habits, on the other hand, are divided into trophic categories (Walker and Bambach, 1974). A trophic category is defined by the character and location of food resources and the selection and acquisition of food. A very complete classification of trophic categories is given by Walker and Bambach (1974, Table 6). A comparable classification is given by Valentine (1973, Table 5-1). The major categories are (1) suspension feeders, (2) deposit feeders, (3) browsers, (4) carnivores, (5) scavengers, and (6) parasites. Imbrie (1964) proposed a similar classification of species that also included mode of movement or attachment. Because trophic structure depends upon the type of energy resources available and the partitioning of these resources, trophic structure is sensitive to factors controlling energy resources, such as environment, geography, and evolution.

Energy within a trophic system is measured either by standing crop per unit area or by energy fixed per unit area per unit time at each trophic level (Odum, 1971, p. 79). Trophic pyramids consist of (1) numbers of individuals, (2) biomass based on dry weight or caloric value, or (3) rate of energy flow. None of these factors is readily or accurately measured in ancient communities. Although numbers of individuals may be faithfully preserved in in-place communities, their generation in the community may not be obvious. Walker's biovolume measures may be related to biomass, but the relation is not always direct. Feeding habit structure is the most direct aspect of the trophic structure to be studied by the paleoecologist, but its problems are discussed later.

Controlling Factors

Theories that relate species diversity and trophic structure to environmental processes are speculative at this time. Many additional data are needed to test these theories before they can be used with any degree of confidence in interpreting paleocommunities. Paleocommunity studies also serve to test these theories by integrating results from autecologic and sedimentologic analyses. The following discussion is simply a summary of the present theoretical framework, which needs much testing.

Trophic structure and community diversity are closely related to the major community-controlling factors of stability of the community and of the environment, spatial variability, and resource level (Fig. 2). Slobodkin and Sanders (1969) have suggested that species diversity is closely regulated by environmental stability. By this they mean the regularity, predictability, and range or severeness of environmental changes. High species diversity is maintained in predictable environments where tolerance limits are rarely reached. Low diversity characterizes highly unpredictable environments; as the limits of tolerance are approached, diversity is suppressed even further. This relation between diversity and environmental predictability or stability seems to be reflected in some Paleozoic communities (Bretsky and Lorenz, 1970) and in some Early Cretaceous communities (Scott, 1974), as well as in Holocene communities (Johnson, 1973). Environmental stability must be distinguished from community stability. A community is stable as long as population structure, rank dominance, density, diversity, and trophic structure are maintained. Community stability may be a result of environmental stability, however. A stable community requires a stable energy flow, which may be partitioned through several trophic levels and result in increased diversity (Valentine, 1973, p. 292). In some environments, complex food webs appear to be directly related to high diversity (Bretsky, 1969b). Therefore, stable communities in stable environments are

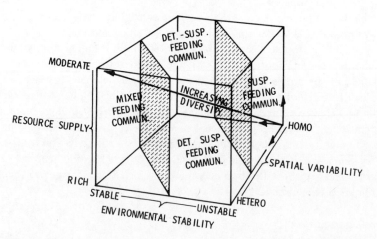

Figure 2
Relations of community diversity and trophic structure with environmental stability, spatial variability, and resource supply.

diverse and have a complex trophic structure. In unstable environments some communities are unstable and have low diversity, but not necessarily simple trophic relations. However, Bretsky (1969b) has suggested that some nearshore communities in unstable environments are stable and persist through several geologic periods because no other species adapt to the rigorous environment.

Level and stability of resource supplies also affect community diversity and trophic structure. Where food supplies are stable but low, K-selection (for carrying capacity of the environment) produces a trophic structure efficiently utilizing available energy (MacArthur and Wilson, 1967). Unpredictable food supplies, on the other hand, although occasionally rich, result in r-selection or selection of species capable of rapid reproduction during optimum times. If resource levels are very rich, such as in eutrophic lakes or upwelling currents, a few species tend to reproduce rapidly at the expense of others and diminish overall diversity (Valentine, 1971). Environments having stable and relatively low amounts of resources sustain the most diverse, highly specialized communities, so that complex trophic structures develop. Unstable and periodically rich resources sustain very low diversity and generalized populations, and the trophic structure is simple. These relationships have been investigated in terrestrial communities and are inferred rather tenuously for marine environments. Experiments are badly needed.

The third controlling factor is spatial variability. The uniformity or variability of substrate conditions, sunlight penetration, wave current energies, and rates of change in water depth are a few factors that characterize spatial variability on a macroscale. Homogeneous environments usually contain less diverse communities than heterogeneous environments (Bretsky and Lorenz, 1970). Valentine (1973, Fig. 7-4) has combined spatial variability, resource level, and stability into a block diagram and has contoured this three-dimensional space by diversity. He associates lowest diversities with lowest spatial variability, resource stability, and resource levels. Highest diversities correspond with highest space variance and resource stabilities and intermediate resource levels. If trophic structure can be correlated with these same factors and with diversity, the most complex structures would develop under conditions of greatest spatial variability and resource stability, as well as intermediate resource levels. Less developed trophic relations, then, would match the same conditions as lowest diversity. These relations are shown in Fig. 2.

Rather simple trophic relations are also a function of the evolutionary stage of the biosphere (Tappan, 1971) and the stage of biotic succession (Sanders, 1968). The first organisms to immigrate into a new environment will have feeding habits near the base of the food chain and will be on the lower trophic levels. The first organisms to appear in the fossil record of the biosphere are producers, browsers, and detritus feeders. Only with time do species evolve feeding habits enabling them to occupy higher trophic levels. Finally, substrate type strongly influences the type of food chain and also the trophic structure that can develop. Hard substrates and coarser-grained substrates tend to support food chains based on green plants, and muddy substrates tend to sustain detritus-based chains (Purdy, 1964; Sanders, 1958).

In summary, community trophic structure appears to be closely related to the dynamic properties of the environment, just as species diversity is; it is also related to substrate conditions. Consequently, trophic structure can be used as both a predictor of the environment and as a basis for classification of communities.

Fidelity of the Fossil Record

Several important studies have shown that significant parts of the community species composition are lost between the moments of death and discovery (Johnson, 1964; Lawrence, 1968; Herm, 1972). Less than 25 percent of the species in Oligocene oyster communities is preserved (Lawrence, 1968). And as Stanton (this volume) shows, this loss can alter the reconstruction of trophic structure significantly. This necessitates the distinction between actual or realized trophic structure and preserved trophic structure. The *actual trophic structure* exists while all species populations actively occupy the habitat together. Each species has attained its realized niche (Valentine, 1973). The *preserved trophic structure* consists only of those species found as fossils or represented by traces. In paleocommunities only relic properties can be studied: diversity, substrate relations, population dynamics, and density. It is only when these properties are interpretable in light of ecological theory that they take on ecological meaning. If so little of the community is preserved that it is not interpretable, synecological analysis is virtually impractical. One cannot assume that a given fossil deposit is even representative of a community, much less that the preserved trophic structure accurately reflects the actual structure. Each conclusion must be tested along the lines suggested by Kauffman and Scott (this volume). Once a suite of species is related to an environment by their recurrence with a specific lithofacies, the feeding habits can be analyzed and the preserved trophic structure described. A reliable comparison of trophic structures can be made of communities comprised of similar taxonomic groups.

Feeding Habit–Substrate Niche Classification

Samples of fossil deposits as well as species of taxa sets (association or biofacies) can be classified by the percentages of species in each feeding habit and substrate niche (Scott, 1972). The methodology of reconstructing feeding habits is in a primitive state and much work is needed. However, feeding habits of many fossils are known in general terms because they possess typical morphologies analogous to feeding paradigms (Rudwick, 1968; Stanley, 1970) and because of inferences drawn from Holocene relatives. Many details about the feeding of some modern organisms are yet unknown, and will never be known about some fossils. However, generalized feeding habits of some groups are known (Valentine, 1973; Walker and Bambach, 1974).

By selection of feeding categories and combinations of others, three main feeding-habit groups have proved useful in the classification of ancient communities: (1) suspension feeders, (2) detritus feeders, and (3) predators. *Suspension feeders* remove from the water small particles such as phytoplankton and zooplankton. As such, many are first- or second-level consumers of the plant-based food chain in the marine ecosystem. These species function as the interface between the planktic and benthic communities. Usual feeding mechanisms are flagellae, ciliated lophophores and ctenidia, and tentacles. Common ancient suspension feeders were sponges, many smaller anthozoans, hydrozoans, and stromatoporoids, bryozoans, brachiopods, many bivalves, some gastropods, some annelids and crustaceans, pelmatozoans, and graptoloids. Scott (1972) combined suspension feeders with herbivores to form first-level consumers, but this is an inaccurate designation because many herbivores also feed on detritus (Walker and Bambach, 1974).

Detritus feeders consist of deposit feeders, which swallow or scrape smaller organic particles and organic-rich sediment grains from within the sediment, and scavengers, which eat larger particles and dead animals upon or within the sediment. Both serve to recycle organic matter. This category grades into the predator or carnivore category in some crustaceans and echinoids that feed on both dead and live organisms. Other detritus feeders common as fossils are some gastropods and bivalves, scaphopods, some annelids, ophiuroids, and holothurians. Detritus feeders also grade into browsers that feed upon both detritus and live plants. Browsers or herbivores are first-level consumers that mainly scrape, rasp, or chew live algae and other plants. Browsers can be grouped with the detritus feeders to form the second major feeding-habit group because they feed on organic matter upon the sediment and some also feed upon detritus. Another reason to ally browsers with detritus feeders is because many Paleozoic archeogastropods may have fed in one or the other or both ways, and their precise food resources may never be known. Common fossil herbivores are Amphineura and gastropods.

The third major feeding habit is the *predator* or *carnivore* class. These organisms capture live prey by active search and seize techniques or by passively waiting for prey to pass within seizing range. The prey normally are nektic or benthic low-level consumers within either food chain, rather than plankton or larvae. Some common fossil predators are larger anthozoans, cephalopods, and many gastropods, some annelids and crustaceans, asteroids, ophiuroids, and some echinoids. Parasitic organisms are important in the trophic system but are uncommon or unrecognizable in the fossil record.

The three major feeding-habit categories form the apices of the feeding habit triangle, which is the basis of a trophic classification of ancient communities (Fig. 3). Other combinations of feeding categories are possible, as is the separation of browsers from detritus feeders. The choice depends upon the type of organisms studied and the needs of the classification. Likewise, the decision to divide the area into fields and the location of the boundaries are based on empirical data. The suggested fields have worked well with Cretaceous communities. A *suspension-feeding community* consists of more than 80 percent suspension-feeding species. A *detritus-suspension community* is composed of between 50 to 80 percent suspension feeders, between 10 to 50 percent detritus-browsers, and fewer than 10 percent predators. A *mixed community* differs by having subequal numbers of detritus-browsers and predators. A *predator-suspension community* contains fewer than 10 percent detritus browsers and between 10 and 50 percent predators. A *predator community* is comprised of no more than 50 percent of either suspension or detritus-browser feeders. A *detritus community* consists of no more than 50 percent of either suspension or predator feeders.

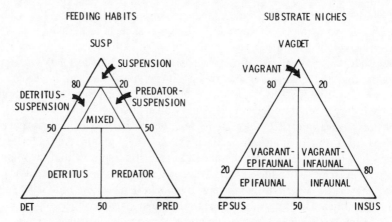

Figure 3
Classification of trophic categories. Susp, suspension feeders; det, detritus feeders; pred, predators; vagdet, vagrant detritus feeders; epsus, epifaunal suspension feeders; and insus, infaunal suspension feeders.

The substrate niche of a species should be included in a trophic classification because the relation of an organism to the substrate influences the food resources available to it (Walker and Bambach, 1974). In fact, some trophic groups are defined by relation to substrate as well as feeding mechanism (Walker, 1972a). Again, three substrate relations are most commonly found in the fossil record (Fig. 3). Sessile infaunal organisms live within the sediment, and some are mobile chiefly when disturbed; sessile epifaunal organisms are upon the substrate; and vagrant organisms actively peruse the substrate both within and upon it. Most suspension feeders are sessile and dwell within or upon the substrate. These categories can be subdivided further if other distinctions need to be made, such as rapid and slow, or shallow and deep burrowers, nestlers, spiny epifauna, cemented, byssal, rooted, or swimmers. These distinctions may be useful in dealing with communities dominated by sessile forms, and can be displayed by means of separate graphs or ternary diagrams.

Field boundaries are empirical in the substrate-niche diagram, as with the feeding-habit diagram. Five categories have proved useful. An *infaunal community* contains fewer than 20 percent vagrant organisms and no more than 50 percent epifauna. An *epifaunal community* is similar but contains no more than 50 percent infauna. A *vagrant-infaunal* group is comprised of 20 to 80 percent vagrant animals and more infauna than epifauna; a *vagrant-epifaunal community* has more epifauna than infauna. A *vagrant community* consists of more than 80 percent vagrant animals. These substrate-niche class names are used as adjectives to precede the feeding-habit class names; for example, a vagrant-epifaunal, detritus-suspension feeding community. Conceivably, these names could be contracted to "vagepdetsusp," or some equally undecipherable word. Let us not follow the carbonate petrographers into a limbo of words.

Sources of Data

Different methods of handling data in the classification system would produce different results. A choice between plotting numbers of species or numbers of specimens must be made; the source of data may be samples of fossil deposits, a complete list of species in all deposits of a given taxa set, a list of characteristic species, or the average number of specimens in all samples of a given taxa set.

Plots of species rather than individuals in given feeding habit-substrate niche categories have proved most satisfactory for Cretaceous studies. Advantages of giving each species equal weight are (1) the vagaries of preservation and sampling are less likely to affect total number of species than relative or absolute abundances of individuals; (2) rare species have an impact upon the classification outcome so that a more complete picture of trophic structure is achieved; and (3) samples tend to be less clustered and clumped within a ternary diagram so that differences are more apparent. Disadvantages of plotting species are (1) unusual feeding habits may be given more importance than they originally had in the community, and (2) species whose feeding habit is unknown must be omitted, resulting in a bias. Plotting numbers of individuals seems to be more consistent with trophic theory. Biomass is measured in numbers of individuals, and in some communities one species is typically more abundant than others at the same trophic level (Walker, 1972a). However, because the number of species in a sample is statistically more reliable than the relative abundance of the species (Scott, 1970a; Chang, 1967), a plot of species is preferred to a plot of individuals.

Species lists can be generated from samples of fossil deposits, from all samples in a given taxa set or community, or from lists of characteristic species of a community. The plotting of all samples will define a field or area within the ternary diagrams representative of a given formation, facies, or geographic locale. Normally, samples of specific taxa sets will be represented by distinctive symbols; these will show whether or not several taxa sets do, indeed, possess different preserved trophic structures. To select a class name for the trophic structure, one point can be plotted to represent all cumulative species of that taxa set. This point represents an averaged vector point for the community structure and should fall within the field outlined by the plot of all samples of the taxa set.

Many fossil deposits have a mixed species constituency resulting from postmortem winnowing and transport. Hence, the species list from a single sample may actually represent species from more than one taxa set. When this is the case, the species composition of a taxa set must be constructed from all samples, and individual species in a mixed deposit can be allocated to its respective group. In this manner, eclectic species lists of taxa sets can be constructed and plotted. This, however, introduces a subjective factor into the analysis. In some studies, it may not be possible to eliminate this factor, and such samples need to be clearly designated as such.

TROPHIC STRUCTURE AND ENVIRONMENTAL STABILITY

Environmental stability may be defined as the constancy and uniformity of physical conditions for a long duration (Sanders and Hessler, 1969, p. 1420).

This type of stability implies either no changes in conditions or a steady state in which conditions will return to the original point following a short-term perturbation of one factor (Margalef, 1969). Under these circumstances the environment is relatively changeless.

A second type of stability may be defined as the predictability of environmental fluctuations (Slobodkin and Sanders, 1969). This type of stability implies that changes are regular, periodic, and predictable. In most environments the fluctuations are well within the tolerance levels of most species. Only where tolerance limits are regularly surpassed does the environment become rigorous.

Species diversity of a community is positively correlated with environmental constancy and predictability (Slobodkin and Sanders, 1969; Bretsky and Lorenz, 1970). Insofar as diversity reflects species interactions, complexity of community structure should also increase with increasing constancy and predictability, as long as the regular fluctuations are within the tolerance limits of most species. As tolerance limits of successive species are exceeded, both diversity and complexity of community structure should decrease. Diversity tends to decrease also as the climax stage of community succession is approached. The following examples will attempt to demonstrate these relations in ancient communities.

Carbonate Shelf to Shoreface: Main Street Limestone

Structural modifications resulting in the transition from stable, predictable to unstable, unpredictable environments are illustrated by the communities preserved in the lower Cenomanian Main Street Limestone in north-central Texas. The change from a stable and predictable carbonate shelf below normal wave base to unstable shoreface carbonate sands within the zone of wave action can be established independently by petrographic evidence. Associated diversity trends support an interpretation of change from environmental stability to instability proceeding northward along the outcrop belt. Corresponding alterations in community structures can be assumed to be the result of environmental rather than paleogeographic factors or evolutionary factors. The Main Street outcrop between Fort Worth, Texas, and Durant, Oklahoma, of about 160 km (Fig. 4) does not cross provincial boundaries; the unit is entirely within the *Plesioturrilites brazoensis* zone of Young (1967, p. 68).

The basal contact of the Main Street with the underlying Pawpaw Formation is gradational and intertongued from Fort Worth southward (McGill, 1967). North of Denton, Texas, the contact is disconformable (Stephenson, 1918). A ravinement disconformity resulting from transgression is indicated by a sharp basal contact, clay pebbles of Pawpaw lithology, collophane nodules, reworked Pawpaw fossils, and abundant quartz. The Pawpaw facies north of Denton represent deltaic-estuarine environments (Fee, 1974) that have been buried by transgressing shelf-shoreface sands of the basal Main Street. A transgressing environment is rigorous and features wave action and currents that produce an unstable substrate. Deposition of the Main Street was occasionally interrupted with bored hard-ground surfaces formed most commonly in the northern part of the outcrop area. Main Street deposition terminated when clay muds of the overlying Grayson Formation began

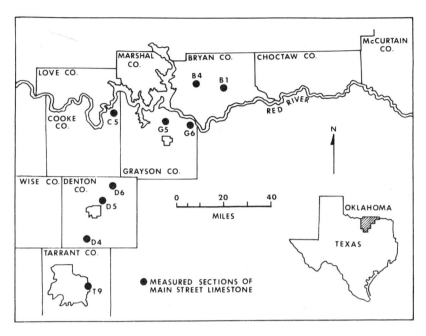

Figure 4
Location of measured sections of the Main Street Limestone in north-central Texas and southern Oklahoma.

to accumulate in response to deepening of the basin (Albritton et al., 1954). In summary, the stratigraphic setting of the Main Street suggests transgression of shelf carbonates into the zone of wave action followed by progressive deepening from shelf to basin depths.

Lateral progression of carbonate facies supports the northward change from shelf to shoreface (Fig. 5A). The southern shelf environment is represented by bioturbated mollusk-echinoid wackestone. It grades northward into lower shoreface oyster packstone comprised mainly of *Ilmatogyra* [*Exogyra*] *arietina*, which, in turn, is replaced by and overlies sparry mollusk-echinoid packstone of the middle shoreface. Trends in petrographic components further substantiate the shelf-to-shoreface progression (Fig. 5B and C). Micrite percentage decreases from an average of 70 percent in the south to 40 in the north. Spar cement is nil in the southern shelf facies but averages about 6 percent in the northern shoreface facies. Quartz and glauconite increase into the shoreface facies from nil to 8 percent; intraclasts and peloids are absent in the wackestone facies but range from 2 to 9 percent in the mollusk-echinoid packstone facies. Each of these trends coincides with facies patterns to support the interpretation that northern environments of the Main Street were deposited in more unstable and rigorous conditions than the southern facies. The shoreface environment is unpredictable as well as unstable and rigorous. Storms alter depth of wave base at irregular intervals, and salinity is sporadically reduced by high runoff or increased by drought.

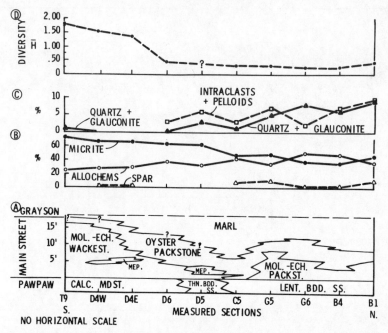

Figure 5
Relations among sedimentary properties and diversity in the Main Street. MEP, mollusk–echinoid packstone. Pawpaw facies after Fee (1974).

The unpredictability of the Main Street shoreface environment is further substantiated by diversity trends (Fig. 5D). The mean number of taxa per section in the southern, stable area is 11, and in the northern, unstable area it is 6. The mean Shannon diversity coefficient, $\bar{H}$, per section in the southern area is 1.51, and in the northern area it is 0.33. Trends in abundances of other organisms also coincide with these trends in diversity. Echinoids are up to 5 percent in the southern sections and present in trace amounts in the north. The scallop, *Neithea texana*, and the oyster, *Texigryphaea* sp., also decrease in abundance northward. The brachiopod, *Kingena wacoensis*, forms a large colony in one of the southern sections and is almost absent in the north. On the other hand, oysters such as *Ilmatogyra arietina* and *Lopha quadriplicata* become more abundant northward (Fig. 6). Calcispheres and planktic foraminifers decrease in abundance in the northern outcrops, and benthic foraminifers, particularly the arenaceous *Cribratina* [*Haplostiche*] *texana*, increase northward. Each of these trends suggests a northward approach to shoreline accompanied by an increase in the rigorousness and instability of the environment.

Four communities are recognized in the Main Street: *Neithea–Plicatula–Planolites*, *Ilmatogyra arietina*, *Kingena*, and oyster-boring bivalves. The first two are distinct on Jaccard coefficient *R*-mode dendrograms; *Kingena* as well as *Ilmatogyra* are recognized also by relative abundances; the last community has a distinctive field occurrence. Each group consists of an empirical taxa set that occurs as disturbed-

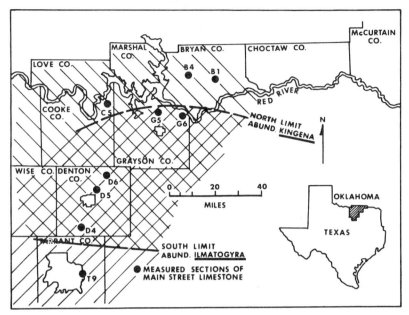

Figure 6
Overlapping distribution of stenohaline taxa represented by *Kingena wacoensis* extending from south to north and euryhaline taxa such as *Ilmatogyra arietina* extending from the north southward.

neighborhood deposits. The suites of taxa have distinctive feeding habit-substrate niche structures (Fig. 7) and diversity patterns. This trophic structure is related to the stability, predictability, and rigor of the environment.

The southernmost community is the *Neithea-Plicatula-Planolites* community preserved in thin, nodular-bedded, mollusk-echinoid wackestone. Diversity is relatively high (6 to 17 taxa) and $H = 1.1$ to 1.8. Characteristic species are *Neithea texana*, *N. bellula*, *N. sexcostata*, *Plicatula dentonensis*, *P. incongrua*, and *Planolites* burrows. Dominant taxa are *Texigryphaea washitaensis* and various echinoids; associated taxa are *Chondrites* (a burrow), *Lima wacoensis*, *Turritella* cf. *T. leonensis*, *Serpula*, *Thalassinoides* (a burrow), *Spondylus*, and pteriids. Faunal density in disturbed neighborhood deposits is low to moderate (8 to 95 specimens per 100 cm^2). Feeding-habit structure of this carbonate shelf community is relatively complex and well balanced with suspension feeders dominant, detritus and deposit feeders generally more than 20 percent of the species, and predators around 10 percent. Epifaunal taxa are the most abundant of the substrate niche types, and vagrant organisms are more common than infaunal taxa. Consequently, the structure of the *Neithea-Plicatula-Planolites* community is diverse, vagrant-epifaunal, detritus-suspension feeding (Fig. 7).

The *Kingena wacoensis* community grades out of the *Neithea-Plicatula-Planolites* community as the relative abundance of *Kingena* becomes more than about 50 per-

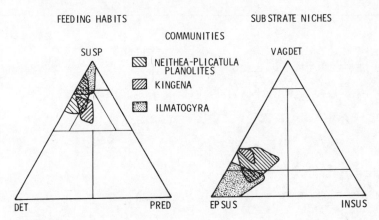

Figure 7
Trophic structure of three Main Street communities.

cent. Associated taxa are *Neithea, Texigryphaea, Serpula, Ilmatogyra, Chondrites, Thalassinoides, Planolites, Plicatula, Turritella*, and echinoids. Species diversity is high (8 to 10 taxa) and $H = 1.1$ to 2.2. Density of the deposits is moderate. Community structure is relatively complex and comparable with that of the *Neithea* community (i.e., diverse, vagrant-epifaunal, detritus-suspension feeding). The community is preserved in the carbonate shelf, mollusk-echinoid wackestone facies. Sandidge (1928) suggested that *Kingena* colonies developed in clear, nutrient-rich waters following marine transgressions. These data tend to support that conclusion.

The *Ilmatogyra arietina* community occurs in the oyster packstone facies from Denton County northward. This cemented to nestling, corkscrew-shaped oyster comprises more than 50 percent of the individuals in deposits of this community. In the southern range of this community, taxa of the *Neithea* and *Kingena* communities overlap; additional associated taxa are *Texigryphaea, Plicatula, Serpula, Chondrites, Planolites, Cliona*, and echinoids. Species diversity is low to moderate (3 to 11 taxa) and H ranges from 0.06 to 1.3. Northward in the Red River area *Lopha quadriplicata* replaces *Kingena* in the community, and local associates are *Texigryphaea, Neithea, Lima, Botula, Serpula, Planolites, Chondrites, Thalassinoides*, and *Turritella*. Diversity is low to moderate (4 to 8 taxa) and H ranges from 0.06 to 0.78. Community structure is generally species dominant, epifaunal suspension feeding, particularly if measured by numbers of individuals rather than by species. This community developed in the lower shoreface transition zone between the carbonate shelf with its *Neithea* and *Kingena* communities and the middle to upper shoreface environments. In the mid shoreface, mollusk-echinoid packstone facies *Planolites* and *Thalassinoides* are the only in-place taxa; *Ilmatogyra* and other taxa are in transported deposits.

The fourth community consists of gastrochaenid boring bivalves and encrusting juvenile oysters. These specialized organisms colonized local hardground surfaces on mollusk-echinoid wackestone and packstone mainly in the northern outcrop area.

Density of individuals is moderate to high and diversity is low, about four species. Other encrusting species are *Serpula* and *Lopha quadriplicata*. The community structure is epifaunal suspension feeding, similar to the spatially associated *Ilmatogyra* community. Similar submarine bored surfaces in the Lower Cretaceous Edwards Formation, Edwards Plateau, formed during periods of slow sedimentation prior to transgression (Rose, 1970).

In summary, the coincidence between facies patterns, petrographic trends, diversity patterns, and community feeding habit-substrate niche structure supports the conclusion that environmental stability and predictability, among other factors, influence the complexity of community structure. More complex communities develop under stable and predictable conditions, and complexity decreases with diminished stability and predictability.

Terrigenous Bay to Shoreface: Tucumcari-Kiowa Formations

The mainly Upper Albian Tucumcari and Kiowa Formations crop out in the southern part of the Western Interior Province (Kansas, Oklahoma, New Mexico, and Colorado). Analyses of stratigraphic relations (Scott, 1970b) and lithofacies (Scott, 1974) indicate that open-sea or bay muds accumulated in the center of the seaway and graded both eastward and westward into sea-margin, shoreface deposits. These nearshore deposits grade into coastal plain sediments that are largely unfossiliferous.

Eight different faunal communities are preserved in this facies complex. Distribution of the communities is controlled mainly by substrate, water turbulence, depth, and salinity. The feeding structure of these communities is controlled by substrate and environmental stability and predictability. Probably the most stable and predictable environment was the open sea where dark gray muds accumulated from pelagic suspension. The open-sea substrates were below normal wave base and probably below the depth affected by most storms. Only infrequently was water turbulence great enough to disturb the bottom by storm-generated density currents. That salinity fluctuations were not extreme is suggested by the relatively diverse fauna. The *Nucula-Nuculana* community occupied this environment. Associated taxa are *Yoldia, Pholadomya, "Breviarca", Turritella belviderei, Drepanochilus*, and *Lingula*. Diversity is moderate (8 to 9 taxa per 100 cm^2); density is low to moderate, and few taxa are abundant, most comprising less than 30 percent of a given fossil deposit. The feeding structure of this community is marked by a diversity of infaunal detritus feeders, and the community classification is vagrant-infaunal, detritus-suspension feeding (Fig. 8). The prominence of infaunal taxa reflects the muddy substrate, and the mixture of feeding types reflects the relatively stable and predictable environment. Studies of modern benthic communities also suggest that deposit feeders characterize stable and predictable environments (Levinton, 1972).

The lower shoreface is a transition zone between the deeper, stable open-sea substrate and the shallower, unstable middle shoreface environment. The distal end of tidal and storm-generated currents commonly deposits thin layers of silt and sand upon the lower shoreface, resulting in interbedded sands and clays. Infaunal orga-

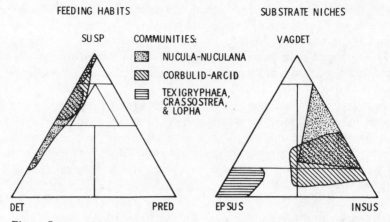

Figure 8
Trophic structure of five offshore communities in the Albian, Southern Western Interior. One symbol represents three separate oyster communities.

nisms rework the sediments, producing discrete burrows and mottled textures. The Corbulid-Arcid community (previously known as *Corbula-Breviarca*) is a complex community consisting of overlapping populations in which one or two of several taxa are dominant: *Corbula ? smolanensis*, "*Breviarca*" spp., *Crassinella semicostata*, *Protocardia multistriata*, and *Turritella belviderei*. Associated taxa are *Syncylonema inconspicuus Inoceramus* sp., *Lucina*? sp., and other species characteristic of the *Nucula-Nuculana* community. Diversity is moderate to high (8 to 11 taxa per 100 cm^2); and density is low to moderate. The Corbulid-Arcid community is dominated by infaunal suspension feeders (Fig. 8) where it is not mixed with members of the *Texigryphaea-Lopha* community. Samples from the Tucumcari are classed as an infaunal, suspension-feeding community; Kiowa samples reflect a greater overlapping of species of the *Nucula-Nuculana* community and fall within the vagrant-infaunal, detritus-suspension feeding field. In the lower shoreface environment detritus feeders tend to be reduced in diversity and infaunal-suspension feeders tend to be accentuated in importance. This may be a response to fluctuating conditions of sand substrates alternating with mud substrates. The introduction of each new substrate required a new colonization by some species, and only a few residual species were capable of burrowing out of the blanket sediments.

Localized oyster biostromes were scattered across this lower shoreface environment (Scott, 1974). In the Tucumcari Formation of New Mexico and the Kiowa Formation of southern Kansas, *Texigryphaea* spp. dominated this community and are associated with *Neithea* spp., *Plicatula* sp., and *Syncyclonema inconspicuus*. Epizoans, such as bryozoans, clionid sponges, lithoglyptid barnacles, *Serpula*, *Botula*, and oyster spat, were attached to many of the oyster shells. Diversity in most disturbed-neighborhood deposits is low (4 to 6 taxa); and density is relatively high (40 or more specimens per 100 cm^2). Transported deposits of the *Texigryphaea* community are common and usually have a higher diversity because of mixing from several communities. In the Tucumcari, *Lopha quadriplicata* is associated with *Texigryphaea* and dominates some deposits. In the Kiowa of Kansas, however, *Lopha quadriplicata* is associated with *Crassostrea kiowana* or forms its own community. The replacement of *Texigryphaea* by *Crassostrea* in the upper Kiowa of southern Kansas and in the central Kansas outcrops of the Kiowa, together with reduced diversity by the loss of stenohaline species (bryozoans, sponges, and barnacles), indicates that the *Crassotrea* and *Lopha* communities formed biostromes in the nearshore brackish parts of the Western Interior seaway. These latter two communities characterize more rigorous, brackish environments than *Texigryphaea*. All three communities developed in the relatively unstable and unpredictable lower shoreface environment, and the feeding structure of each is simple, epifaunal, suspension feeding (Fig. 8).

The middle shoreface environment is within the zone of wave action, the sand substrate is constantly shifting, and burrowing organisms rework much of the sediment. Turbulence level and currents can change regularly or unpredictably. Three communities are preserved in this bioturbated sandstone facies: *Trachycardium-Turritella, Scabrotrigonia-Turritella,* and Pteriid-Mytilid. The *Scabrotrigonia emoryi-Turritella seriatim-granulata* community occupied the mid shoreface of the west side of the seaway in New Mexico (Scott, 1974). Associated species are *Protocardia texana, Homomya* sp., *Cucullaea recedens, Flaventia belviderensis, Cyprimeria kiowana, Cardita?* sp., and *Lima utahensis.* Diversity is moderate (9 taxa per 100 cm^2), and density is highly variable. No species is particularly dominant. In Kansas the *Trachycardium kansasense-Turritella belviderei* community characterizes an analogous community. *Scabrotrigonia emoryi, Protocardia* spp., and *Flaventia belviderensis* are associates. Diversity is moderate and density is variable. The feeding structure of these two communities is also comparable and is classed as infaunal, suspension feeding (Fig. 9). The Pteriid-Mytilid community occurs in the same lithofacies as the previous two communities. Characteristically, *Gervillia mudgeana, Phelopteria salinensis,* and *Modiolus* sp. are associated with *Ostrea rugosa* and taxa of the *Trachycardium-Turritella* community. No single species usually dominates; diversity is relatively high (10 to 20 species per collection), and density is variable. Most deposits of this community are transported mixtures of several communities. Because of this, the feeding structure spans several categories, and a combination of the samples falls in the field of infaunal, detritus-suspension feeding (Scott, 1972). However, the majority of samples plot in the suspension feeding field, which is consistent with other nearshore communities (Fig. 9).

In summary, community feeding structure in the relatively stable, predictable open-sea environment is complex and well developed. As the shore and unstable conditions are approached, community structure tends to become simplified and dominated by infaunal suspension feeders. This change is like that in carbonate shelf-to-shore substrates.

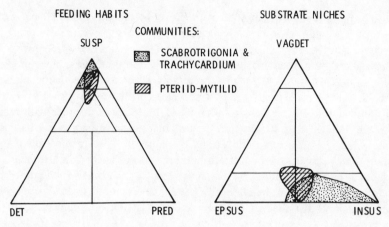

Figure 9
Trophic structure of three nearshore communities in the Albian, Southern Western Interior. Two communities are represented by one symbol.

FEEDING STRUCTURE THROUGH TIME

General

Community feeding structure is bound to have changed through geologic time. As the biosphere evolved, new feeding adaptations appeared within many phyla with the effect of altering community structure. Evolutionary effects upon Paleozoic communities have been analyzed by Bretsky (1969b) and Boucot (1970). Environmental fluctuations have also controlled community structure. Ormiston (1974, pers. comm.) points out that species of communities in unstable environments seem to evolve more slowly. Valentine (1973) has reviewed diversity patterns and trends of both extinction and new appearances of higher taxonomic categories through geologic time. He attributes trends of increased diversity and community complexity to the interrelated factors of continental fragmentation, marine transgression, climatic stability, increasing provinciality, trophic resource stability, and spatial heterogeneity. Times of such diversification and community complexity were Late Cambrian-Ordovician, Medial Ordovician-Early Devonian, Carboniferous-Early Permian, and Late Cretaceous-Cenozoic. On the other hand, species diversity and community structure deteriorated because of a plexus of factors, among which are continental assembly, regression of the seas, environmental and trophic resource instability, reduced provinciality, and environmental homogeneity. Times of maximum deterioration were Late Precambrian, Late Devonian, and Permo-Triassic. Detailed community studies are needed to verify these changes. Effects of biosphere evolution are seen in the contrast between Early Cambrian and Ordovician communities. Cambrian communities apparently consisted of detritus feeders and suspension feeders (Valentine, 1973, p. 452). The apparent absence of species in certain feeding categories may be the result either of their nonpreservability or their nonexistence. During the Ordovician, suspension-feeding brachiopods, bryozoans, eocrinoids, and predaceous cephalopods diversified tremendously. The presence of these organisms probably resulted in many biologically accommodated communities whose feeding structures contrasted with the dominantly physically controlled Cambrian communities. At this point, separation of effects of environment and evolution becomes difficult. However, a valid comparison of community structures of different ages must take these factors into consideration, and only communities in comparable times of diversity and stability can be compared. Furthermore, different substrate niche categories must be used to accommodate the diverse Paleozoic epifauna. Attached suspension feeders are stratified into low- and high-level (Walker, 1972b); bryozoans are ramose or encrusting (Walker and Ferrigno, 1973). Various modes of attachment of brachiopods are described by Rudwick (1965).

A standardized classification scheme of benthic communities would enable the recognition of the combined effects of environmental changes and evolution. A comparison of communities from similar environments occupied during times of environmental stability and equitability should show similar feeding structures, provided that evolution had not produced either new groups having new feeding habits or unfossilizable taxa having a feeding type previously preserved. One community may have well-skeletonized detritus feeders and in another the detritus feeders may be without a preservable skeleton. Also, the evolutionary expansion of carnivorous gastropods during the Late Cretaceous and Cenozoic will produce different feeding structures for those communities than early Mesozoic communities. Aware of these problems, we shall attempt comparisons of nonsynchronous communities from the same environments. The source of data is detailed and comprehensive studies in which large samples from accurate locations were censused for the entire fauna.

Shoreface Environments

Shoreface environments consist of sandy substrates extending in depth from mean low tide to just below normal wave base (40 ft ±). Distinctive sedimentary structures are cross-stratification and bioturbation. The shoreface grades seaward into the inner shelf environment. Other environmental names that include shoreface facies are "nearshore," "littoral," "inner sublittoral," and "infrasublittoral." Reliable faunal studies from this environment range in age from Cretaceous to Ordovician (Rhoads et al., 1972; Scott, 1970a, b, 1974; Wright, 1974; Ziegler et al., 1968; Bretsky, 1969a). The Cretaceous, Silurian, and Ordovician studies represent times of stability, and the Jurassic fauna developed during a time of incipient continental fragmentation. The general features of the communities are compared in Table 2 and feeding structures are compared in Fig. 10.

The upper shoreface communities of Jurassic and Cretaceous ages in the Western Interior Province compare quite well. Comparable communities from older periods

Table 2

Comparison of some shoreface and nearshore communities through time. Asterisk means trophic structure based on reported characteristic taxa; total faunal list not published. Diversity is number of taxa.

	Environments			
		Shoreface		
Age and reference	Upper	Middle	Lower	
Upper Cretaceous (Rhoads et al., 1972)				
Dominant taxa	1. *Tancredia* *Ophiomorpha*	2. *Protocardia* *Phelopteria*	3. *Protocardia* *Oxytoma*	4. *Limopsis* *Corbulamella*
Diversity	~5	~9	~5	~5
Trophic class	Inf Susp*	Vag-Inf, Det-Susp*	Vag-Epi, Det-Susp*	Epi Susp*
Stability	Unstable Unpredictable	Unstable Predictable	Stable Predictable	Stable Predictable

	Environments		
	Shoreface		
Age and reference	Middle	Lower	
Lower Cretaceous (Scott, 1970a)			
Dominant taxa	5. *Trachycardium* *Turritella*	6. Pterid Mytilid	7. Corbulid Arcid
Diversity	17	29	34
Trophic class	Inf Susp	Inf, Det-Susp	Vag-Inf, Det-Susp

	Unstable Unpredictable	Unstable Unpredictable	Stable Predictable
	Upper	*Middle*	*Lower*
(Scott, 1974)	8. *Arenicolites Skolithos Ophiomorpha* 3+ Inf Susp Unstable Unpredictable	9. *Scabrotrigonia Turritella* 9 Inf Susp Unstable Unpredictable	10. Corbulid Arcid 8–11 Inf Susp Stable Predictable
	Upper	*Middle*	
Jurassic (Wright, 1974)	11. *Tancredia* 3 Inf Susp Unstable Unpredictable	12. *Trigonia Pronoella* 6 Inf Susp Unstable Unpredictable	

Table 2 continued

Age and reference		Nearshore Environments	
Devonian (Johnson, 1974; Ormiston, pers. comm.)	*Lingula* 6 Inf Susp	Acrospiriferid Leptocoeliid 42 Epi Susp	
Silurian (Ziegler et al., 1968)	13. *Lingula* 15 Vag-Epi, Det-Susp Unstable Unpredictable	14. *Eocoelia* 29 Vag-Epi, Det-Susp Unstable Unpredictable?	
Ordovician (Bretsky, 1969a)	15. Linguloid 5+ Vag-Inf, Det* Unstable Unpredictable?	16. Rhynchonellid 6+ Epi Susp* Unstable Unpredictable?	

have not been well described or consist of organisms or traces for which feeding habits are unknown. Cretaceous and Jurassic upper shoreface communities (1 and 11, Fig. 10) are dominated by an abundance of infaunal suspension feeders such as *Tancredia, Dosinopsis* (bivalves), *Arenicolites,* and *Skolithos; Ophiomorpha* burrows are common, as well, and represent scavenging and detritus-feeding crustaceans. The faunal compositions of Late Cretaceous and Jurassic communities are quite similar (Wright, 1974) and differ from the Early Cretaceous communities, which normally do not contain body fossils. The feeding structures of these upper shoreface communities are infaunal suspension feeding, and this structure correlates well with the low diversity, suggesting restrictive, unstable, and unpredictable conditions. This environment is most affected by changes in sea level, turbulence, salinity, and temperature, among other limiting factors.

Most middle shoreface communities of the Cretaceous and Jurassic are similarly classified as infaunal suspension feeding. These communities (5, 9, and 12, Table 2 and Fig. 10) are characterized by infaunal bivalves, such as *Trigonia, Scabrotrigonia, Protocardia,* and *Trachycardium,* and turritellid snails. Associated taxa are infaunal arcticids, corbulids, and mactrids and epifaunal mytilids. One other middle shoreface community (6, Table 2 and Fig. 10) consists, in addition to cardiids and mytilids, of pteriids and several detritus-feeding mollusks such as *Nuculana, Linearia,* and *Drepanochilus.* Consequently, the community structure is classed as infaunal, detritus-suspension feeding. Part of this complexity in the Kiowa may be the result of transport mixing by taxa from the *Trachycardium-Turritella* community with byssally attached bivalves (Scott, 1970a, p. 47). Although the feeding structure of mid shoreface communities is similar to that of upper shoreface communities, increased diversity of infaunal mollusks distinguishes the middle shoreface communities.

Most lower shoreface communities of the Cretaceous are classed as vagrant-infaunal, detritus-suspension feeding (2, 7, 10); one is vagrant-epifaunal, detritus-suspension

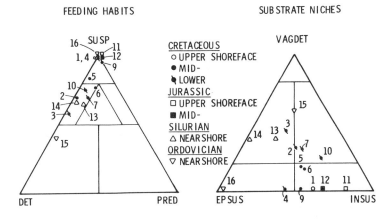

Figure 10
Trophic structure of selected nearshore communities through time.

(3), and one is simply epifaunal suspension feeding (4). These communities are taxonomically diverse, consisting of species from both the middle shoreface and the offshore open-sea muds. The lower shoreface is a zone of ecotonal overlap of several species populations. In the Upper Cretaceous Fox Hills Formation of South Dakota, Rhoads et al. (1972) recognized three communities in clay-silt substrates, which illustrate the variety of adaptations and structure on the lower shoreface. The *Protocardia-Phelopteria* community (2) is dominated by *Protocardia* associated with low abundances of epifaunal *Phelopteria*, *Ostrea*, and *Oxytoma*, and infaunal *Nuculana*, *Malletia*, *Tellinimera*, *Cucullaea*, and *Cymbophora*. The feeding structure of this community is vagrant-infaunal, detritus-suspension. A selected set of these taxa comprises the localized *Protocardia-Oxytoma* community characterized by an abundance of those two taxa with lesser numbers of the byssate *Tenuipteria* and infaunal *Nuculana* with *Malletia*. Structurally, the community is vagrant-epifaunal, detritus-suspension. It may represent colonies of byssal bivalves clustered across the lower shoreface and localized by favorable substrate conditions. The third community, *Limopsis-Corbulamella*, has an epifaunal suspension feeding structure because of diverse byssal bivalves associated with the dominant infaunal suspension feeders. This community may also represent local clumps of byssate colonies upon the lower shoreface, which is dominated by a more widely distributed infaunal population of *Protocardia* and *Corbulamella* among others.

In the Kiowa and Tucumcari formations the Corbulid-Arcid community (7, 8, 10) has a vagrant-infaunal, detritus-suspension feeding structure. The dominant infaunal suspension feeders are associated with less abundant vagrant and infaunal detritus feeders, such as *Nuculana*, *Nucula*, *Drepanochilus*, and scaphopods. This community is associated with oyster biostromes or banks of *Texigryphaea*. Only locally do pteriid bivalves occur. None of Wright's Jurassic communities are similar.

Paleozoic lower shoreface communities, as such, are not reported. However, several "nearshore" communities have been described. The structure of Silurian and Ordovician nearshore communities in analogous sand lithofacies (13, 14, 15, and 16, Table 2 and Fig. 10) are somewhat different from Cretaceous lower shoreface communities. The Silurian *Lingula* community (Ziegler et al., 1968) is dominated by *"Camarotoechia" decemplicata*, with a lesser abundance of *Lingula* spp., *Palaeoneilo rhomboidea*, and *Pteronitella* cf. *P. retroflexa*. Based on numbers of taxa, the feeding structure is classed as vagrant-epifaunal, detritus-suspension because of the diversity of browsing gastropods and deposit-feeding bivalves. Of the seven epifaunal organisms, the two brachiopods and *Cornulites* sp. reflect the Paleozoic character of the community; the epifaunal bryozoans and bivalves are analogous to younger taxa. The much more diverse *Eocoelia* community is dominated by *Eocoelia curtisi* and associated with *"Camarotoechia"* sp., *Dalejina* sp., *Strophochonetes* sp., and a leptostrophid. The vagrant-epifaunal, detritus-suspension feeding classification reflects the diversity of both epifaunal brachiopods and bryozoans and browsing snails and grubbing trilobites. These taxa are typically Paleozoic in character; however, these niches were repopulated during the Mesozoic following the Permo-Triassic extinction. The structural difference of the *Eocoelia* community from Cretaceous nearshore communities may be due in part to real environmental differences as well as to the decreased diversity of epifaunal brachiopods. An additional factor influencing community structure is the possible mixing of species from other communities. Ziegler et al. (1968) described evidence that specimens of the *Eocoelia* community have been transported and possibly mixed with taxa from other communities. The similarity in feeding structure of the Silurian *Lingula* and *Eocoelia* communities suggests comparable substrates and environmental stability and predictability. However, the higher diversity of the *Eocoelia* community (31 species versus 16) suggests either that stability and predictability were greater or that several communities are mixed together. Without additional data the question cannot be resolved.

The Ordovician *Lingula* community of Bretsky (1969a) has a vagrant-infaunal, detritus feeding structure. The detritus-feeding habit is emphasized by the number of browsing snails and by the overall low sample diversity (five taxa). The community occupied a sandy substrate close to shore or a barrier beach. The closely associated Rhynchonellid community not only shares some common taxa, but also occurs on a nearshore silty substrate downslope from the *Lingula* community. Its feeding structure is epifaunal suspension because of the apparent absence of browsers and detritus feeders. Perhaps detailed collections and species lists would modify these conclusions. In summary, lower shoreface and nearshore community structures change from Paleozoic epifaunal-dominated to Cretaceous infaunal-dominated, and feeding types remain approximately constant, detritus-suspension feeding.

SUMMARY

These few samples of somewhat comparable communities do not provide an adequate basis for establishing structural trends through time or in relation to environmental gradients. It would be useful to make detailed studies of contemporaneous communities in the deltaic to shelf transition to test structural trends such as decrease in detritus feeders and increase in suspension feeders from the delta front to shelf environments. Such a change is not unexpected in view of the seaward increase in stability and predictability and decrease in turbidity.

The preceding examples have been designed primarily to test the applicability of the feeding habit-substrate niche technique to faunas of different environments and different ages. The approach appears to be useful in discriminating among feeding structures of near-contemporaneous communities from a related set of environments. Also, it appears to aid in defining not only the direction but also the size of structural changes through geologic time within analogous environments. The technique deserves more application, testing, and refinement.

ACKNOWLEDGMENTS

Many of these ideas were stimulated by conversations with Norman Sohl and Curt Teichert. Discussions with Robert Stanton, Ronald West, Jerry Vincent, Thomas Hellier, Donald Reaser, Robert Parker, and

Allan Ormiston contributed significantly in the maturation of ideas. Thanks are due Charles McNulty and Allan Ormiston for carefully reviewing the manuscript. However, I take full responsibility for any errors of misinterpretations of other's ideas. Financial support for the collection of data was received from the University of Kansas (NSF Grant GB-4446X, G. W. Byers, Director), Waynesburg College, and the University of Texas at Arlington.

REFERENCES

Albritton, C. C., Jr., et al. 1954. Foraminiferal populations in the Grayson Marl. Geol. Soc. Amer. Bull., 64:327-336.

Bandy, O. L., and R. E. Arnal. 1960. Concepts of foraminiferal paleoecology. Amer. Assoc. Petrol. Geol. Bull., 44:1921-1932.

Beerbower, J. R., and Dianne Jordan. 1969. Application of information theory to paleontologic problems: taxonomic diversity. Jour. Paleont. 43:1184-1198.

Bloom, S. A., J. L. Simon, and V. D. Hunter. 1972. Animal-sediment relations and community analysis of a Florida estuary. Marine Biol. 13:43-56.

Boucot, A. J. 1970. Practical taxonomy, zoogeography, paleoecology, paleogeography and stratigraphy for Silurian and Devonian brachiopods. North Amer. Paleont. Convention, Proc., F:566-611.

Braun-Blanquet, J. 1913. Die vegetationsverhnältnisse der Schneestufe in den Rätisch-Lepontischen Alpen. Ein Bild des Pflanzenlebens and seinen aussersten Grenzen. Schweiz. Naturf. Ges., Neue Denkschr., 48:1-347.

Bretsky, P. W. 1969a. Central Appalachian Late Ordovician communities. Geol. Soc. Amer. Bull., 80:193-212.

———. 1969b. Evolution of Paleozoic benthic marine invertebrate communities. Palaeogeog. Palaeoclimat. Palaeoecol., 6:45-59.

———, and D. M. Lorenz. 1970. An essay on genetic-adaptive strategies and mass extinctions. Geol. Soc. Amer. Bull., 81:2449-2456.

———, K. W. Flessa, and S. S. Bretsky. 1969. Brachiopod ecology in the Ordovician of eastern Pennsylvania. Jour. Paleont., 43:312-321.

Chang, Yi-Maw. 1967. Accuracy of fossil percentage estimation. Jour. Paleont., 41:500-502.

Clements, F. E. 1916. Plant succession: analysis of the development of vegetation. Publ. Carnegie Inst., Wash., 242:1-512.

———, and V. E. Shelford. 1939. Bioecology. John Wiley & Sons, Inc., New York. 425p.

Crocker, R. L., and J. G. Wood. 1947. Some historical influences on the development of the South Australian vegetation communities and their bearing on concepts and classification in ecology. Roy. Soc. South Australia Trans. 71:91-136.

Deevey, E. S., Jr. 1969. Specific diversity in fossil assemblages. In G. M. Woodwell and H. H. Smith (eds.), Diversity and stability in ecological systems. Brookhaven Symp. Biol., 22:224-242.

Elton, C. S., and R. S. Miller. 1954. The ecological survey of animal communities: with a practical system of classifying habitats by structural characters. Jour. Ecol., 42:460-496.

Fee, D. W. 1974. Lithofacies and depositional environments of the Weno and Pawpaw Formations (Lower Cretaceous) of north-central Texas and south-central Oklahoma. Unpublished M.S. thesis, University of Texas at Arlington, 153p.

Forbes, E. 1844. Report on the Mollusca and Radiata of the Aegean Sea, and on

their distribution considered as bearing on geology. Report of the British Association for the Advancement of Science for 1843, p. 130-193.
Fox, W. T. 1962. Stratigraphy and paleoecology of the Richmond Group in southeastern Indiana. Geol. Soc. Amer. Bull., 73:621-642.
———. 1968. Quantitative paleoecologic analysis of fossil communities in the Richmond Group. Jour. Geol., 76:613-640.
Gleason, H. A. 1926. The individualistic concept of the plant association. Torrey Bot. Club Bull., 53:7-26.
Halleck, M. S. 1973. Crinoids, hardgrounds, and community succession: the Silurian Laurel-Waldron contact in southern Indiana. Lethaia, 6:239-252.
Hedgpeth, J. W. 1957a. Classification of marine environments. In J. W. Hedgpeth (ed.), Treatise on marine ecology and paleoecology, v. 1, Ecology. Geol. Soc. Amer. Mem. 67:17-28.
———. 1957b. Concepts of marine ecology. In J. W. Hedgpeth (ed.), Treatise on marine ecology and paleoecology, v. 1, Ecology. Geol. Soc. Amer. Mem., 67:29-52.
Herm, D. 1972. Pitfalls in paleoecologic interpretation—an integrated approach to avoid the major pits. 24th Intern. Geol. Cong. Proc., Sec. 7:82-88.
Humboldt, Alexander von. 1805. Essai sur la géographie des plantes; accompagné d'un tableau physique des régions equinoxides. Par Al. de Humbolt et A. Bonpland, rédigé par Al. de Humbolt. Paris, Levrault, Schoell et Cie. 155p.
Imbrie, J. 1964. General ecological classification of marine fossils (abst.). Geol. Soc. Amer. Spec. Paper, 82:99.
Jago, J. B. 1973. Cambrian agnostid communities in Tasmania. Lethaia, 6:405-422.
Johnson, J. G. 1974. Early Devonian brachiopod biofacies of western and Arctic North America. Jour. Paleont., 48:809-819.
Johnson, R. G. 1962. Interspecific associations in Pennsylvanian fossil assemblages. Jour. Geol., 70:32-55.
———. 1964. The community approach to paleoecology. In John Imbrie and N. D. Newell (eds.), Approaches to paleoecology. John Wiley & Sons, Inc., New York, p. 107-134.
———. 1973. Conceptual models of benthic marine communities. In T. J. M. Schopf (ed.), Models in paleobiology. Freeman, Cooper & Company, San Francisco, p. 148-159.
Kaesler, R. L. 1966. Quantitative reevaluation of ecology and distribution of Recent Foraminifera and Ostracoda of Todos Santos Bay, Baja California, Mexico. Kansas Univ. Paleont. Contrib. Paper 10, 50p.
Kauffman, E. G., and R. W. Scott. 1974. Basic concepts in community paleoecology. Geol. Soc. Amer. Abst., 6:815-816.
———, and N. F. Sohl. 1974. Structure and evolution of Caribbean Cretaceous rudist frameworks. Festschrift für Hans Kugler. Verhandl. Naturf. Ges. Basel, 84:399-467.
Lawrence, D. R. 1968. Taphonomy and information losses in fossil communities. Geol. Soc. Amer. Bull., 79:1315-1330.
Levinton, J. S. 1972. Stability and trophic structure in deposit-feeding and suspension-feeding communities. Amer. Naturalist, 106:472-486.
Lewontin, R. C. 1969. The meaning of stability. In G. M. Woodwell and H. H. Smith (eds.), Diversity and stability in ecological systems. Brookhaven Symp. Biol., 22:13-24.
Lowenstam, H. A. 1957. Niagaran reefs in the Great Lakes area. In J. W. Hedgpeth

(ed.), Treatise on marine ecology and paleoecology, v. 2, Paleoecology. Geol. Soc. Amer. Mem., 67:215-248.

Loya, Y. 1972. Community structure and species diversity of hermatypic corals at Eilat, Red Sea. Marine Biol., 13:100-123.

MacArthur, R. H., and E. O. Wilson. 1967. The theory of island biogeography. Princeton University Press, Princeton, N.J., 203p.

Margalef, R. 1960. Temporal succession and spatial heterogeneity in phytoplankton. *In* A. A. Buzzati-Traverso (ed.), Perspectives in marine biology. University of California Press, Berkeley, Calif., p. 323-349.

———. 1968. Perspectives in ecological theory. University of Chicago Press, Chicago, 111p.

———. 1969. Diversity and stability: a practical proposal and a model of interdependence. *In* G. M. Woodwell and H. H. Smith (eds.), Diversity and stability in ecological systems. Brookhaven Symp. Biol., 22:25-37.

McGill, D. W. 1967. Washita Formations, north Texas, correlated to Georgetown Limestone, central Texas. Soc. Econ. Paleont. Mineral., Permian Basin Sec. Publ., 67-8:218-239.

Mello, J. R., and M. A. Buzas. 1968. An application of cluster analysis as a method of determining biofacies. Jour. Paleont., 42:747-758.

Nelson, H. F. 1973. The Edwards Reef complex and associated sedimentation in central Texas. Geol. Soc. Amer. Guidebook No. 15, Bureau Econ. Geol., Univ. Texas at Austin, 35p.

Newell, N. D., J. K. Rigby, A. G. Fischer, A. J. Whiteman, J. E. Hickox, and J. S. Bradley. 1953. The Permian reef complex of the Guadalupe Mountains region, Texas and New Mexico: a study in paleoecology. W. H. Freeman and Company, San Francisco, 236p.

———, J. Imbrie, E. G. Purdy, and D. L. Thurber. 1959. Organism communities and bottom facies, Great Bahama Bank. Amer. Museum Natural Hist. Bull., 117: 177-228.

Odum, E. P. 1971. Fundamentals of ecology, 3rd ed. W. B. Saunders Company, Philadelphia, 574p.

Parker, R. H. 1956. Macro-invertebrate assemblages as indicators of sedimentary environments in east Mississippi delta region. Amer. Assoc. Petrol. Geol. Bull., 40: 295-376.

———. 1960. Ecology and distributional patterns of marine macro-invertebrates, northern Gulf of Mexico. *In* F. B. Phleger and T. H. van Andel (eds.), Recent sediments, northwest Gulf of Mexico. Amer. Assoc. Petrol. Geol., Tulsa, Okla., p. 302-337.

———. 1964. Zoogeography and ecology of some macro-invertebrates, particularly mollusks, in the Gulf of California and the continental slope of Mexico. Vidensk. Medd. Dansk Naturh. Foren., 126:178p.

Petersen, C. G. J. 1913. Valuation of the sea. II. The animal communities of the sea bottom and their importance for marine zoogeography. Rept. Danish Biol. Sta., 21:1-44.

Phleger, F. G. 1960. Ecology and distribution of Recent Foraminifera. Johns Hopkins Press, Baltimore, Md., 297p.

Purdy, E. G. 1964. Sediments as substrates. *In* John Imbrie and N. D. Newell (eds.), Approaches to paleoecology, John Wiley & Sons, Inc., New York, p. 238-271.

Ramensky, L. G. 1924. Die Grundgesetzmässigkeiten in Aufbau der vegetationsdecke. [Russian] Wjestn. opytn. djela Woronesch. 37 pp. (Bot. Centralbl., N.F., 7:453–455, 1926).

Rhoads, D. C., I. G. Speden, and K. M. Waagé. 1972. Trophic group analysis of Upper Cretaceous (Maestrichtian) bivalve assemblages from South Dakota. Amer. Assoc. Petrol. Geol. Bull., 56:1100–1113.

Rose, P. R. 1970. Stratigraphic interpretations of submarine versus subaerial discontinuity surfaces: an example from the Cretaceous of Texas. Geol. Soc. Amer. Bull., 81:2787–2798.

Rudwick, M. J. S. 1965. Ecology and paleoecology. In R. C. Moore (ed.), Treatise on invertebrate paleontology, Pt. H, Brachiopoda, v. 1. Geological Society of America and Kansas University Press, p. H199–H213.

———. 1968. Some analytic methods in the study of ontogeny in fossils with accretionary skeletons. In D. B. Macurda, Jr. (ed.), Paleobiological Aspects of Growth and Development, a Symposium. Paleont. Soc. Mem. 2, Jour. Paleont., 42:35–49.

Sanders, H. L. 1958. Benthic studies in Buzzards Bay. I. Animal-sediment relationships. Limnol. Oceanog., 3:245–258.

———. 1968. Marine benthic diversity: a comparative study. Amer. Naturalist, 102:243–282.

———, and R. R. Hessler. 1969. Ecology of the deep-sea benthos. Science, 163:1419–1424.

Sandidge, J. R. 1928. The recurrent brachiopods of the Lower Cretaceous of northern Texas. Amer. Jour. Sci., 5th Ser., 15:314–318.

Scott, R. W. 1970a. Paleoecology and paleontology of the Lower Cretaceous Kiowa Formation, Kansas. Univ. Kansas Paleont. Contrib., Art. 52 (Cretaceous 1), 94p.

———. 1970b. Stratigraphy and sedimentary environments of Lower Cretaceous rocks, southern Western Interior. Amer. Assoc. Petrol. Geol. Bull., 54:1225–1244.

———. 1971. Early Cretaceous benthic communities in Washita rocks. Geol. Soc. Amer. Abst., 3:244–245.

———. 1972. Preliminary ecological classification of ancient benthic communities. 24th Intern. Geol. Congr. Proc., Sec. 7:103–110.

———. 1974. Bay and shoreface benthic communities, Lower Cretaceous, southern Western Interior. Lethaia, 7:315–330.

———, J. F. Wallace, R. L. Bowers, M. R. Box, D. W. Fee, and S. A. Root. 1973. Benthic community succession in a Cretaceous carbonate to shale transition. Geol. Soc. Amer. Abst., 5:801.

Shaffer, B. L., and S. C. Wilke. 1965. The ordination of fossil communities; an approach to the study of species interrelationships and communal structure. Michigan Acad. Sci. Arts Letters Papers, 1:199–214.

Slobodkin, L. B., and H. L. Sanders. 1969. On the contribution of environmental predictability to species diversity. In G. M. Woodwell and H. H. Smith (eds.), Diversity and stability in ecological systems. Brookhaven Symp. Biol., 22:82–95.

Speden, I. G. 1966. Paleoecology and the study of fossil benthonic assemblages and communities. New Zealand Jour. Geol. Geophys., 9:408–423.

Stanley, S. M., 1970. Relation of shell form to life habits in the Bivalvia (mollusca). Geol. Soc. Amer. Mem. 125, 296p.

Stanton, R. J., Jr., and Ian Evans. 1972. Community structure and sampling requirements in paleoecology. Jour. Paleont., 46:845–858.

Stephenson, L. W. 1918. Contribution to the geology of northeastern Texas and southern Oklahoma. U.S. Geol. Surv. Profess. Paper, 120-H:129-163.

Stephenson, T. A., and Ann Stephenson. 1950. Life between the tide marks in North America, I. Florida Keys. Jour. Ecol., 39:354-402.

Stephenson, W. 1973. The validity of the community concept in marine biology. Roy. Soc. Queensland Proc., 84:73-86.

Sutton, R. G., Z. P. Bowen, and A. L. McAlester. 1970. Marine shelf environments of the Upper Devonian Sonyea Group of New York. Geol. Soc. Amer. Bull., 81:2975-2992.

Tappan, H. 1971. Microplankton, ecological succession and evolution. North Amer. Paleont. Conv., Proc., H:1058-1103.

Thorson, G. 1957. Bottom communities. *In* J. W. Hedgpeth (ed.), Treatise on marine ecology and paleoecology, v. 1, Ecology. Geol. Soc. Amer. Mem. 67:461-534.

Valentine, J. W. 1971. Resource supply and species diversity patterns. Lethaia, 4:51-62.

———. 1973. Evolutionary paleoecology of the marine biosphere. Prentice-Hall, Inc., Englewood Cliffs, N.J., 511p.

———, and Bob Mallory. 1965. Recurrent groups of bonded species in mixed death assemblages. Jour. Geol., 73:683-701.

———, and R. G. Peddicord. 1967. Evaluation of fossil assemblages by cluster analysis. Jour. Paleont., 41:502-507.

Walker, K. R. 1972a. Trophic analysis: a method for studying the function of ancient communities. Jour. Paleont., 46:82-93.

———. 1972b. Community ecology of the Middle Ordovician Black River Group of New York State. Geol. Soc. Amer. Bull., 83:2499-2524.

———, and L. P. Alberstadt. 1974. Ecological succession as an aspect of structure in fossil communities. Geol. Soc. Amer. Abst., 6:997-998.

———, and R. K. Bambach. 1974. Analysis of communities. Principles of benthic community analysis. Sedimenta IV, Comp. Sed. Lab., Univ. Miami, 2.1-2.20.

———, and K. F. Ferrigno. 1973. Major Middle Ordovician reef tract in east Tennessee. Amer. Jour. Sci., 273-A:294-325.

Warme, J. E. 1969. Live and dead mollusca in a coastal lagoon. Jour. Paleont., 43:141-150.

———. 1971. Paleoecological aspects of a modern coastal lagoon. Univ. Calif. Publ. Geol. Sci., 87:131p.

West, R. R. 1972. Relationship between community analysis and depositional environments: an example from the North American Carboniferous. 24th Intern. Geol. Congr. Proc., Sec. 7:130-146.

Whittaker, R. H. 1956. Vegetation of the Great Smoky Mountains. Ecol. Monog., 26:1-80.

———. 1962. Classification of natural communities. Botanical Rev., 28:1-239.

Wright, R. P. 1974. Jurassic bivalves from Wyoming and South Dakota: a study of feeding relationships. Jour. Paleont., 48:425-433.

Young, K. 1967. Ammonite zonations, Texas Comanchean (Lower Cretaceous). Soc. Econ. Paleont. Mineral., Permian Basin Section Publ., 67-8:65-70.

Ziegler, A. M. 1965. Silurian marine communities and their environmental significance. Nature, 207:270-272.

———, L. R. M. Cocks, and R. K. Bambach. 1968. The composition and structure of Lower Silurian marine communities. Lethaia, 1:1-27.

Classification of Communities Based on Geomorphology and Energy Levels in the Ecosystem

Robert H. Parker   Coastal Ecosystems Management, Inc.

ABSTRACT

Most biologists and many paleontologists describe and categorize biotic communities from groupings of species bound by one or two predominant species that lend groups their names (e.g., *Nucula–Ampelisca*, *Turritella communis*, oak–hickory, and willow–sedge). Use of scientific or common names to designate communities can be misleading, as faunal and floral names of many identical forms differ throughout the world. The *Arca* of European terminology equals *Anadara* in America. Malacologists recognize *Arca* as an epifaunal filter feeder, whereas *Anadara* is an infaunal deposit feeder. Moreover, predominant species of many communities may be entirely soft-bodied, and thus leave no fossil record and are of no use to paleoecologists.

An overall scheme for classifying both aquatic and terrestrial communities is suggested by grouping natural aggregations of plants and animals that are confined by terrain features and controlled by radiant, kinetic, and chemical energy input and utilization. Primary classification and terminology are based on geomorphology; secondary categories can be formed on the basis of ecosystem type and energy utilization. Typical primary categories are (1) *terrestrial:* tundra, high plain, river basin, coastal plain, mountain, barrier spit, and (2) *aquatic:* upper estuary, bay center, bay margin, reef, inlet, surf zone. Some secondary categories are (1) *coastal plain:* hummock, marsh (fresh), marsh (salt), prairie, mott, and (2) *bay center:* clayey bottom, muddy sand, low salinity, hypersaline, normal salinity.

Secondary groupings contain communities or organisms influenced by energy levels (amount of radiant energy, turbulence, and chemical degradation). All can be recognized in past sedimentary records by faunal remains and lithology.

INTRODUCTION

The existence of discrete groupings of organisms (depending upon the scale of observation) is generally recognized visually by homogeneity of composition within the group and statistically by homogeneity of distributional patterns, abundances, and diversities. Boundaries of organism groupings are more easily discerned in terrestrial environments and in waters where visibility permits a clearer view of the bottom than in most aquatic areas. These small groupings have been termed "communities" since the beginnings of the study of ecology. Communities of humans, characterized by similar ethnic traits and socioeconomic goals for communal existence, have been recognized since humans became a fully social animal. Before that time, their place within natural organismal communities is still somewhat conjectural, depending upon whether they were aboriginal predators, scavengers, or grazers.

Human communities are organized around the need for cooperation in meeting environmental needs. The larger the group, the greater the need for delegating various tasks to individuals with differing skills. This division of labor leads to a more efficient utilization of resources. Natural communities (excluding man) must operate on the same basis (Odum, 1975). Resources available to natural functioning communities are energy in the form of the sun's radiant energy, chemical energy (additional heat, light, and electron capture or release from natural oxidation and reduction reactions), and kinetic energy (derived from the motion of water, air, or gravity). These energy resources were originally contained in the atmosphere, hydrosphere, or lithosphere in the form of gases, water, chemical liquids, and rocks, which were later supplemented by organic elements (biosphere) added to the environment through the biological conversion of energy.

For plants and animals to make the most efficient use of their resources, a division of labor usually takes place. With few exceptions, numerous species of plants comprise a typical plant community. Altogether, these species utilize available radiant energy with optimum efficiency, although singly they may absorb light at slightly different spectral levels and intensity. Other resources available to the individual species of plants within a community may be shared also through different modes of uptake. Some plants have root hair systems that can absorb minute quantities of growth substances from surface soil layers only; other members of the community may use large tap root systems to get nutrients from deeper subsoil layers. When single sources of nutrients and energy are available at stable and predictable rates, few species of plants are needed to furnish the basis for a terrestrial community. On the other hand, when there are multiple sources of energy, soil types, and nutrients also available at a predictable level, many species of plants will eventually become established for optimal utilization of these resources within any habitat.

Examples of opposite ends of this ecosystem spectrum can be given. For instance, a mature grassland biome (an aggregation of plant communities) where only a few species of large grasses predominate, each attaining ascendancy by season governed by light intensity and temperature, is an example of a stable but low-diversity association (Küchler, 1974). Nutrient sources are few, but resources high. American prairie climates ranging from unseasonable frosts to searing droughts are relatively unpredictable, an additional pressure leading to low plant diversity. Overall diversity of animal life dependent upon the prairie is low, reflecting the low diversity of plant life. On the opposite side of the spectrum is the classic example of the rain forest, where light intensity is high, and nutrient sources are many and involve numerous chemical and reproductive cycles (Janzen, 1973; Fleming, 1973). Because of the nearly solid high canopy that controls light to plants underneath, light sources are variable, promoting high habitat internal community diversity. In this habitat, predictability is high, because of constant seasonal light, temperature, and, to some extent, rainfall. Animal diversity is extremely high (so high that limits have not yet been discovered), resulting from a nearly limitless number of available habitats and niches for colonization.

Animal communities dependent upon the basic plant communities within a specific ecosystem vary proportionally to the diversity of plants (Fleming, 1973). A prairie grass community is inhabited by a few species of large grazing animals, rodents, and predators, which prey on both grazers and rodents. Bird diversity is low, and insect populations are geared to a few kinds of plants. In a tropical jungle, animal diversity is high because of the enormous variety of plants supplying different forms of food (fruits, seeds, bark, roots, etc.). Selection produces specialized herbivores and even more specialized animals to prey on the herbivores.

In general, these principles of community organization apply to marine or aquatic communities as well. Light as the prime energy source is of course limited to the shallower waters or euphotic zone. In deeper waters, this energy source is replaced by chemical and kinetic energy through chemical degradation of biotic substances in the sunlit layers, and through the often slow but always persistent movement of waters. In deeper waters, energy sources are produced more through the biosphere than from the atmosphere. Niche diversification and resultant organismal diversity again depend upon energy and nutrient resources modified by predictability of the environment. It is well known that tropical coral reefs produce the highest diversity and numbers of animals in the sea; highly variable, low-light-energy estuaries have the lowest diversity and abundance of animal life. The major difference between terrestrial and aquatic communities is that most of the world's aquatic bottom communities contain no benthic plant life, and must subsist on highly decomposed plant debris from the surface layers as the base of the food pyramid. The base of terrestrial communities, on the other hand, is formed almost entirely of living plant.

How can these broad generalizations as to energy restrictions of community organization be used to classify communities, and can these communities be recognized in the fossil record? It is my belief that the energy restrictions are intimately tied to surface geomorphology, latitudinally and altitudinally. Predominant plants and animals of any habitat are there because of the particular resources presented for utilization, first, and specialization for feeding on certain early colonizers second, names of communities should be more geomorphic than biologic. Biologists tend to give names to communities that emphasize the scientific names of predominant *visible* members. For example, Thorson and his followers designated worldwide shelf communities as *"Venus-Syndosmya"* or *"Tellina-Ophiura"* or *"Ampelisca-Nereis"* communities (Thorson, 1957). *Venus* is a fossil genus, *Syndosyma* is a synonym for *Abra*, *Ampelisca* forms species complexes; each species has different food preferences and means of feeding. The last community, for instance, is named after soft-bodied entities that would not leave a fossil record. Names are given to communities based on a bias of screen mesh size. Most benthic biologists and virtually all environmental-impact census takers use screens with large mesh, which result in fewer animals to count. Because only large animals are seen, their names become fixed in community literature.

More recent studies, including those by Sanders (1956, 1958, 1960), Parker (1969, 1975), and Wade (1972), indicated that the bulk of bottom animal life is composed of animals less than 1 cm and certainly less than 1 mm in size. Whereas there may be one large clam in a square meter of bottom, the same area may contain 150,000 nematodes, amphipods, harpacticoid copepods, cumaceans, and other animals still larger than 0.50 mm. Because scientific species names do not adequately describe a community, it is preferable to use names that describe the physical habitat of each respective community. Geomorphic or physiographic names also may indicate energy levels and thus give a hint as to the type of ecosystem involved.

More important is the fact that, if real parallels can be drawn between modern communities and assemblages of shells from ancient sedimentary horizons, the knowledge of their habitats as geography can be used to reconstruct shorelines, depths of the sea, and other topographic features. More definitive knowledge of ancient topography at the habitat level can lead to more efficient exploration for fossil fuels and toward a more accurate reconstruction of the earth's past. For this reason, Parker (1956) emphasized all organisms within the community, including the shell remains. The use of the word "assemblage" by Parker (1956, 1959, 1960) indicated that both the living animals and their undecomposed remains were used to describe the biological elements of a particular habitat. Parker (1975) has used the term "habitat" more often than "community" to describe different small ecosystems. For instance, a small area encompassing a series of tidal channels yielded a number of well-defined animal communities that lived within relatively narrow habitat boundaries. Figure 1 shows the boundaries of the Hadley Harbor community aggregations, but in the published caption the areas circumscribing these communities (as habitats) are emphasized. These well-defined communities/habitats were designated as inner harbor, sand flats, channels, and eel-grass beds. All have their equivalents in marine and freshwater areas around the world. The habitat names may be slightly different, such as inner sound or bay, outer sound or open bay, inlets rather than channels, and shallow sandy shelf rather than sand flats. Each of these habitats has distinctive animal and plant groupings (communities) composed of various feeding types occupying various trophic levels that make the most efficient use of the available resources.

Community Classifications Based on the Ecosystem 71

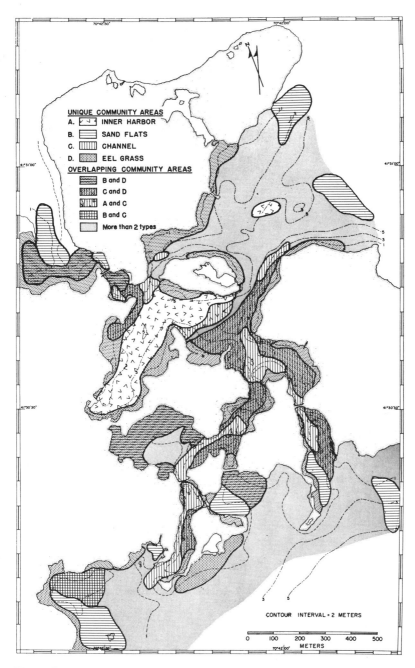

Figure 1
Community groups arranged within habitats of the Hadley Harbor tidal channels, near Woods Hole, Massachusetts. Supporting data for groupings can be found in Parker (1975).

GENERAL CLASSIFICATION OF COMMUNITIES

Having set the stage for a general classification of communities, let us look at a few of today's common terrestrial and aquatic community types—as neither time nor space are available for a detailed examination of all major and subcommunity types. Taking the knowledge of existing communities and their limiting factors, we can then select those with the most preservable remains for detailed studies of the food chains and trophic levels that operate within these ecosystems. Knowledge of all organisms and their biology as they relate to the life and death of the preservable forms can then be used to reconstruct the paleocommunities. If these same modern and fossil assemblages also reflect geomorphology, ancient landscapes can be reconstructed.

Terrestrial Communities

Land plant communities can be separated into four basic types, all of which should leave some remains in the fossil record. These basic associations are forests, grasslands, desert vegetation, and tundra (Table 1). Each basic community type of terrestrial vegetative association can be divided into five to seven communities, all of which are of larger extent than the assemblages designated as communities in the marine environment. Table 1 lists each major community under what terrestrial ecologists might call *biomes*. These aggregations cover hundreds of thousands of square miles and have no real equivalent in the marine world, with the possible exception of the abyssal community.

1. Low-diversity forest communities, in general, produce a low diversity of animal life with perhaps only one species of large mammal, few reptiles and amphibians, and relatively few species of perching birds (Fleming, 1973). Lack of a variety of fruits and nuts or annual forbs restricts the kinds of animals that can be supported. Birch, aspen, and tamarack groves have few rodents (no nuts), and thus few predators. Oak thickets and pine forests, on the other hand, support a fair variety of small mammals, thus encouraging a few large predators (Fleming, 1973).

2. High-diversity forests, such as mixed temperate hardwoods and tropical rain forests, support a rich variety of mammals, birds, reptiles, and amphibians. The greater the variety of plant food sources and types, the greater the diversity of animal species. Arboreal species often predominate in rain forests (Janzen, 1973); in mixed hardwood forests, large omnivores, carnivores, and herbivores are more evident.

3. Prairie communities are marked by the predominance of very large and relatively small herbivores. Ungulates (hoofed mammals) are the characteristic animal of the prairies on all continents, with small grain-eating rodents the next most abundant animal. Large predators follow the grazers on African prairies and apparently were common on the prehuman prairies of the Americas. Large scavengers and predaceous birds predominate on prairies of all types, almost to the exclusion of perching birds. Reptiles are common, whereas amphibians are restricted to infrequent permanent watercourses.

4. Contrary to normal expectations, animal communities in deserts and within xerophytic plant communities are diverse and abundant. To survive under arid

Table 1
Major plant communities for typical but not restrictive habitats.[a]

Forest Biome	
Low-Diversity Communities	*High-Diversity Communities*
Taiga (birch) forest: flat glacial plains (Pleistocene)	Temperate rain forest: maritime climate, igneous soil
Pine barrens: old strand lines (subclimax), late Tertiary, Pleistocene	Mixed eastern hardwoods: maritime, weathered acid soils
Palm groves: coral atolls, Holocene beaches	Coastal thickets: semitropical, floodplain soils
Mountain pine forest (yellow pine, spruce): unweathered soils	Tropical rain forest: lateritic soils, floodplains, tropics
Live oak groves: Pleistocene strand lines, sandy ridges	Coastal chaparral: moderately weathered coastal soils
Tamarack forest: subarctic glacial plain	Bottom-land hardwoods: alluvium clay and silt, streambeds
Aspen groves: subclimax, burned over pine forests	

Grasslands Biome	Desert Biome
Prairies: great central plains alkaline soils, former inland seas	Tropical xerophytic forest: Baja California (cactus)
Coastal and tropical savannahs	Desert chaparral
Temperate-zone meadows: subclimax to temperate forests	Salt falts: dry lake beds, mostly below sea level
Freshwater marshes: cattails, papyrus, reeds (subclimax)	Desert grasslands: cactus and yucca
Salt marshes: sedges, alligator weed, salt grass (subclimax)	Mixed cactus–chaparral: desert shrubs
	Dune communities

Tundra and Alpine Biome
Arctic tundra: forbs and short grass, soil limited by permafrost
Alpine meadows above tree line: forbs and short grass, soil thin

[a] More explicit data on diversity and composition of biomes and communities can be found in Fleming (1973), Janzen (1973), Küchler (1973, 1974), and Shreve and Wiggins (1964).

conditions, desert plants assume many shapes and methods for assuring reproduction, thus furnishing many small niches for animals (Beatley, 1974; Wallace et al., 1974). Browsers rather than grazers predominate, followed by reptiles and small rodents. Predators are common but of smaller size (for energy conservation) than those in the more mesophytic prairies and forests.

5. Tundra communities plus the alpine meadows above the tree line in most mountain areas are limited more by temperature than any other physical factor. Usable soils are perhaps limiting as well. In the Arctic, soils are permanently frozen; on high mountains, soils are thin because weathering processes have not created a thick soil horizon. Here plant diversity is relatively high because of an abundance of flowering plant species, mosses, and lichens. Rodent populations are high; thus, predaceous mammals and birds are also common. Tundra grasses and flowers support large herds of herbivores, thus encouraging large mammal predators, although diversity is low (Fleming, 1973).

A schematic depiction of typical Edwards Plateau communities as influenced by various outcrops of Lower Cretaceous formations is shown in Fig. 2 (Bryant et al., 1973), which demonstrates how broad plant-community associations are influenced by geology. At the top of the section, a thick limestone facies exists which supports a mixed semiarid forest and grassland. The Walnut Formation below is somewhat steeper and more porous and supports only grasses, small shrubs, and cactus; the Glen Rose Formation supports relatively large trees. Finally, the Quaternary alluvium at the bottom of the active arroyos supports a good mixed hardwood forest. With a little imagination, this cross section can be

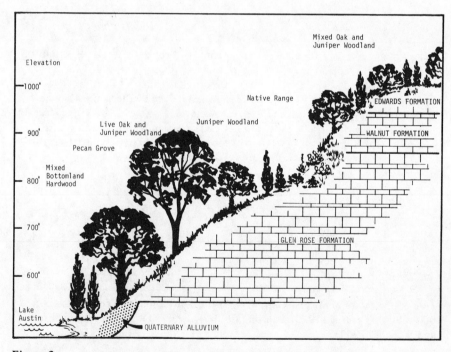

Figure 2
Generalized cross section of lower Cretaceous outcrops on the hills west of Austin, Texas. Note correspondence of vegetational changes with changes in geological strata. (Geological cross section compiled by R. W. Scott; in Bryant et al., 1973.)

expanded to the size of a continent. Shrubby vegetation occupies high elevations below the timberline. This is followed by large coniferous forests, which, in turn, phase into grasslands at gentler elevations. Grasslands interphase with forests on the more acidic soils of the river valleys. Similar geological-soil-community relationships are discussed in Küchler (1973).

Only a few of the terrestrial communities, most of which may have existed from Mesozoic times, have been discussed here. Plant remains are not as well preserved or as obvious as animal remains in both terrestrial and aquatic environments.

Aquatic Communities

Paleoecologists are on more familiar ground with the discussion of aquatic or marine animal communities, primarily because members of these communities leave abundant and permanent remains. The majority of contributors to this symposium will deal with specific modern and ancient communities and the factors that control them. For this reason, only some of the principles that govern diversity and abundance of organisms in the more evident aquatic communities will be discussed. Some of these principles have been presented before in different form; others are being tossed to the predators for digestion.

Aquatic ecosystems, although more familiar to most paleoecologists, in reality are more complex and less understood than terrestrial ecosystems. Terrestrial cycles are relatively simple, and all members of the living portions of the ecosystem are easily visible. On the other hand, many steps of aquatic ecosystems are only guessed at, inferred from animal or plant remains or by capture of animals by blind groping. Only a fraction of 1 percent of the plant and animal communities of the sea bottom has been mapped visually, whereas nearly the whole terrestrial environment has been made visible by satellite mapping. Most marine communities have been mapped on the basis of distantly spaced, nonseasonal, bottom faunal samples. Only coral reefs, oyster and mussel beds, and kelp beds are easily mapped by eye.

Freshwater communities. These communities are usually easily distinguished from marine communities. However, where speciation and selection have been extreme, such as in Lake Tanganyika and Lake Baikal, differences based on molluscan fauna are not so great. Although not generally realized, two families of freshwater mollusks occupying Lake Tanganyika for about 5 million years have produced look-alike species resembling every major marine gastropod family. What is more amazing is that each freshwater form lives in the exact same habitat or niche that its look-alike counterpart lives in within the marine environment. Knowledge of this phenomenon, gained some 20 years ago, prompted me to consider geomorphology and the physical environment that influence habitats as prime forces in governing community composition. Examples of Lake Tanganyika species and their marine counterpart are shown in Fig. 3. On the whole, however, freshwater communities are relatively simple, of low diversity, and easily recognized at the phylum and order level. Mollusk types are limited to only two or three families of pelecypods and about the same number of gastropod families. A discussion of diversity and sediment relationships within lacustrine mollusk communities can be found in

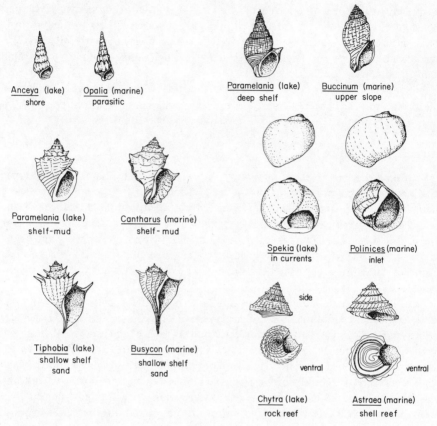

Figure 3
Examples of various species of freshwater gastropods endemic to Lake Tanganyika. Look-alike genera of marine families of gastropods are displayed with their counterparts.

Harman (1972). Shell structure and composition are different, and both trace element and isotopic geochemistry easily separates marine from freshwater shells. A lack of large crustaceans, echinoderms, polychaetes, stony corals, and other preservable marine forms characterizes most freshwater habitats. An excellent discussion of freshwater communities and their older counterparts can be found in Lister (1974).

Estuarine communities. The estuarine transition between freshwater and marine communities, on the other hand, produces distinct and recognizable animal and plant communities, most of which are governed primarily by the physical environment surrounding them. Controlling factors in the estuary are (1) salinity (as a function of river discharge or runoff), (2) sediment source (silts and clays from a low-energy coastal plain or sands and gravel from a high-energy steep coast), (3) tidal forces (controlling circulation), (4) nutrient sources and abundances, and (5) shape of the estuary (governed by physical forces and geological processes).

Community Classifications Based on the Ecosystem   77

As set forth many times in the past I recognize five main estuarine communities controlled by the general shape and physicochemical character of the estuary (Parker, 1956, 1959, 1960, 1969, 1975). These are (1) low salinity, river influenced, (2) mollusk reef, (3) interreef, (4) bay margin, and (5) bay center communities. Each is controlled or influenced primarily by the shape of the shore, circulation, and sediment type. Figure 4 illustrates the various geomorphic types and some of the biological details for each. The inlet habitat is related to both estuaries and

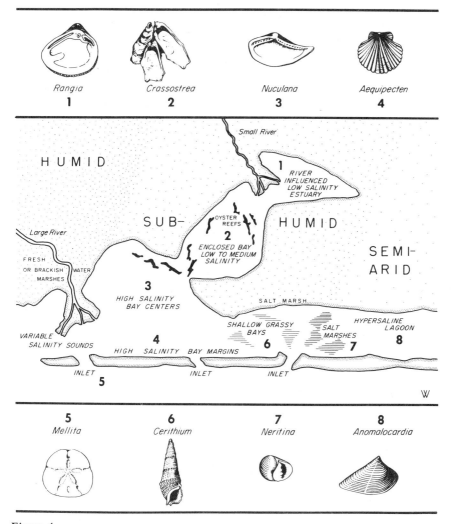

Figure 4
Typical Gulf coast estuarine habitats and representative members of the invertebrate communities that inhabit them. (From Parker, 1959; by permission of the American Association of Petroleum Geologists.)

open sea, since it is the interface between the two. It, too, is governed by circulation, sediment type, and shape.

Important considerations to remember concerning the regulation of estuarine communities include the following:

1. Areas with great extremes of climate in which changes occur with low predictability produce communities with very low diversity and low populations. The greater the age of the region, regardless of predictability, the more diverse the faunas become. Recently formed estuaries in areas with great ranges of physical factors have lowest diversity and population density; ancient estuaries located in areas of extreme ranges of physical factors, but greater predictability, have low diversity but higher populations.

2. There appears to be a direct correlation between water movement (vertical and horizontal) and organism abundance. Areas of both high wave motion and strong currents produce either dense beds of organisms or diverse communities of both epifaunal and infaunal animals and plants. Water movement in estuaries is controlled by geomorphology (or may also control geomorphology), that is, size of inlets, kinds of deltas, and shape of beaches, bays, and tidal islands.

3. Sediment types govern diversity and abundances of bottom organisms. Fine, well-sorted sands and well-sorted clays produce low diversity, through lack of a variety of niches, and low abundance of organisms as a result of providing little or no interstitial space for the accumulation of detrital organic matter (the food source for "primary" meiofaunal production). Poorly sorted coarse sediments with high detrital organic matter contents result in the production of higher animal populations and provide more niches and thus higher diversity.

4. Shallow bays with clay bottoms in which sediment is easily thrown into suspension usually support benthic communities based on a production cycle of bacterial degradation of refractory organic matter rather than on a phytoplankton production of organic matter. Turbid waters inhibit plant growth and enhance bacterial growth in the bottom.

Knowledge of these basic principles can be used to interpret the shape of ancient landforms at the seashore interface. The presence of dense shell reefs in living attitudes indicates the presence of strong currents in shallow water. Most mollusk species of groups that form reefs grow across the prevailing currents so as to derive the greatest benefit when filtering out microscopic life from the water. Low diversity and small populations of organisms with hard parts are indicative of fine sediments in settling basins, whereas high diversity of organisms with stout hard parts usually indicates proximity to the wave base in a high-energy estuarine area.

Shallow-water or nearshore marine communities. Shallow-water marine communities in the open sea are influenced primarily by depth, distance from shore, sediment type, and prevailing wave base. There is no question that bottom faunal composition changes with depth from shore to continental shelf edge. These changes occur regardless of sediment type, although sediment type is often dependent upon depth. Studies in European waters by Thorson (1957) and countless other benthic ecologists since have demonstrated the existence of a relationship between depth and organism composition. Parker (1960) revealed the existence of parallel and deepening communities along the Texas coast. Similar zonations were observed by this author in the Gulf of California (Parker, 1964a), along the western African coast by Longhurst (1958), and by Samuel (1944) on the Madras coast of India. A diagrammatic version of this banding of communities by depth is shown in Fig. 5. Note that in the high-energy surf zone, a special community of relatively low diversity but extremely high organism abundance exists. Characteristic of high-energy areas are mollusk communities such as oyster reefs, mussel reefs, and, in the surf zone, *Donax* (wedge clam) or *Spisula* (surf clam) beds dominated by single species. An extremely diverse and abundant community lives beyond the breakers, but still within sandy bottoms. All forms of energy are high (light, kinetic, and chemical), and sediments are less well sorted. Because wave energy is still high, animals often have stout, wave-resistant shells and are easily preserved. This community probably provides the most extensive and most easily recognized marine community in the fossil record. High production is possible because this zone is close enough to shore to receive a high level of nutrients to support high phytoplankton populations, which in turn form the base of a rich benthic food pyramid. The shallow or nearshore shelf community extends to depths of about 26 m (Fig. 5).

The next deepest community is the intermediate shelf community, which generally lives in water from 27 to 65 m deep. As waves are no longer providing the dominant energy for sediment deposition, bottom types are more often a result of other geological processes or events, such as transgressive and regressive shorelines, long shore currents and extreme tides, tidal waves, or storm waves. Although certain invertebrate animals (few benthic plants live this deep) provide an overall matrix of uniformity, this zone may be separated into various subcommunities that are dependent upon sediment types. Fine-sediment bottoms, being off major rivers, provide a niche for deposit feeders and few colorful mollusks; sandy bottoms furnish a substrate that supports epifaunal animals and large numbers of echinoderms and showy mollusks. Some seasonality of fauna may occur, because seasonal temperature changes are effective to these depths.

The deepest shelf community exists from about 66 m to the break in the slope of the shelf (120 to 160 m). In most areas, these depths are beyond wave base and are typically covered by clay or silty clay. Communities are usually infaunal and contain a large percentage of soft-bodied animals. Mollusks are generally small and scarce and diversity is low. There is little difficulty in separating nearshore fossil beds from offshore or deep-shelf fossil beds. One peculiarity of Holocene continental shelves is the fact that the shelf edge in many areas is marked by relict sands, calcareous prominences, or shell beds. Lowered sea levels have left their mark, characterized by the remains of the highly abundant high-energy communities that once thrived in a surf zone. Former eustatic and tectonic changes in sea levels around the continental nuclei have been marked by such communities in the past and should be recognizable from mixtures of shallow- and deep-water organism remains.

Continental slope and deep-sea marine communities. A final set of recognizable marine communities can be found on the sea bottom from the top of the continental slope to the greatest depths of the sea. Examples of the depth zones in the deep

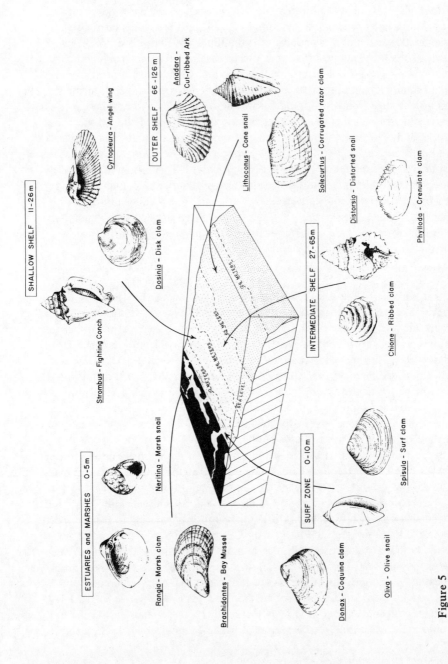

Figure 5
Location of typical shelf and coastal habitats, with representative (diagnostic) mollusk species (typical of subgeneric groups with ecological restrictions) for each depth zone. Community trophic organization within depth zones is partially controlled by sediment types. (Parker, 1964b, p. 362-363.)

sea can be seen in Fig. 6. Note that parallel zones can be discerned according to depth, but that the deepest zone is not the farthest from land. Most deep-sea trenches are relatively close to land, and their faunas are quite different from those on the adjacent abyssal sea floor. Until the work of Danish biologists on the *Galathea Expedition*, most of our knowledge of deeper animal communities was gained from the *Challenger Expedition* and several smaller European and American voyages. Diversity and abundance of deep-sea animals are now well established by Sanders (1968) and many others since then. I have taken an enormous number of different kinds of sea animals from depths between 1,000 and 4,000 m in areas thought to be relatively barren. What was more remarkable was that this diversity has been little changed since upper Miocene times. Olsson (1942) discovered Plio-Miocene deposits on the Burica Peninsula of Panama and Costa Rica (Fig. 7), which proved to be richly fossiliferous. Examination of these fossils by me revealed that the

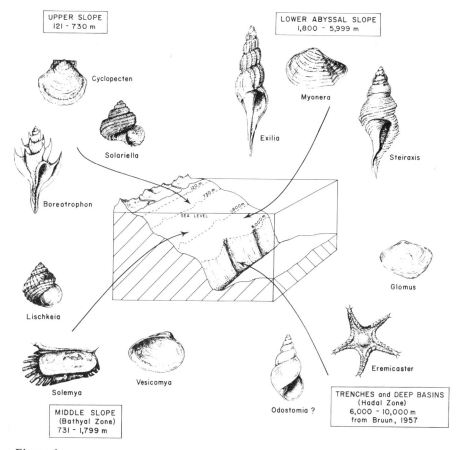

Figure 6
Generalized depth zones or habitats in the deep sea, with typical invertebrate genera usually found at these depths.

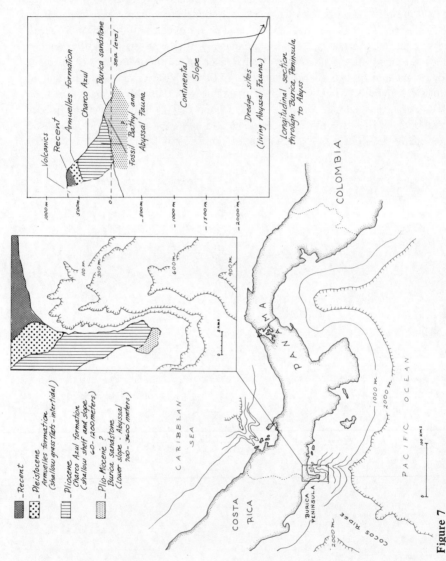

Figure 7
Location and generalized cross section of Plio-Miocene to Holocene geological formations on the Burica Peninsula of Panama and Costa Rica. (Geological information from Olsson, 1942.)

shells were of deep-sea forms exactly like those trawled in depths of 3,000 m off the same coast. This is one of the few places in the world where deep-sea deposits have become uplifted to sea level rather than becoming a part of the mantle again. If sea-floor spreading and continental drift is a valid theory, it is not likely that many deep-sea deposits will be found at the surface on continental margins. Even so, the fact that some deep-sea fossils do exist suggests that the criteria for deep-sea community organization should be included in this discussion.

1. The closer the community is to the shelf edge, the higher the population density. Organic matter slumps off the shelf edge and there are still sufficient nutrients to support high plankton populations whose remains fall to the bottom. The farther away the community is from shore, the fewer the animals, but not necessarily the lower the diversity.

2. Diversity in the abyssal areas may approach that of the continental shelf, but for different reasons. Scarcity of food (based on the rain of decomposing organic matter) promotes the selection of many unique animal forms, each equipped to digest organic matter at a different chemical base of degradation. Whereas shallow-water infaunal animals may ingest a wide range of decomposed organic matter, deep-sea animals may only eat very specific compounds or combinations of compounds. Likewise, because of the high diversity of infauna and particularly scavengers, many kinds of predators have evolved to make use of this diversity.

3. Deepest living animals appear to be descended directly from animals living in slightly shoaler depths, which suggests that great depths are transient in terms of animal evolution. Lack of high diversity and the presence of small populations in present-day trenches suggest that the trenches are not very old, geologically.

4. There appears to be little difference in composition between benthic communities living on different sediments on the abyssal plain. Greatest differences in diversity and abundance among animal communities are evident among those living on rock bottom of the steeper continental slopes and on the igneous slopes of submarine volcanos.

CONCLUSIONS

In summary, a number of basic principles can be presented that appear to govern the composition and location of both terrestrial and aquatic plant and animal communities. It is probable that these principles have been in operation since the arrival of most of the dominant plant and animal phyla on earth.

1. Terrestrial communities are controlled by latitude, altitude, availability of moisture, and soil types. Soil types are dependent upon the source and age of rocks from which they are derived. Sunlight and water, along with nutrients, are also a function of geological processes. Position of continents upon which plant communities grow is a function of tectonic forces. The amount of sunlight that plants receive depends upon latitude and altitude, both governed by crustal deformation and continental drift. Water, soils, and nutrients depend upon geomorphology and geological processes, which reflect latitude, altitude, and past geological environments.

2. Plant communities controlled by geological processes directly are plains (grasslands), primarily developed from weathering of carbonates derived from epeiric or epicontinental seas; pine forests, more often growing on ancient strandline deposits; and hardwood forests, mostly found growing on highly weathered and acidic soils of primarily Paleozoic age.

3. Diversity of plants and consequent diversity of terrestrial animal life depend upon predictability of climate, porosity of soil, and availability of water. Areas of even and abundant sunlight, high rainfall, and abundant nutrients produce highly diverse plant and animal communities. Regions with widely fluctuating temperatures and rainfall, and consequently varying nutrient supplies, yield low-diversity communities.

4. Community boundaries on land are easily discerned; aquatic community organization and boundaries are much more hypothetical. Almost all aquatic communities lack attached plant life; almost all terrestrial communities are based on rooted plants. Different criteria must be used to discern terrestrial community groupings than are used for grouping aquatic communities. Overall terrestrial and aquatic community composition is dependent upon energy availability for all communities, but radiant energy becomes a less important direct controlling factor in driving aquatic ecosystems.

5. Diversity and abundance of all organisms in all communities are governed primarily by availability of radiant energy (to support the biosphere), kinetic energy of motion, and chemical energy of compound oxidation or reduction. Ecosystems are organized through the utilization and conservation of these energy sources through various trophic levels. Internal community composition depends upon how animals and plants divide up the energy and pass it through each system for maximum utilization, plus the stability and predictability of these energy resources.

6. Of all aquatic communities, freshwater communities are the simplest, least diversified, and support the lowest number of organisms. The primary reasons for low diversity and abundance in freshwater communities are their transitory nature and geological youthfulness. Secondary reasons for low diversity relate to low kinetic energy levels and lack of diverse chemical niches.

7. Estuarine community diversity, composition, and organism abundances are governed by stability and predictability of salinity (at various ranges), temperature, sediment sources, and nutrient sources. Arrangement of communities within estuaries depends upon circulation, which in turn is governed primarily by river discharge, runoff, tides, and shaping of the estuaries by previous geological process.

8. Trophic organization of shallow shelf communities is governed by available amounts of wave energy, light penetration, vertical and horizontal water circulation, sediment source, and distance from shore. Depth facies are related to energy levels. Subcommunity groupings within depth facies depend upon previous geological events, which result in patchy sediment groupings.

9. Deep-sea communities reflect depth zones and distance from shore or continental shelf edge. Diversity and abundance of deep-sea faunas relate geologically to age of the sea floor (least diverse communities live in trenches, which are geo-

logically young) and to availability of organic matter. Most abundant populations of deep-sea animals live in zones where sources of organic matter are close by, adjacent to the slope, or under areas of continual upwelling.

10. The majority of fossil assemblages can be used to interpret not only environments of deposition, but also the geomorphology of land or sea-bottom habitats in which they formerly lived.

REFERENCES

Beatley, J. C. 1974. Phenological events and their environmental triggers in Mojave Desert ecosystems. Ecology, 55:856–863.

Bryant, H. H., L. E. Alderson, D. J. Arnold, R. H. Parker, and R. W. Scott. 1973. Environmental inventory and assessment of East Ranch. *For* Southern Living and Leisure Corporation, Wimberley, Texas: Coastal Ecosystems Management, Inc., Fort Worth, Texas. 120p.

Fleming, T. H. 1973. Numbers of mammal species in North and Central American forest communities. Ecology, 54:555–563.

Harman, W. H. 1972. Benthic substrates: their effect on fresh-water mollusca. Ecology, 53:271–277.

Janzen, D. H. 1973. Sweep samples of tropical foliage insects: description of study sites, with data on species abundances and size distributions. Ecology, 54:659–686.

Küchler, A. W. 1973. Problems in classifying and mapping vegetation for ecological regionalization. Ecology, 54:512–523.

———. 1974. A new vegetation map of Kansas. Ecology, 55:586–603.

Lister, K. H. 1974. The significance of temporal changes in a lacustrine ostracode community from the Great Salt Lake Basin, Utah. Geol. Soc. Amer. Abst., 6:847–848.

Longhurst, A. R. 1958. An ecological survey of the West African marine benthos. Colonial Office (Br.) Fishery Publ. 11, 102p.

Odum, H. T. 1975. Marine ecosystems with energy circuit diagrams. *In* J. C. J. Nihoul, (ed.), Modelling of marine systems. Elsevier Scientific Publishing Company, Amsterdam. p. 127–151.

Olsson, A. A. 1942. Tertiary and Quaternary fossils from the Burica Peninsula of Panama and Costa Rica. Bull. Amer. Paleont., 27:1–82.

Parker, R. H. 1956. Macro-invertebrate assemblages as indicators of sedimentary environments in east Mississippi Delta region. Amer. Assoc. Petrol. Geol. Bull., 40:295–376.

———. 1959. Macro-invertebrate assemblages of central Texas coastal bays and Laguna Madre. Amer. Assoc. Petrol. Geol. Bull., 43:2100–2166.

———. 1960. Ecology and distributional patterns of marine macro-invertebrates, northern Gulf of Mexico. *In* F. P. Shepard, F. B. Phleger, and T. H. van Andel (eds.), Recent sediments, northwest Gulf of Mexico. American Association of Petroleum Geology, Tulsa, Okla., p. 302–337.

———. 1964a. Zoogeography and ecology of some macro-invertebrates, particularly mollusks, in the Gulf of California and the continental slope off Mexico. Vidensk. Medd. Dansk naturh. Foren., 126:178.

———. 1964b. Zoogeography and ecology of macro-invertebrates, Gulf of California and continental slope of western Mexico. *In* T. H. van Andel, and G. G. Shor, Jr.

(ed.), Marine geology of the Gulf of California. Amer. Assoc. Petrol. Geol. Mem. 3:331–376.

———. 1969. Benthic invertebrates in tidal estuaries and coastal lagoons. *In* A. A. Castañares, and F. B. Phleger (ed.), Lagunas costeras, un simposio. Mem. Simp. Intern. Lagunas Costeras, UNAM-UNESCO, p. 563–590.

———. 1975. The study of benthic communities—a model and a review. Elsevier Scientific Publishing Company, Amsterdam. 279p.

Samuel, M. 1944. Preliminary observations on the animal communities of the level sea-bottom of the Madras coast. Jour. Madras Univ., 15:45–71.

Sanders, H. L. 1956. Oceanography of Long Island Sound, 1952–1954. X. Biology of marine bottom communities. Bull. Bingham Oceanog., 15:345–414.

———. 1958. Benthic studies in Buzzards Bay. I. Animal-sediment relationships. Limnol. Oceanog., 3:245–258.

———. 1960. Benthic studies in Buzzards Bay. III. The structure of the soft-bottom community. Limnol. Oceanog., 5:138–153.

———. 1968. Marine benthic diversity: a comparative study. Amer. Naturalist, 102:243–282.

Shreve, F., and I. Wiggins. 1964. Vegetation and flora of the Sonoran Desert, Vols. 1 and 2. Stanford University Press, Stanford, Calif. 1740p.

Thorson, G. 1957. Bottom communities. *In* J. W. Hedgpeth (ed.), Treatise on marine ecology and paleoecology, v. 1, Ecology. Geol. Soc. Amer. Mem., 67:461–534.

Wade, B. A. 1972. A description of a highly diverse soft-bottom community in Kingston Harbour, Jamaica. Marine Biol., 13:57–69.

Wallace, A., S. A. Bamberg, and J. W. Cha. 1974. Quantitative studies of roots of perennial plants in the Mojave Desert. Ecology, 55:1160–1162.

Paleocommunities: Toward Some Confidence Limits

Keith B. Macdonald   University of California,
                                      Santa Barbara

ABSTRACT

Quantitative studies of present-day life and death assemblages, representing different community types developed under various environmental conditions, provide an important empirical approach for establishing "confidence limits" on the accuracy and significance of paleocommunity reconstructions and studies of community evolution.

Qualitative community characteristics, such as environmental setting, spatial–temporal distribution, and taxonomic composition, are the most likely to be accurately preserved in the fossil record. Species presence–absence patterns will usually be adequate for paleocommunity identification. Quantitative "structural" community characteristics—taxonomic diversity, equitability, homogeneity, and trophic relationships—are adequately preserved in low-turbulence environments but not in higher-turbulence settings. Measures of community dynamics such as productivity and energy flow are unlikely to be retrieved from the fossil record. In all cases, community reconstructions restricted to "shelled" taxa will be more reliable than those attempting to include the record of soft-bodied organisms.

INTRODUCTION

Paleocommunity research is rapidly expanding in several different directions (Watkins et al., 1973; West and Scott, 1974; Ziegler et al., 1974). The use of community patterns for refining paleoenvironmental and paleogeographic interpretations has been studied by several workers (Ziegler, 1965; Fox, 1970; Sutton et al., 1970; Anderson, 1971; Stevens, 1971; Donahue and Rollins, 1974; Thayer, 1974; and others); others are using the "community technique" (Ziegler, 1974) to refine biostratigraphic correlations (Boucot, 1970; Ziegler and Boucot, 1970; Kauffman, 1974). Research objectives with a greater biological emphasis include the study of paleocommunities as a context for phylogenetic evolution (Olson, 1951, 1966; Sloan, 1970; Berry, 1974; Boucot, 1974) and evolutionary studies of intrinsic community properties (Sorgenfrei, 1962; Hedgpeth, 1964; Ziegler et al., 1968; Bretsky, 1969, 1970; Goulden, 1969; Bretsky and Lorenz, 1970; Rhoads et al., 1972; Scott, 1972; Valentine, 1972; Bowen et al., 1974; and others).

Of these different research objectives, the last—community evolution—probably offers the greatest challenge. Many aspects of community dynamics are clearly more readily studied in Recent environments than in the geologic record. The added perspective of geologic time, however, provides paleoecologists with a unique opportunity to study both the origins and evolution of community structure and trophic relationships and their response to long-period environmental changes. Such studies might significantly expand general ecological theory.

The advantages provided by geologic time are partially offset by the incompleteness of the fossil record, which often makes detailed interpretation difficult. Although paleotonologists have long been aware of both the bias and incompleteness of the record (Newell, 1959; Chave, 1964; Speden, 1966; Durham, 1967; Lawrence, 1968; Raup, 1972; Schäfer, 1972), this awareness remains largely qualitative. Quantitative data concerning the modes and mechanisms of formation of fossil assemblages, of their relationships to original living communities, and of their potential limitations remain scarce. Comparative quantitative studies of present-day life and death assemblages, representing different community types developed under various environmental regimes, provide one of the few empirical approaches available for establishing "confidence limits" on the accuracy and significance of paleocommunity reconstructions.

This paper reviews data describing the life and death assemblages of present-day *marine benthic macrofaunal communities* and outlines tentative answers to the following questions:

1. What basic community properties should be examined in order to describe adequately present-day community structure and function?

2. Which of these community properties are likely to be preserved in the fossil record?

3. How accurately do presently accumulating marine death assemblages reflect these preservable community attributes?

COMMUNITY ATTRIBUTES

Communities must be reasonably objectively identified before their structure, function, and evolution can be adequately studied. Fager's (1963) operational definition of a community as "a group of species which are often found living together" is accepted here. What is sufficiently "often" remains a subjective decision, although several widely used explicit grouping procedures (Buzas, 1970; Macdonald, 1975) can promote reproducibility and consistency. The determination that species are "living together" will depend upon the mobility, activity, and sphere of influence of individuals and the interspersion of populations (Fager, 1963).

Basic community properties that may be examined once a community is satisfactorily identified include *taxonomic composition*—usually established during the community identification procedure, *taxonomic diversity*—the number of species present, and *equitability* or *evenness*—a measure of the equality with which individuals are distributed among species. "Species diversity" measures have also been widely used in community studies, but they are less informative than taxonomic diversity and equitability for they combine both variables into a single statistic (Pielou, 1969; Buzas, 1972). Differences in community "pattern" can be measured by *homogeneity* indexes (Murdoch et al., 1972; Peterson, 1972, 1974; Driscoll and Swanson, 1973). Suppose, for example, that two communities share the same species with the same relative-abundance relationships; in one community all individuals are randomly distributed, but in the second the species are restricted to separate, nonoverlapping patches of substrate. The homogeneity of the two communities is clearly quite different, yet by definition their taxonomic diversity and equitability are identical. Characteristic patterns of *species succession* within a community might also be documented if successive replicate sample sets are available (Nicol, 1962; Johnson, 1972; Peterson, 1972; Halleck, 1973; Kauffman, 1974).

Considerable insight into community function can be gained from examination of the *trophic structure* (i.e., nature and relative abundance of different feeding types; Walker, 1972; Walker and Bambach, 1974) as well as the *standing crop* and *productivity* of successive trophic levels. The spatial and temporal distribution and environmental setting (inferred from the sedimentary record) of the community are also of prime importance.

Besides the intrinsic properties of multispecies assemblages, the following characteristics of each component species can be measured: *frequency*—the proportion of community samples containing a given species; *absolute, relative,* and *rank abundance* (Fager, 1957, 1963; Walker and Bambach, 1974); and *dispersion*—an index of the aggregation, random distribution, or even spacing of the individuals (Greig-Smith, 1964). *Fidelity*—a measure of the restriction of a species to a single community, (Johnson, 1971), *vitality*—the degree to which a species completes its life cycle within a single community, and *periodicity*—temporal, particularly seasonal, changes in species abundance, are also informative statistics but remain hard to quantify (Fager, 1963). Data summarizing the ecological life histories, environmental requirements, and adaptive strategies of component species would round out the information required to fully describe community structure and function.

CONFIDENCE LIMITS

Adequate description of all the parameters listed above remains out of reach for all but the simplest present-day communities, and, as in paleoecology, most ecological studies are limited to arbitrarily defined parts of communities—particular taxa, certain faunal size categories, or distinctive ecological groups. The choice of such categories usually reflects the availability or limitations of particular field-sampling or laboratory procedures or the ease of identification of the sampled organisms. It cannot be emphasized too strongly that the taxonomic composition of the communities identified will reflect the sampling techniques, sample size, and analytical methods chosen, as well as the distribution and behavior of the organisms being sampled. Furthermore, several community parameters—taxonomic diversity, equitability, and dispersion, for example—may change significantly as the number or size of samples is altered. Obviously, careful consideration must be given to the relationship between sampling and organisms before analytical results can be equated with real community properties (Cochran et al., 1954; Fager, 1963; Longhurst, 1964; Dennison and Hay, 1967; Stanton and Evans, 1972).

Before real evolutionary changes in community structure and function can be assessed, the paleoecologist must also be able to separate intracommunity variability from real intercommunity differences. The statistical concept of "confidence limits" is valuable here. A single sample of any community parameter is likely to prove unreliable. As additional samples are collected, both the average value and the variability associated with the parameter can be increasingly accurately assessed. Confidence intervals provide a precise statistical method of stating both how close the measured value of a statistic is likely to be to the "true" value of a parameter and the probability of its being that close (Tate and Clelland, 1957; Dixon and Massey, 1969; Reyment, 1971). Paleoecologists should be encouraged to collect sufficient data to assess both the variability and confidence limits of basic community attributes for the ancient communities that they describe. Comparative data describing present-day community variability are also needed: How do comparable communities vary between widely spaced localities? Do all communities exhibit similar levels of variability or are some more uniform in their properties than others? Which community attributes exhibit the greatest variability, which the least? While presently available data are insufficient to provide statistical confidence limits for most community attributes, the studies outlined next permit a general assessment of their reliability and provide a starting point for more rigorous future studies.

TOTAL VERSUS PRESERVABLE FAUNAS

The most widely recognized bias of the marine fossil record remains the nonpreservation of soft-bodied organisms. A recent survey of southern California shelf benthos (0 to 22 km offshore, water depths to 180 m; Jones, 1969) provides an excellent example of the consequences of such nonpreservation. Analyses of 176 quantitative samples yielded 1,473 macroinvertebrate species. Of this total, 523 species were polychaetes, 419 were crustaceans, 408 were mollusks, 64 were echinoderms, and the remaining 59 taxa represented several other phyla. Because few species other than the mollusks possessed resistant hard parts, the potential fossil assemblage would contain less than one third of the total living species. When the species were ranked in order of decreasing frequency of occurrence, the most widespread mollusk ranked only twenty-third. When ranked in order of decreasing mean population density, the most abundant mollusk ranked seventh.

In a comparable survey of Georgia's shelf benthos (0 to 20 km offshore, water depths to 15 m), Dörjes (1972) identified 268 macroinvertebrate species. Of these polychaetes accounted for 76 species and 56 percent of the total individuals, crustaceans for 67 species and 15 percent of the individuals. With 90 species, the mollusks were the most diverse phylum, yet they accounted for only 3.5 percent of the total individuals. Two species (*Donax variabilis* and *Tellina texana*) alone accounted for 88 percent of the total living mollusks collected! Comparable surveys of other marine level-bottom communities (Johnson, 1964) indicate that 33 to 93 percent of the total species and 13 to 99 percent of the total individuals of such communities do not possess hard parts and are therefore unlikely to be preserved. Similar problems of nonpreservation are also characteristic of other types of marine communities (Lawrence, 1968; Driscoll and Swanson, 1973).

In the absence of body fossils, trace fossils may confirm the former presence of soft-bodied organisms. It is apparent, however, that relatively few species of the total soft-bodied macrofauna leave unique, preservable traces that can be accurately identified. Hertweck (1972), for example, found that of the 268 macroinvertebrates identified from the Georgia shelf (Dörjes, 1972) only 40 produced distinctive lebensspuren, less than half of which were found preserved in the shelf sediments.

These examples indicate that nonpreservation of soft-bodied animals alone will often produce a death assemblage with a substantially different quantitative properties and trophic structure than the living community from which it was derived. Unless—or until—it can be shown that the role of soft-bodied organisms has changed substantially through geologic time we should assume that this conclusion also applies to ancient marine communities.

PRESERVABLE FAUNAS: LIVE-DEAD RELATIONSHIPS

Although studies of the evolution of ancient communities *in toto* are likely to remain out of reach, lack of preservation of soft-bodied animals does not prevent the study of changing community relationships among "shelled" taxa or otherwise preservable fossil groups. Studies of the accuracy and completeness with which presently accumulating assemblages of such "shelled" taxa reflect the attributes of the living "shelled communities" from which they were derived should provide valuable insight into the possible significance of community reconstructions developed from comparable fossil shelly faunas. Examples of such studies include those by Miyadi and Habe (1947), Habe (1956), van Straaten (1956, 1960, 1967), Valentine (1961), Johnson (1965), Wilson (1967), Cadée (1968), Evans (1968), Macdonald (1969a, b), Warme (1969, 1971), Hertweck (1971, 1972), Ekdale (1972, 1974), Peterson (1972, 1974), and Driscoll and Swanson (1973).

Johnson (1965) compared bivalve life and death assemblages in Tomales Bay, California. Transported shell material was readily identified, and most death assemblages appeared to have formed by the gradual accumulation of bivalves that lived at or near the site of burial. The taxonomic composition of the live faunas was well represented in the death assemblages. Absolute- and relative-abundance relationships were less accurately preserved: the same species occupied top abundance rank among both live and dead shells in 44 percent of the samples; the second ranked species corresponded in 23 percent of the samples; the third in only 13 percent of the samples.

Warme (1969, 1971), examining molluscan samples from Mugu Lagoon, southern California, found the live faunas adequately reflected by the death assemblages, whether compared on the basis of individual species, whole communities (defined from cluster analysis of presence-absence data), or species relative abundances within communities. He concluded that postmortem transport of shells within the lagoon was insignificant for most paleoecological purposes. The correspondence between live and dead shells from the lagoon's subtidal sandy channels (Table 1) is particularly impressive, for this habitat was characterized by coarse well-sorted sands subject to strong (>2.5 m/s) tidal currents.

Peterson (1972, 1974) reexamined Mugu's sand channel faunas in greater detail and compared them with similar faunas from Tijuana Slough, another lagoonal complex 260 km farther south. At both sites he compared a single sampling of the death assemblage with 10 replicate samplings of the living community, collected over a 3-year period. At Mugu, 74 percent of the individuals contained in the life-death assemblage sample set belonged to species that were represented by both live and dead shells in the same sample; at Tijuana this figure was 62.7 percent. (Equivalent figures for pooled sample sets from *all* environments, of 50.4 and 68.8 percent, are cited by Warme, 1969, and Johnson, 1965, respectively.) This confirms Warme's conclusion that live-dead correspondence is high and postmortem transport minimal. This same sample set yielded 27 live species and empty shells of 47 species; the Tijuana live-dead assemblage sample set also contained 27

Table 1
Mugu Lagoon sand channel fauna: total abundance (and frequency) relationships among live and dead macroinvertebrates collected in nine sand channel samples (Warme, 1969). B—bivalve, E—echinoid, G—gastropod.

Species	Live	Dead
Sanguinolaria nuttallii B	676 (9)	454 (7)
Cryptomya californica B	203 (6)	294 (8)
Dendraster excentricus E	42 (7)	54 (9)
Diplodonta orbella B	15 (5)	5 (4)
Olivella biplicata G	3 (1)	16 (8)
Chione californiensis B	2 (2)	6 (4)
Spisula dolabriformis B	1 (1)	2 (1)
Nassarius fossatus G	1 (1)	1 (1)
Lunatia lewisi G	1 (1)	1 (1)
Polinices reclusianus G	1 (1)	1 (1)

live species but only 38 "dead" species. Pooling the 10 replicate sample sets from the living communities still underestimated the taxonomic diversity of the death assemblages. Although this discrepancy could be partly explained by species rarity, patchy species distributions, and shifting habitats (Warme, 1969), most of the difference apparently reflected temporal variability of species populations. Calculation of an index of proportionate similarity (Peterson, 1974) showed that pooling of successive sample sets steadily increased the similarity of species relative-abundance relationships between the live and dead assemblages. Spatial homogeneity values (cf. pattern diversity; Murdoch et al., 1972) confirmed that the dead assemblages were significantly ($p > 0.01$) more homogeneous than the living communities. Significantly different homogeneity values obtained for the live communities at Mugu and Tijuana were accurately reflected by corresponding but smaller differences in the death assemblage values.

In my study (Macdonald, 1969a, b) of the restricted molluscan faunas of Pacific Coast salt marshes and tidal creeks, I examined the variability of live-dead assemblage relationships within single community types (Table 2). Recurrent group analysis (Fager, 1957) of quantitative data from 11 widely spaced sites (27 to 47° N), representing two molluscan provinces, readily separated and identified the distinctive molluscan communities characteristic of the salt marshes and tidal creeks. When described in terms of the identity and number of species present and their respective relative abundances and size-frequency distributions, the living communities exhibited a characteristic structure that remained fairly uniform between different sites within the same faunal province. This structure remained essentially the same despite differences in the latitude and area of the sites and in the species composition and density of the faunas. Within both environments, the community structure remained similar in the number of types of niches occupied; however, in the creek faunas the specific niche type changed with latitude (Table 2).

Twenty-eight of the 30 species collected alive in the quantitative samples were also represented by empty shells. The death assemblages contained an additional 34 species, 31 of which were represented by less than 10 individuals each and were clearly rare accidentals from adjacent habitats. When these rarer species were excluded, the death assemblages adequately represented the species composition and relative-abundance relationships of the living communities. The mean densities of living marsh and creek mollusks were positively correlated (Table 2; Kendall Tau, rank correlation = 0.46, $p > 0.06$, one-tailed), which suggests that some sites were biologically richer than others. The relative sizes of living and dead populations, however, varied erratically between sites (Table 2). On the average, species relative-abundance relationships corresponded better between live and dead assemblages than described by Johnson (1965) but not as closely as found by Peterson (1974). For 14 of 24 species examined, size-frequency distributions in the death assemblages were not different from those in the living material. In the remaining 10 species, either living or dead material was so much less common that comparisons could not be made.

Of the models currently available in the ecological literature, only the logarithmic curve gave an adequate representation of the individuals-species relationships ob-

Table 2

Salt marsh and tidal creek molluscan assemblages: number of live and dead species, mean density of live individuals, ratio of total live to total dead individuals and tidal creek trophic structure, recorded from various Pacific Coast localities. Feeding types: CS–ciliary suspension, D–deposit, R–rasper, P/Sc–predators and scavengers (after Macdonald, 1969a, b).

	Salt Marsh				Tidal Creeks				Feeding Types (%)			
Locality	Live Sp.	Density	Dead Sp.	L:D	Live Sp.	Density	Dead Sp.	L:D	CS	D	R	P/Sc
Grays Harbor	2	0.6[a]	2	0.21[b]	5	0.1[a]	5	0.32[b]	78	21	<1	0
Coos Bay	4	0.3	3	0.06	3	<0.1	3	1.35	95	5	0	0
Humboldt Bay	3	0.7	3	0.09	0	–	4	–	–	–	–	–
Tomales Bay (A)	2	0.2	3	0.05	2	0.9	11	0.22 }	22	0	78	0
Tomales Bay (B)	3	0.8	3	0.04	8	3.5	11	2.29				
Elkhorn Slough	2	5.8	2	1.23	4	7.8	11	2.54	10	<1	89	0
Mugu Lagoon	4	1.8	10	2.30	7	0.7	16	3.17	23	1	75	<1
Mission Bay	7	1.2	8	0.60	11	1.1	20	0.86	24	0	66	10
San Quintin Bay (A)	3	0.9	5	1.77	3	1.1	7	3.96 }	0	0	90	10
San Quintin Bay (B)	5	0.9	7	0.67	2	5.9	5	0.46				
Black Warrior Lagoon	5	0.8	19	7.36	5	0.3	16	2.09	1	0	91	8

[a] Mean density live mollusks, $1{,}000/m^2$.
[b] Ratio total live mollusks: total dead mollusks.

served in the living faunas. The fit to observation was far poorer in the case of the death assemblages but became quite adequate when all the species represented by single occurrences were eliminated. This suggests that the lack of fit to the raw data was largely due to sampling variability involving both rare and accidental species.

Cadée (1968) provided an excellent account of living and dead mollusk assemblages from the Ria de Arosa (northwest Spain), a considerably larger, deeper, and environmentally more diverse bay than those cited above. Marginal-bay, central-bay and oceanic zones were each characterized by different species. Estuaries within the bay were occupied by euryhaline species from the marginal zone. This same overall zonation pattern (cf. Bay of Arcachon, France; van Straaten, 1956) was readily recognizable in the molluscan death assemblages. Postmortem shell transport was essentially restricted to the narrow marginal-bay zone (depths of 0 to 3 m in the protected inner bay, and 0 to 20 m in the wave-exposed outer bay) and the exposed rocky coast of the oceanic zone. Within the marginal-bay zone, transport included (1) lateral mixing, (2) shoreward transport of deeper-water species, apparently confined to a belt from the tidal zone to depths of about 5 m, and (3) offshore transport of shallow-water mollusks. Although most marginal-bay species were recorded among the central-bay samples, they rarely accounted for more than 1 percent of the total individuals collected. Death assemblages from the oceanic zone (>60 m) evidenced both lateral mixing of local shelly-gravel infaunas with adjacent rocky-reef epifaunas and offshore transport of shallow-water species.

Whereas the taxonomic composition of the live molluscan communities were reasonably well represented in their death assemblages, quantitative relationships were less well preserved. After pooling all available data from each bay zone, Cadée compared the following characterisitcs of the "total" life and death assemblages: species relative-abundance relationships (Fig. 1); bivalve : gastropod : scaphopod ratio; epifauna : infauna ratio; and relative abundance of major trophic groups (Fig. 2). The observed relationships are clearly complex and vary considerably between bay zones.

Additional insight into the mechanisms and results of shell transport and faunal reworking, such as described from the "marginal-bay zone" by Cadée, are provided by Wilson's (1967) study of intertidal shell beds bordering Solway Firth, Scotland. Empty shells were more widely distributed than live animals of the same species. "Exotic" species from other habitats were widespread but rarely abundant; certain *Mytilus edulis* valves were shown to have originated more than 6 km from the study site. Like Cadée (1968), Wilson found predation to be a major cause of shell breakage (Driscoll, 1970; Trewin and Welsh, 1972). Shell beds accumulating on the tide-flat surface differed from channel-floor shell beds in taxonomic composition, species relative abundance, and species size-frequency distributions. Both types of shell beds also differed significantly in all three criteria from the local live populations.

Quantitative studies of continental shelf life and death assemblages remain scarce. Ekdale (1972, 1974) examined shallow-water (0 to 60 m), reef-associated molluscan assemblages from eastern Yucatan, Mexico. Fifty samples yielded 290 species, of which only 82 were represented by live animals. Seventy-five species were recorded (alive or dead) from single samples only. Cluster analysis of presence-absence

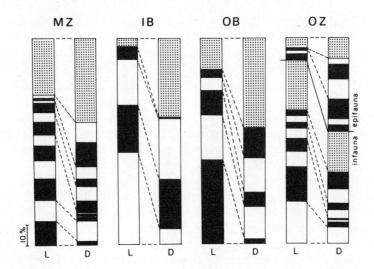

Figure 1
Ria de Arosa, Northwest Spain: Quantitative comparisons of species relative abundances in life (L) and death (D) assemblages. Alternating black and white bars indicate percentages of corresponding species; stippled bars represent additional rare species. [MZ—marginal zone, IB—inner bay, OB—outer bay, OZ—oceanic zone (see text and Fig. 2; modified after Cadée, 1968).]

and relative-abundance data for 66 common species distributed among 25 samples both yielded the same general associations, whether live mollusks or empty shells were considered. The three communities identified occupied lagoonal settings with restricted circulation, open marine settings with good circulation, and current-swept straits, respectively. Ekdale concluded that postmortem transport between habitats was insignificant and noted, like Warme (1969), that coarse-grained sediments supported and preserved the most distinctive and easily identified molluscan communities.

Evans (1968) analyzed 22 dredge samples from the western Puerto Rico shelf (0 to 36 m) at Anasco Bay. A fauna of 55 macroinvertebrate species was recorded, but only 12 of these were represented by live individuals. Four communities were identified; however, the richest of these in terms of live animals, a littoral fauna, was virtually absent from the death assemblage. Evans also noted downslope transport of skeletal grains from shallow, turbulent, back-reef habitats into deeper-water muddy-shelf environments. The two remaining communities, characteristic of the Anasco delta and mud-shelf environments, respectively, yielded no evidence of postmortem transport, and Evans concluded that their abundant skeletal remains had slowly accumulated in situ.

Dörjes (1971), following Petersen's (1918) classical approach to community definition, recognized four animal communities occupying parallel depth zones (>50 m) across the Gulf of Gaeta shelf, Italy. Hertweck (1971), analyzing death

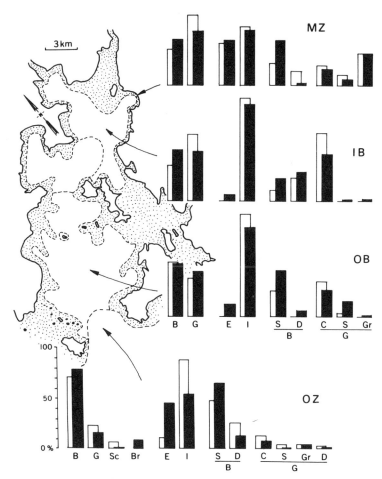

Figure 2
Ria de Arosa, Northwest Spain: Histograms showing composition (% total individuals) of shelled faunas within life (white) and death (black) assemblages collected from different bay zones. [B—bivalves, G—gastropods, Sc—scaphopods, Br—brachiopods, E—epifauna, I—infauna. Trophic categories: S—suspension feeders, D—deposit feeders, C—carnivores, Gr—grazers (see text and Fig. 1; modified after Cadée, 1968).]

assemblages from the same samples, recognized a "definite boundary" between in situ and transported shells, marked by wave base at about 6-m water depth. At greater depths, death assemblages "conformed well" with the living communities. In shallower water, shoreward shell transport and lateral mixing produced a mixed death assemblage (Cadée, 1968). Direct comparisons of Dörjes's (1971) and Hertweck's (1971) published data are difficult; however, it appears that distributions of species were broader among empty shells than among live animals. Land and freshwater snail shells occurred regularly in sample to 15 m water depth. Live *Abra alba* were restricted to water depths of more than 8 m, and live *Turritella communis* to depths greater than 15 m;—yet empty shells of both species were present in most shallower samples (van Straaten, 1960). The more abundant live mollusks also appear to have been more common in the death assemblages and, except above wave base, transported species apparently accounted for a relatively small percentage of the total dead individuals.

Qualitative continental shelf studies by other authors (Cadée, 1968) also have suggested that significant postmortem shell transport and faunal mixing by waves and currents is generally restricted to shallow waters (Table 3). Additional support for this view comes both from direct observations (Clifton, 1971) and from the fact that, despite their submergence, relict shallow-water molluscan assemblages on the continental shelves of both the eastern United States (Pilkey et al., 1969) and New Zealand (Norris, 1972) have retained their identity. Qualitative glimpses of deeper (16 to 600 m) molluscan faunas (Valentine, 1961; Emery and Hülsemann, 1962; van Straaten, 1967) also suggest in situ accumulations of death assemblages. It must be stressed, however, that recent research (Powers and Kinsman, 1953; Dobson et al., 1971; Brenner and Davis, 1973; Swift et al., 1973) has shown continental shelf environments to be considerably more dynamic than previously believed; thus, additional quantitative studies of shelf assemblages are certainly warranted.

Kranz (1974a, b) has demonstrated that computer simulation techniques can also be of great value in assessing differences between the attributes of death assemblages and their parent communities. Simulating interactions between rapid burial and bivalve escape behavior, Kranz found that, whereas deep (50 to 100 cm) burial faithfully preserved community properties, shallower burial favored some life habitat

Table 3
Maximum water depths at which shells are significantly transported by waves and currents.

Locality (Reference)	Depth (m)
Ria de Arosa, Spain,	
Entrance and outer shelf (Cadée, 1968)	∿ 60
Rhone Delta, France (van Straaten, 1960)	10–20
Wester Schelde, Netherlands (Wilderom, 1966)[a]	18
W. Jutland, Faeroes and Iceland (Johansen, 1902)[a]	15
Georgia Shelf, U.S.A. (Hertweck, 1972)	∿ 12–15
Kattagat, Netherlands (Johansen, 1901)[a]	12
Bohuslän, S. Sweden (Antevs, 1928)[a]	10
Gulf of Gaeta, Italy (Hertweck, 1971)	6
Ria de Arosa, Spain	
Marginal bay zone (Cadée, 1968)	5

[a]Full reference in Cadée (1968).

groups (e.g., infaunal mucus tube feeders) at the expense of others (e.g., epifaunal suspension feeders). The resulting death assemblages therefore provided a biased view of the species composition, relative-abundance relationships, and trophic structure of the original community. Because the survival value of bivalve mobility and escape behavior varies between habitats, Kranz concluded that the potential bias produced by rapid burial would be greater among quiet-water, stable-bottom communities than for communities typical of higher-energy, unstable bottom conditions.

DISCUSSION AND CONCLUSIONS

The results previously outlined clearly confirm that quantitative comparisons of present-day marine macroinvertebrate communities and their accumulating death assemblages can provide valuable data on the potential fidelity of the fossil record. Few such studies are presently available, but those that have been published lend support to several general conclusions.

It is a general observation (Eltringham, 1971; Thorson, 1971; and others) that many present-day marine benthic communities are dominated by soft-bodied animals, both in terms of numbers of species and total individuals. Body fossils of such groups are rare and, although trace fossils may reflect the presence of certain species, it is doubtful that they can provide quantitative data on either the true taxonomic diversity or abundance of the total soft-bodied fauna. Because the relative roles of soft-bodied versus shelled invertebrates in ancient communities remain unknown, paleoecological reconstructions of "total" communities are likely to prove far less reliable than studies of community evolution within single shelled phyla (e.g., molluscan communities, brachiopod communities) or co-occurring groups of such phyla.

Even when restricted to shelled phyla, it is already clear that both the scope for paleocommunity studies and their probable reliability change significantly between different environmental settings. The different levels of turbulence, sedimentation rates, and climatic regimes (frequency and intensity of storms, runoff patterns) characteristic of different environments in different geographic settings will directly influence the degree of postmortem transport, breakage, or dilution of shell deposits. These same variables can also create second-order effects. A particular sedimentary environment may favor epifaunal species over infaunal species, or mobile shallow burrowers over sedentary deep-burrowing forms. Because these different life habitat groups have differing propabilities of in situ preservation (Purdy, 1964; Craig and Jones, 1966; Cadée, 1968; Kranz, 1974a, b), the communities of which they are a part will exhibit varying levels of agreement between life and death assemblages.

In protected low-turbulence, low-runoff, shallow-water settings, such as studied by Johnson (1965), Warme (1969), and Peterson (1974), postmortem faunal mixing is negligible. Within-habitat species presence–absence data for live and dead mollusks exhibited close agreement, and comparable analytical procedures (recurrent group or cluster analysis of presence–absence data; Macdonald, 1975) identified broadly similar communities whether living animals or empty shells only were considered. Species relative-abundance and rank-abundance relationships also showed fairly good correspondence, but were less reliable than the presence-absence data. Living and dead material shared parallel patterns of spatial heterogeneity; however, as would be expected, absolute differences were less distinct among the death assemblages than the living communities. The relatively accurate preservation of community structure within these death assemblages would also ensure meaningful reconstruction of trophic relationships and spatial distribution of the live community, while appropriate analysis of sedimentary structures (Warme, 1971) should permit correct identification of the environmental setting. Comparable environments in the geologic record should provide excellent opportunities for paleocommunity analysis.

The situation on exposed tidal flats and along beaches and bay margins, above wave base, is much less satisfactory. Wave and current action tend to mix shells from adjacent varied microhabitats into a single death assemblage. Longshore transport, frequent sediment reworking, and selective mechanical and biological destruction of shells will all modify the species composition of the mixed assemblage. Species relative-abundance relationships and size-frequency distributions bear little resemblance to those of the original living communities. Homogeneity, trophic structure, and community spatial distribution patterns are all severely disrupted. Independent geologic evidence of a nearby shoreline, distinctive sedimentary structures, and the general aspect of the fauna might still permit correct identification of the environmental setting, but most faunal data would be inadequate for paleocommunity studies.

Quantitative data comparing life and death assemblages from offshore settings are very scarce. Cadée (1968) and Hertweck (1971) both found that species presence-absence data for communities developed below wave base were well reflected in associated death assemblages. Hertweck's data suggest that species relative-abundance and trophic relationships were also reasonably well preserved in the death assemblages, although the latter were more homogeneous than the corresponding live communities. Cadée found little overall quantitative correlation between life and death assemblages, but noted that the best correspondence occurred well inside the Ria de Arosa; it declined markedly on the exposed outer shelf. These apparently conflicting results may reflect differences in exposure, wave base, and tidal action between the two study sites. Because shallow shelf communities rank high among the more commonly preserved fossil assemblages, additional quantitative studies of life and death assemblages from such shelf settings would be instructive.

Death assemblages may result from sudden catastrophic burial (Kranz, 1974a, b) or the slow, gradual accumulation and reworking of approximately in situ skeletal remains (Warme, 1969). The second process, subject to the vagaries of shifting habitats, temporal variability of species populations, and patchy species distributions, creates time-averaged assemblages. These assemblages and the paleocommunities identified from them are likely to differ from their parent communities in several predictable ways. Taxonomic diversity, equitability, and homogeneity will be greater in the time-averaged assemblages, and their trophic structure may be modified through the addition of many rather rare species. The spatial distribution and environmental tolerance of time-averaged communities may appear broader than was really the case, because of temporal fluctuations in limiting parameters. Individual component species may also exhibit higher frequencies and more random dispersion patterns in the time-averaged assemblage than the original living community. These differences should be kept in mind as paleoecologists attempt to unravel possible changes in community structure and function through time.

The studies reviewed warrant the following general conclusions regarding the general reliability of different community attributes: qualitative community attributes such as the environmental setting, spatial and temporal distribution, and taxonomic composition are the most likely to be accurately preserved in the fossil record. Species presence–absence data will usually be adequate for correct paleocommunity identification. Quantitative "structural" community characteristics, such as taxonomic diversity, equitability, homogeneity, and trophic relationships, will be adequately preserved in some low-turbulence environments, but will become progressively less reliable under increasingly turbulent conditions. Measures of community dynamics, such as productivity and energy flow, are least likely to be retrievable from the fossil record. Limitation of paleocommunity studies to readily preservable "shelled" taxa will considerably enhance their overall reliability. Because soft-bodied organisms may have played significant roles in ancient communities, it must be remembered that structural and trophic relationships among such shelled taxa may be quite different from those of the "total community" to which they once belonged.

The reliability with which the mode of formation and preservation of fossil assemblages can be interpreted has increased considerably in recent years (Johnson, 1960; McAlester et al., 1964; Fagerstrom, 1964; Boyd and Newell, 1972; Seilacher, 1973; Bowen et al., 1974; and others). Along with knowledge of the approximate reliability with which various community attributes can be retrieved from the fossil record of different environments, this reliability should increase both the accuracy and potential significance of future paleocommunity studies.

REFERENCES

Anderson, E. J. 1971. Environmental models for Paleozoic communities. Lethaia, 4:287–302.

Berry, W. B. N. 1974. Types of early Paleozoic faunal replacements in North America: their relationship to environmental change. Jour. Geol., 82:371–382.

Boucot, A. J. 1970. Practical taxonomy, zoogeography, paleoecology, paleogeography and stratigraphy for Silurian and Devonian brachiopods. North Amer. Paleont. Convention, 1969, Proc., F:566–611.

———. 1974. Evolution and extinction rate controls. American Elsevier Publishing Company, New York, 370 p.

Bowen, Z. P., D. C. Rhoads, and A. L. McAlester. 1974. Marine benthic communities in the Upper Devonian of New York. Lethaia, 7:93–120.

Boyd, D. W., and N. D. Newell. 1972. Taphonomy and diagenesis of a Permian fossil assemblage from Wyoming. Jour. Paleont., 46:1–14.

Brenner, R. L., and D. K. Davis. 1973. Storm-generated coquinoid sandstone: genesis of high-energy marine sediments from the Upper Jurassic of Wyoming and Montana. Geol. Soc. Amer. Bull., 84:1685-1698.

Bretsky, P. W. 1969. Evolution of Paleozoic benthic marine invertebrate communities. Palaeogeog. Palaeoclimat. Palaeoecol., 6:45-59.

———. 1970. Late Ordovician benthic marine communities in north-central New York. N.Y. State Museum Sci. Ser. Bull., 414:1-34.

———, and D. M. Lorenz. 1970. Adaptive response to environmental stability: a unifying concept in paleoecology. North Amer. Paleont. Convention, 1969, Proc., E:522-550.

Bruun, A. 1957. Deep sea and abyssal depths. In J. W. Hedpeth (ed.), Treatise on marine ecology and paleoecology, v. 1, Ecology. Geol. Soc. Amer. Mem., 67:641-672.

Buzas, M. A. 1970. On the quantification of biofacies. North Amer. Paleont. Convention, 1969, Proc., B:101-116.

———. 1972. Patterns of species diversity and their explanation. Taxon, 21:275-286.

Cadée, G. C. 1968. Molluscan biocoenoses and thanatocoenoses in the Ria de Arosa, Galicia, Spain. E. J. Brill, Leiden, The Netherlands, 121p.

Chave, K. E. 1964. Skeletal durability and preservation. In J. Imbrie and N. D. Newell (eds.), Approaches to paleoecology. John Wiley & Sons, Inc., New York, p. 377-387.

Clifton, H. E. 1971. Orientation of empty pelecypod shells and shell fragments in quiet water. Jour. Sed. Petrol., 41:671-682.

Cochran, W. G., F. Mosteller, and J. W. Tukey. 1954. Principles of sampling. Jour. Amer. Statistical Assoc., 49:13-35.

Craig, G. Y., and N. S. Jones. 1966. Marine benthos, substrate and palaeoecology. Palaeontology, 9:30-38.

Dennison, J. M., and W. W. Hay. 1967. Estimating the needed sampling area for subaquatic ecologic studies. Jour. Paleont., 41:706-708.

Dixon, W. J., and F. J. Massey. 1969. Introduction to statistical analysis (3rd ed.). McGraw-Hill Book Company, New York, 683p.

Dobson, M. R., W. E. Evans, and K. H. James. 1971. The sediment on the floor of the southern Irish Sea. Marine Geol., 11:27-69.

Donahue, J., and H. B. Rollins. 1974. Paleoecological anatomy of a Conemaugh (Pennsylvania) marine event. In G. Briggs (ed.), Carboniferous of the southeastern United States. Geol. Soc. Amer. Spec. Paper 148, 153-170.

Dörjes, J. 1971. De Golf von Gaeta (Tyrrhenisches Meer). IV. Das makrobenthos und seine küstenparallele zonierung. Senckenbergiana Maritama, 3:203-246.

———. 1972. Distribution and zonation of macrobenthic animals. Senckenbergiana Maritama, 4:183-216.

Driscoll, E. G. 1970. Selective bivalve shell destruction in marine environments, a field study. Jour. Sed. Petrol., 40:898-905.

———, and R. A. Swanson. 1973. Diversity and structure of epifaunal communities on mollusc valves, Buzzards Bay, Massachusetts. Palaeogeog. Palaeoclimat. Palaeoecol., 14:229-247.

Durham, J. W. 1967. The incompleteness of our knowledge of the fossil record. Jour. Paleont., 41:559-565.

Ekdale, A. A. 1972. Ecology and paleoecology of marine invertebrate communities in calcerous substrates, northeast Quintana Roo, Mexico. Unpub. M.A. thesis. Rice University, Houston, Texas, 159p.

———. 1974. Recent marine molluscs from northeastern Quintana Roo, Mexico. *In* A. E. Weidie (ed.), Field Seminar on water and carbonate rocks of the Yucatán Peninsula, Mexico. Geol. Soc. Amer. Guidebook No. 2, Field Trip to Yucatan Peninsula, p. 199-218.

Eltringham, S. K. 1971. Life in mud and sand. Crane, Russak & Co., New York, 218p.

Emery, K. O., and J. Hülsemann. 1962. The relationships of sediments, life and water in a marine basin. Deep-Sea Res., 8:165-180.

Evans, I. 1968. The post-mortem history of the skeletal material of benthic invertebrate fauna in Anasco Bay, Puerto Rico. Unpub. M.S. thesis, University of South Carolina, Columbia, S.C., 50p.

Fager, E. W. 1957. Determination and analysis of recurrent groups. Ecology, 38: 586-595.

———. 1963. Communities of organisms. *In* M. H. Hill (ed.), The sea, Vol. 2. John Wiley & Sons Inc., (Interscience Division), New York, p. 415-437.

Fagerstrom, J. A. 1964. Fossil communities in paleoecology: their recognition and significance. Geol. Soc. Amer. Bull., 75:1197-1216.

Fox, W. T. 1970. Analysis and simulation of paleoecologic communities through time. North Amer. Paleont. Convention, 1969, Proc., B:117-135.

Goulden, C. E. 1969. Developmental phases of the biocoenosis. Proc. Natl. Acad. Sci., 62:1066-1073.

Greig-Smith, P. 1964. Quantitative plant ecology. Butterworth & Company (Publishers) Ltd., London, 256p.

Habe, T. 1956. Studies on the shell remains in bays. Contrib. Physiol. Ecol. Kyoto Univ., 77:28-31.

Halleck, M. S. 1973. Crinoids, hardgrounds, and community succession: the Silurian Laurel-Waldron contact in southern Indiana. Lethaia, 6:239-252.

Hedgpeth, J. W. 1964. Evolution of community structure. *In* J. Imbrie and N. D. Newell (eds.), Approaches to paleoecology. John Wiley & Sons, Inc., New York, p. 11-18.

Hertweck, G. 1971. Der Golf von Gaeta (Tyrrhenisches Meer) V. Abfolge der biofaziesbereiche in den vorstrand-und schelfsedimenten. Senckenbergiana Maritima, 3:247-276.

———. 1972. Georgia coastal region, Sapelo Island, U.S.A.: sedimentation and biology. V. Distribution and environmental significance of lebensspuren and in situ skeletal remains. Senckenbergiana Maritima, 4:125-168.

Johnson, R. G. 1960. Models and methods for analysis of the mode of formation of fossil assemblages. Geol. Soc. Amer. Bull., 71:1075-1086.

———. 1964. The community approach to paleoecology. *In* J. Imbrie and N. D. Newell (eds.), Approaches to paleoecology. John Wiley & Sons, Inc., New York, p. 107-134.

———. 1965. Pelecypod death assemblages in Tomales Bay, California. Jour. Paleont., 39:80-85.

———. 1971. Animal-sediment relations in shallow water benthic communities. Marine Geol., 11:93-104.

———. 1972. Conceptual models of benthic marine communities. *In* T. J. M. Schopf (ed.), Models in paleobiology. Freeman, Cooper & Company, San Francisco, p. 148-159.

Jones, G. F. 1969. The benthic macrofauna of the mainland shelf of southern California. Allan Hancock Monog. Marine Biol. 4, 219p.

Kauffman, E. G. 1974. Cretaceous assemblages, communities and associations, Western Interior United States and Caribbean Islands. Principles of benthic community analysis. Sedimenta IV Comp. Sed. Lab. Univ. Miami, 12.1–12.27.

Kranz, P. M. 1974a. The anastrophic burial of bivalves and its paleoecological significance. Jour. Geol., 82:237–265.

———. 1974b. Computer simulation of fossil assemblage formation under conditions of anastrophic burial. Jour. Paleont., 48:800–808.

Lawrence, D. R. 1968. Taphonomy and information losses in fossil communities. Geol. Soc. Amer. Bull., 79:1315–1330.

Longhurst, A. R. 1964. A review of the present situation in benthic ecology. Bull. Inst. Océanog. Monaco, 63:1–54.

Macdonald, K. B. 1969a. Quantitative studies of salt marsh mollusc faunas from the North American Pacific Coast. Ecol. Monog., 39:33–60.

———. 1969b. Molluscan faunas of Pacific Coast salt marshes and tidal creeks. Veliger, 11:399–405.

———. 1975. Quantitative community analysis: recurrent group and cluster analysis techniques applied to the fauna of the Upper Devonian Sonyea Group, New York. Jour. Geol., 83:473–500.

McAlester, A. L., I. G. Speden, and M. A. Buzas. 1964. Ecology of Pleistocene molluscs from Martha's Vineyard—a reconsideration. Jour. Paleont., 38:985–991.

Miyadi, D., and T. Habe. 1947. On thanatocoenoses of bays. Physiol. Ecol. Tokyo, 1:110–124.

Murdoch, W. F., F. Evans, and C. Peterson. 1972. Diversity and pattern in plants and insects. Ecology, 53:819–829.

Newell, N. D. 1959. The nature of the fossil record. Proc. Amer. Phil. Soc., 103:264–285.

Nicol, D. 1962. The biotic development of some Niagaran reefs—an example of an ecological succession or sere. Jour. Paleont., 36:172–176.

Norris, R. M. 1972. Shell and gravel layers, western continental shelf, New Zealand. New Zealand Jour. Geol. Geophys., 15:572–589.

Olson, E. C. 1951. The evolution of a Permian chronofauna. Evolution, 6:181–196.

———. 1966. Community evolution and the origin of mammals. Ecology, 47:291–302.

Petersen, C. G. J. 1918. The sea bottom and its production of fish food: a survey of the work done in connexion with valuation of Danish waters from 1883–1917. Rept. Danish Biol. Sta., 25:1–62.

Peterson, C. H. 1972. Species diversity, distribution and time in the bivalve communities of some California lagoons. Unpub. Ph.D. dissert., University of California, Santa Barbara, 229p.

———. 1974. Live and dead molluscs of two lagoons. Unpublished ms., 20p.

Pielou, E. C. 1969. An introduction to mathematical ecology. John Wiley & Sons, Inc. (Interscience Division), New York, 286p.

Pilkey, O. H., B. W. Blackwelder, L. J. Doyle, E. Estes, and P. B. Terlecky. 1969. Aspects of carbonate sedimentation off the southern United States. Jour. Sed. Petrol., 39:744–768.

Powers, M. C., and B. Kinsman. 1953. Shell accumulations in underwater sediments and their relation to the thickness of the traction zone. Jour. Sed. Petrol., 23:229–234.

Purdy, E. G. 1964. Sediments as substrates. In J. Imbrie and N. D. Newell (eds.), Approaches to paleoecology. John Wiley & Sons, Inc., New York, p. 238–271.

Raup, D. M. 1972. Taxonomic diversity during the Phanerozoic. Science, 177: 1065-1071.
Reyment, R. A. 1971. Introduction to quantitative paleoecology. American Elsevier Publishing Company, Inc., New York, 226p.
Rhoads, D. C., I. G. Speden, and K. M. Waagé. 1972. Trophic group analysis of Upper Cretaceous (Maestrichtian) bivalve assemblages from South Dakota. Amer. Assoc. Petrol. Geol. Bull., 56:1100-1113.
Schäfer, W. 1972. Ecology and paleoecology of marine environments (translation). C. Y. Craig (ed.), trans. I. Oertel. University of Chicago Press, Chicago, 568p.
Scott, R. W. 1972. Preliminary ecological classification of ancient benthic communities. 24th Intern. Geol. Cong. Proc., Sec. 7:103-110.
Seilacher, A. 1973. Biostratinomy: the sedimentology of biologically standardized particles. *In* R. N. Ginsburg (ed.), Evolving concepts in sedimentology. Johns Hopkins University Press, Baltimore, Md., p. 159-177.
Sloan, R. E. 1970. Cretaceous and Paleocene terrestrial communities of western North America. North Amer. Paleont. Convention, 1969, Proc., E:427-453.
Sorgenfrei, T. 1962. Some trends in the evolution of European molluscan faunas. Proc. First Europ. Malac. Congr., p. 69-78.
Speden, I. G. 1966. Paleoecology and the study of fossil benthic assemblages and communities. New Zealand Jour. Geol. Geophys., 9:408-423.
Stanton, R. J., and I. Evans. 1972. Community structure and sampling requirements in paleoecology. Jour. Paleont., 46:845-858.
Stevens, C. H. 1971. Distribution and diversity of Pennsylvanian marine faunas relative to water depth and distance from shore. Lethaia, 4:403-412.
Sutton, R. G., Z. P. Bowen, and A. L. McAlester. 1970. Marine shelf environments of the Upper Devonian Sonyea Group of New York. Geol. Soc. Amer. Bull., 81:2975-2992.
Swift, D. J. P., D. B. Duane, and O. H. Pilkey. 1973. Shelf sediment transport: process and pattern. Dowden, Hutchinson & Ross, Inc., Stroudsburg, Pa., 656p.
Tate, M. W., and R. C. Clelland. 1957. Nonparametric and shortcut statistics. Interstate Printers & Publishers, Danville, Ill., 171p.
Thayer, C. W. 1974. Marine paleoecology in the Upper Devonian of New York. Lethaia, 7:121-155.
Thorson, G. 1971. Life in the sea. World University Library, McGraw-Hill Book Company, New York, 256p.
Trewin, N. H., and W. Welsh. 1972. Transport, breakage and sorting of the bivalve *Mactra corallina* on Aberdeen Beach, Scotland. Palaeogeog., Palaeoclimat., Palaeoecol., 12:193-204.
Valentine, J. W. 1961. Paleoecologic molluscan geography of the Californian Pleistocene. Univ. Calif. Publ. Geol. Sci., 34:309-442.
———. 1972. Conceptual models of ecosystem evolution. *In* T. J. M. Schopf (ed.), Models in paleobiology. Freeman, Cooper & Company, San Francisco, p. 192-215.
van Straaten, L. M. J. U. 1956. Composition of shell beds formed in tidal flat environment in the Netherlands and in the Bay of Arcachon (France). Geol. Mijnbouw, 18e:209-226.
———. 1960. Marine mollusc shell assemblages of the Rhone Delta. Geol. Mijnbouw, 39e:105-129.
———. 1967. Turbidites, ash layers and shell beds in the bathyal zone of the southeastern Adriatic Sea. Rev. Geog. Phys. Geol. Dyn., 9:219-239.

Walker, K. R. 1972. Trophic analysis: a method for studying the function of ancient communities. Jour. Paleont., 46:82-93.
———, and R. K. Bambach. 1974. Analysis of communities. Principles of benthic community analysis. Sedimenta IV, Comp. Sed. Lab., Univ. Miami, 2.1-2.20.
Warme, J. E. 1969. Live and dead molluscs in a coastal lagoon. Jour. Paleont., 43: 141-150.
———. 1971. Paleoecological aspects of a modern coastal lagoon. Univ. Calif. Publ. Geol. Sci., 87:131p.
Watkins, R., W. B. N. Berry, and A. J. Boucot. 1973. Why "communities?" (Includes replies by P. W. Bretsky and J. W. Valentine.) Geology, 1:55-60.
West, R. R., and R. W. Scott. 1974. The "community" newsletter. Kansas State University, Manhattan, Kan., 22p.
Wilson, J. B. 1967. Palaeoecological studies on shellbeds and associated sediments in the Solway Firth. Scot. Jour. Geol., 3:329-371.
Ziegler, A. M. 1965. Silurian marine communities and their environmental significance. Nature, 207:270-272.
———. 1974. The community technique. Principles of benthic community analysis. Sedimenta IV, Comp. Sed. Lab., Univ. Miami, 1.1-1.10.
———, and A. J. Boucot. 1970. North American Silurian animal communities. *In* W. B. N. Berry and A. J. Boucot (eds.), Correclation of the North American Silurian rocks. Geol. Soc. Amer. Spec. Paper 102:95-106.
———, R. M. Cocks, and R. K. Bambach. 1968. The composition and structure of Lower Silurian marine communities. Lethaia, 1:1-27.
———, K. R. Walker, E. J. Anderson, E. G. Kauffman, R. N. Ginsburg, and N. P. James. 1974. Principles of benthic community analysis. Sedimenta IV, Comp. Sed. Lab., Univ. Miami, 175p.

Relationship of Fossil Communities to Original Communities of Living Organisms

Robert J. Stanton Jr.   Texas A & M University

ABSTRACT

The interpretation of fossil communities depends upon the relationship of the fossil communities to the original communities of living organisms, of which they are only small samples. Two tacit assumptions commonly made in paleontology are (1) that each fossil community represents a sample of a corresponding original community and, consequently, that distribution patterns of the fossil and original communities coincide, and (2) that the structure of the original community is preserved in the fossil community.

The modern communities on the Southern California Shelf have been analyzed to test these assumptions. The modern communities based on the total benthic invertebrate macrofauna, which correspond to the original communities, have been compared with the communities based on the potentially fossilizable taxa, which correspond to fossil communities. It is concluded that the first assumption is valid but that the second is not. Consequently, the community is a useful analytical entity by which modern environmental patterns and information can be related to the fossil record for paleoenvironmental reconstruction, but attempts to describe the paleoenvironment or community evolution on the basis of trophic structure of the fossil community must be made with caution.

INTRODUCTION

The study of fossil assemblages as samples of communities of once-living organisms has become a major topic of interest in paleontology. The community has become an important analytical unit because it is well suited for investigating several types of problems: (1) The comparison of living communities with fossil communities provides a way to compare the corresponding modern and ancient habitats. Thus, the community is a unit by which Holocene environmental data can be applied to the analysis of ancient environments (Dodd and Stanton, 1975). (2) The recognition and analysis of communities of fossils leads to a fuller understanding of ancient life by providing a framework within which organism interactions can be studied. (3) Differences in communities of various ages but from apparently similar environments can be analyzed and interpreted in terms of more subtle and otherwise unrecognized environmental differences, in terms of evolution within the taxonomic lineages present, or in terms of evolution of community structure itself (Walker and Laporte, 1970).

Major steps that have facilitated this work are (1) the analyses by Johnson (1960) and Lawrence (1968) of the processes that alter the composition of a community in the interval of time from its existence as a group of living organisms to its subsequent collection as an assemblage of fossils, (2) the suggestion by Bretsky (1968) that communities inhabiting specific environments have evolved through time as the results of extinction, first appearance, and evolution of component species, and that this community evolution can be described for several Paleozoic examples, and (3), the subsequent application of trophic analysis to the fossil record by Walker (1972) to study the biological structure of ancient communities.

The objectives of this paper are to examine basic assumptions about the nature of fossil communities that determine the extent to which community theory derived from ecology can be applied to paleontology, and to evaluate the adequacy of fossil communities to provide information about the original communities of which they are probably only a small preserved remnant. These objectives have been achieved by comparing communities on the Southern California continental shelf (Fig. 1) based on the total benthic macrofauna with communities based on the shelled and potentially fossilizable components of the same fauna.

The communities based on the shelled taxa are comparable to fossil communities. Those based on the total benthic macrofauna serve as approximations to the corresponding original total communities, or biocoenoses. They are approximations because all the flora, all the pelagic fauna, and the benthic microfauna are completely missing; and benthic macroorganisms that are vagile or deeply burrowing are also underrepresented in the collections. Nevertheless, the total communities are based on unusually complete data resulting from a large survey and coordination of the efforts of a large group of specialists. The two types of communities should be adequate to determine the relation between original and fossil communities.

COMMUNITY CHARACTERISTICS

Community is a valuable term in a general sense, but it has been used in so many ways that in detail it may have a different meaning for each investigator. These dif-

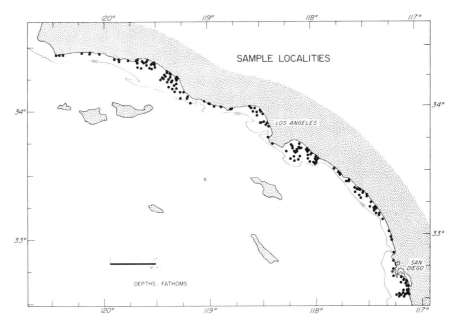

Figure 1
Locations of 195 samples on Southern California shelf from which total and shelled macroinvertebrate communities were constructed.

ferences in meaning reflect differences in method of recognition, scale, and ecologic meaning and significance.

Watkins et al. (1973) described three basic types of organism groupings that are very different in size but which have all been called communities. Valentine (1973), in commenting on their paper, contended that only the most local of their three units, the Peterson animal community, should be called a community. The operational definition of communities as "recurrent organized systems of organisms with similar structure in terms of species presence and abundances" (Fager, 1963, p. 415) reflects similar concepts and is the definition used in this study. In practice, the geographical size of a community will depend upon the magnitude of environmental variations, the amount of biotic variability acceptable within a single community, and the scale of the problem being investigated.

A critical aspect in community definition is the extent to which the total biota is utilized. Newell et al. (1959) have emphasized this by distinguishing the *biocoenosis*, which is based on the total biota, from the *organism community*, which is based on some fraction of the biota, such as mollusks, foraminifers, or shelled macrobenthos. Few ecologic studies have attempted to reconstruct the biocoenosis, and in paleontology this is impossible, for much of the original biota is not preserved. Surveys of modern benthic marine invertebrate faunas have shown that only about 30 percent of the living biocoenosis is potentially preservable to be recognized as a fossil community (Lawrence, 1968, p. 1317).

Communities, whether biocoenoses or organism communities, are defined and mapped on the basis of coincident occurrences of the component taxa. That is, successive samples from within a community have a high degree of taxonomic similarity compared to samples from different communities, and a large proportion of species will have highly correlated distributions. The component species of a community co-occur either because they prefer the same physical conditions or because of interactions (predator-prey, symbiosis, etc.) that link them together and exclude others. The relative importance of these two factors, environment and biologic interaction, has been a matter of continuing controversy among ecologists, and is important to the paleontologist as he attempts to interpret fossil communities.

If the distributions of individual taxa within a biota are determined solely by the physical environment, the compositions and distributions of the constituent biocoenoses should reflect, and be interpretable solely in terms of, the environment; the probability should be high that each biocoenosis would yield to the fossil record a distinctive organism community; and the distribution pattern of communities in the fossil record would correspond to the pattern of the original biocoenoses. This would be the ideal case if we wish to interpret the paleoenvironment from the distribution patterns of fossils and fossil communities.

To the extent, however, that the distributions of individual taxa within a biota are determined by biological factors of organism interactions, the compositions and distributions of the constituent biocoenoses should contain information about organism interactions as well as about the physical environment, and the interpretation of the subsequent fossil community will be meaningful only to the extent that these two components of its information content can be distinguished.

Two tacit assumptions are basic to most studies of communities in paleontology: (1) Each fossil community is a sample drawn from a corresponding original biocoenosis, and boundaries and distribution patterns based on the fossil community coincide with those based on the biocoenosis; (2) the trophic structure of the biocoenosis is preserved in the fossil community. The objective of this research has been to test the validity of these assumptions.

DATA

This report is based on data collected during an oceanographic and biological survey of the Southern California Shelf between Point Conception and the Mexican border by the Allan Hancock Foundation of the University of Southern California (Allan Hancock Foundation, 1965). Discussions of these data in a number of papers by the scientists involved in the survey are cited in the reference above and in Jones (1969). This study utilizes 195 samples that were collected in water less than 70 m deep and for which the fauna has been completely identified (Appendix 1; Figs. 1 and 2). Specific locations of these samples can be determined from lists and maps in Allan Hancock Foundation (1965). The faunal data from these samples were derived for this study from a duplicate of the original card deck from which Allan Hancock Foundation (1965) was prepared. Thus, minor errors of transcription that appear in that publication have been eliminated (personal communication, G. F. Jones, 1968).

The samples were collected with a Hayward Standard Orange Peel grab that sampled an area of 0.25 m^2. They were washed on shipboard through a series of screens, the finest of which had a mesh opening of 0.71 mm, and the organisms were subsequently sorted and identified. Further details of sample collection, preparation, and adequacy are discussed comprehensively by Jones (1969) and in references therein. The richness of major taxonomic groups in the total fauna and in the fauna that would be preserved in the fossil record is tabulated in Appendix 2. The latter fauna is referred to as the shelled fauna, but it does not include taxa with shells that are not likely to be preserved in the fossil record. For example, echinoderms are common in the total fauna, but only the sand dollar is considered likely to be preserved and is included in the shelled fauna. Other echinoderms, such as brittle stars, have shells that are very fragile or that disaggregate after death. Their nonpreservability is confirmed by their extreme rarity in the Tertiary strata of the Pacific Coast.

COMMUNITY DEFINITION AND DISTRIBUTION

Communities were defined in the total fauna and in the shelled and potentially preservable fauna by Q-mode cluster analyses utilizing Jaccard's coefficient and an

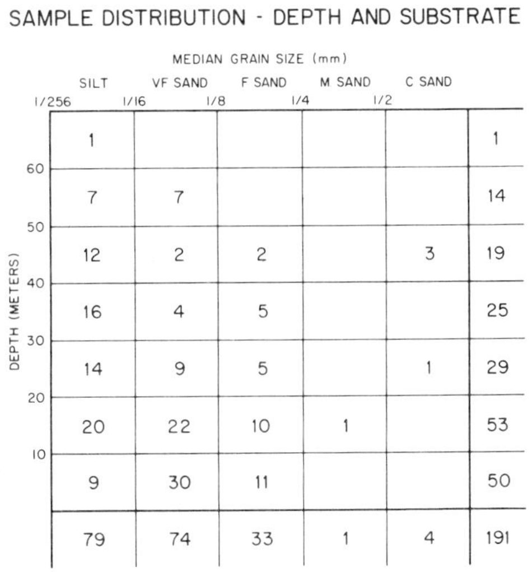

SAMPLE DISTRIBUTION - DEPTH AND SUBSTRATE

DEPTH (METERS)	SILT 1/256	VF SAND 1/16	F SAND 1/8	M SAND 1/4	C SAND 1/2	
60	1					1
50	7	7				14
40	12	2	2		3	19
30	16	4	5			25
20	14	9	5		1	29
10	20	22	10	1		53
	9	30	11			50
	79	74	33	1	4	191

Figure 2
Distribution of samples by depth and by median grain size of substrate. Four samples not charted because one or the other of these parameters is not available.

unweighted pair-group-averaging method. By this procedure, samples are grouped on the basis of similarity of faunal composition. In the past, both Q-mode and R-mode cluster analyses have been used to define paleontologic patterns and groupings. Q-mode analysis was used in this study; it is preferred for geologic or paleontologic problems in general because it is more analogous than R-mode analysis to the usual procedure in paleontology of comparing samples of fossils rather than co-occurrences of species, and because the results are more readily portrayed in mappable units. R-mode analysis, by which taxa are grouped on the basis of similarity of distribution, indicates taxa that are strongly paired or that have disjunct distributions. Thus, it provides the data essential for the study of organism interactions. The species clusters established by R-mode analysis, however, are commonly difficult to relate to actual samples (Lane, 1964).

From the cluster analysis, an area of modern sea floor characterized by relatively similar samples corresponds to the mapped distribution of a community; the community itself is composed of the organisms which occur in that area. Translated into a geologic setting, the area of the sea floor would represent a correlative horizon within a biofacies, and the fauna of that biofacies would be a community. To maintain the connection with paleontology, the areal extents of the modern communities in this study are referred to as biofacies in the following discussion.

SIMILARITY OF SHELLED AND TOTAL BIOFACIES

TOTAL \ SHELLED	1	2	3	4	MISC	
A	74 / 44 / 88	12 / 7 / 17	8 / 5 / 26	3 / 2 / 18	3 / 2 / 7	60
B	4 / 1 / 2	89 / 25 / 60	0	0	7 / 2 / 7	28
C	0	8 / 1 / 2	25 / 9 / 45	0	10 / 2 / 7	12
D	0	2 / 1 / 2	0	88 / 42 / 78	10 / 5 / 17	48
MISC	11 / 5 / 10	17 / 8 / 19	10 / 5 / 26	21 / 10 / 19	41 / 19 / 63	47
	50	42	19	54	30	195

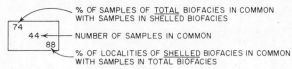

— % OF SAMPLES OF TOTAL BIOFACIES IN COMMON WITH SAMPLES IN SHELLED BIOFACIES
— NUMBER OF SAMPLES IN COMMON
— % OF LOCALITIES OF SHELLED BIOFACIES IN COMMON WITH SAMPLES IN TOTAL BIOFACIES

Figure 3
Similarity matrix of number and percentages of samples in common between shelled and total biofacies.

The 195 samples were grouped on the basis of the total fauna into four biofacies comprising 148 samples. The remaining 47 samples were each so different faunally that they did not fall into any group. Similarly, the samples were grouped on the basis of the fossilizable, shelled fauna into four biofacies comprising 165 samples, with the 30 remaining samples not falling into any group. The total and shelled biofacies to which each sample belongs are indicated in Appendix 1. The strong similarity in sample memberships for the pairs of shelled and total biofacies 1 and A, 2 and B, 3 and C, 4 and D, and shelled miscellaneous and total miscellaneous, evident from Appendix 1, are quantified in the similarity matrix of Fig. 3.

The geographic distributions of the biofacies are distinctive, but are not readily characterized at the map scale possible in this paper. The distinctive environmental parameters of each biofacies are evident, however, from depth-sediment texture plots (Figs. 4 and 5). The high correlation in geographic distributions of the pairs of biofacies noted above is confirmed by the similar depth-texture parameters in these two plots.

The total and shelled biofacies have been identified by letters and numbers respectively for simplicity. The faunal characteristics of the corresponding communities are listed in Appendix 3, and the dominant species of each total community are listed in Appendixes 4 to 7. The communities could be named by the dominant few species of each, but for the purposes of this study, it is more appropriate to use the same letter and number designations as used for the biofacies.

PALEOENVIRONMENTAL INTERPRETATION OF COMMUNITIES

Previous studies have demonstrated that the distribution patterns of communities of living marine benthic macroinvertebrates are strongly correlated with patterns and gradients of the physical environment. In the Gulf of California, for example, faunal patterns are correlated with water depth, temperature, and oxygen content and with substrate texture (Parker, 1964). In the northern Gulf of Mexico, they are correlated with the average values of salinity, temperature, turbidity, current strength, and substrate texture and with the degree of variability of these parameters (Parker, 1960).

Adjacent to the Mississippi Delta, the distribution of communities, which is strongly correlated with geographic position relative to the delta, appears to be primarily controlled by the combined effects of water circulation and energy, as determined by bathymetric and geographic factors, and by gradients in salinity and sedimentation, as determined by amount and proximity of freshwater influx from the Mississippi River. The distribution patterns of fauna and sediment are similar because each is determined by characteristics of the water mass and by the location and nature of freshwater influx (Stanton and Evans, 1971). In San Francisco Bay, the distribution of communities is correlated with variability of the local environment, as determined by the degree of circulation from the open ocean to various

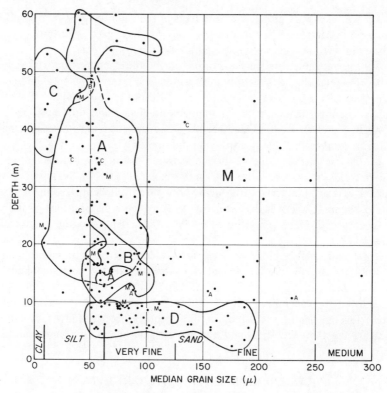

Figure 4
Distribution of samples and total biofacies by depth and by median grain size of substrate. All samples within outlined areas belong to biofacies as indicated by large letter except for the few other samples as indicated.

parts of the bay, and with the location within the bay of freshwater influx (Stanton and Dodd, 1970).

Biofacies patterns in the geologic record are also strongly correlated with environmental trends and patterns. For example, in Pliocene strata of the Kettleman Hills, California, which were deposited within a broad interior sea, community distribution patterns were determined primarily by factors similar to those which are important in San Francisco Bay: the degree of water circulation with the open ocean and localized deltaic sources of freshwater and sediment (Dodd and Stanton, 1975). On a much broader scale, biofacies in the Silurian shelf deposits of Wales paralleled the shoreline and shelf edge and were controlled by water depth and other depth-dependent aspects of the environment (Ziegler, 1965).

These and other studies of fossil and modern organism communities have suggested that the physical environment is the primary determinant of community composition and distribution, and that biological interactions play only a minor

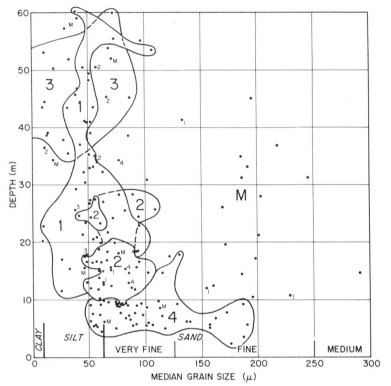

Figure 5
Distribution of samples and shelled biofacies by depth and by median grain size of substrate. All samples within outlined areas belong to biofacies as indicated by large numbers except for the few other samples as indicated.

role in shaping communities within the framework of the physical environment. The geographical and depth-substrate distributions of the total and shelled biofacies discussed herein are clearly in accord with the previous conclusions that biofacies patterns are well correlated with the physical environment. Although past work has shown that boundaries between modern or ancient biofacies are generally perpendicular to environmental gradients, it has not demonstrated that biofacies boundaries recognized in the fossil record should coincide with the boundaries of the original biocoenoses. The correspondence between total and shelled biofacies on the Southern California Shelf, however, does indicate that biofacies boundaries in the fossil record will be approximately in the same position as those defined by the original biocoenoses. The assumption that each fossil community represents a sample drawn from a corresponding biocoenosis, and that the biofacies boundaries and distribution patterns as based on the fossil communities mirror those of the original biocoenosis appears to be valid.

TROPHIC STRUCTURE OF BIOCOENOSES AND ORGANISM COMMUNITIES

Because a community is defined by distinctive taxonomic composition, it may be inferred to have a distinctive structure. That is, in each community, energy would flow through the ecosystem in certain proportions and along certain pathways. Taxonomic composition and proportions of biomass and metabolism in each level of the food pyramid would be distinctive, and interactions between organisms would involve particular species. Thus, community structure is an attribute that potentially might be analyzed and interpreted in terms of both depositional environment and community evolution. The analysis of community structure by mapping energy flow through all the possible pathways of the system has not been possible. Instead, the study of trophic relationships—the kinds and proportions of food resources available and the feeding characteristic of the organisms comprising the community—has been the approach more commonly used.

Our present understanding of these relationships has developed in two directions. One has been the correlation of feeding type with sediment texture and the depositional environment, as exemplified by the results of Sander's (1956) survey of the biology of the marine bottom communities of Long Island Sound. There the relative abundance of deposit feeders on finer-grained sediment and of suspension feeders on coarser-grained sediment is determined by water energy or agitation controlling, concurrently, sediment texture and the proportions of trophic resources that are in the sediment and in suspension. Deposit feeders predominate where water energy is low and food settles onto the sediment; suspension feeders predominate where energy is high and food is maintained suspended in the water column. The same energy conditions result in finer or coarser substrates, respectively. The correlation of feeding type with sediment texture may be modified, however, by other processes, such as the activity of deposit feeders in reworking the substrate and consequently excluding suspension feeders from otherwise potentially habitable areas (Rhoads and Young, 1970; Aller and Dodge, 1974).

Our understanding of trophic relationships has also developed from the work of Turpaeva (1957) and other Russian fisheries scientists. They have shown that a community is generally dominated by organisms of one feeding type (trophic group), that a community is usually dominated by a few most abundant species, and that, if the dominant species are ranked by abundance, successive species are of different feeding types. Furthermore, they recognized that these trophic characteristics of a community were determined by environmental conditions (Savilov, 1957).

The analysis of trophic structure of ancient invertebrate communities was described by Walker (1972). He found that the Silurian and Devonian communities he studied were generally like the communities described by Turpaeva in that each community was dominated by one trophic group, each trophic group was dominated by one species, and the feeding types of the ranked dominants within a community alternated. Subsequently, Rhoads et al. (1972) and Wright (1974) have shown that Cretaceous and Jurassic communities of Wyoming and South Dakota are each characterized by apparently distinctive trophic structure and are each restricted to particular depositional environments. These Mesozoic communities differ from the modern communities described by Turpaeva in two respects, however: (1) suspension feeders are much more common as dominants, and (2) alternation of feeding types among the dominants is much less common. In fact, many of these fossil communities are homogeneous, containing only suspension- or only deposit-feeding organisms, suggesting unusual, extreme environments in which trophic resources were limited to either the water column or the substrate.

For trophic structure to be useful in paleontology, the second of the two assumptions described earlier must be valid. The relation between trophic resources and trophic structure of the biocoenosis must take into account all the resources and, correspondingly, all the organisms as well. Thus, to interpret environmental conditions from the trophic structure of the fossil community, we must assume that the trophic structure of the biocoenosis can be determined from that of the fossil community. This structure can be preserved in the fossil record in three ways: (1) all the species, or at least all the dominant species, in the original biocoenosis are preserved and thus also comprise the fossil community, (2) the abundances and feeding types of soft-bodied or nonpreservable shelled species can be recognized from their traces or other records (Lawrence, 1968), and (3) the trophic characteristics of the soft-bodied organisms cannot be determined, but the shelled fraction preserved will nevertheless reflect the trophic structure of the biocoenosis. In the first case, soft-bodied organisms would have to have been much less abundant or dominant in acient biocoenoses than in modern ones. In the second case, if soft-bodied organisms were abundant and generally dominant as in modern biocoenoses, we must identify the trophic role of the nonpreserved organisms from their traces, as Walker (1972) attempted. The magnitude of this task is indicated by the number of species of soft-bodied organisms on the Southern California Shelf and the great diversity of their feeding types. In the third case, we must assume that the shelled fraction is a good estimate of the original trophic structure of the biocoenosis, including the soft-bodied organisms.

TROPHIC CLASSIFICATION

Analysis of the trophic structure of a community requires accurate information about the feeding methods and food requirements of the component organisms. This information is very incomplete for living organisms and is even scarcer and less precise for fossils. The following analysis is based primarily upon the food resources utilized by organisms at the primary consumer level of the food pyramid and secondarily upon feeding method. The three basic categories of resources are food within the substrate, food on the substrate surface, and food that is in suspension in the water or in the process of settling onto the substrate.

Classification of organisms by the trophic resource utilized is well suited to microphagous organisms, for the proportions of animal or plant material or of partially decomposed organic detritus in their diet are generally not known. The greatest source of error in this classification scheme is the lack of sufficient ecologic data by which to recognize microphagous carnivores, in order to separate them from the primary consumers.

Organisms classified herein as carnivores are either obvious macrophagous carnivores, such as muricid or naticid gastropods, or organisms with particular adaptations for a carnivorous method of feeding, such as the raptorial polychaetes. The category of scavenger is difficult to apply in paleontology and was not used in this study. Whether an animal is a scavenger or carnivore probably depends in many cases upon whether the first thing it encounters is alive or dead. In terms of energy flow through the community, parasitic animals correspond to prey-specific carnivores and, thus, have not been distinguished from carnivores.

Classification by feeding method distinguishes between those organisms which ingest sediment or water in bulk and are not selective (swallowers) and those which feed selectively (collectors). Organisms that feed upon suspended particles are classified as active (those which create water currents to bring food to them, such as most pelecypods and chaetopterid polychaetes) or passive [those with a variety of more passive feeding mechanisms, such as catching sand grains moving past in currents and cleaning the food off them (ampeliscid crustaceans) or laying palps or other structures out on the sea floor and feeding from what settles on them (cirratulid polychaetes)]. In addition, organisms that feed from suspended matter very near the bottom (arbitrarily defined as within 0.5 cm) are distinguished from those which feed from higher levels.

Organisms are difficult to classify by feeding characteristics for several reasons, as described previously. The problems that must be resolved in this process are well illustrated, for example, by the data available for the tellinid pelecypods (Yonge, 1949; Pohlo, 1969). Neither trophic resource nor feeding method is uniform within the family; thus, in general, conclusions about large taxonomic groups based on information about one or a few species may need to be revised as more information is gathered, and conclusions about fossils will inevitably be oversimplified. The inhalent siphon of tellinids ranges from very short, not extending above the sediment surface, to somewhat longer, lying passively on the surface, extending a short distance above the surface, or moving over the surface very much as a vacuum cleaner. Thus, members of the family feed to different degrees from material that is in suspension, lying on the sea floor, or drawn into suspension from the surface; the distinctions between active and passive, selective and nonselective, and suspension and on-sediment feeding are all gradational. The description by Stasek (1965) of some living species of the pelecypod *Yoldia* feeding from suspension as well as from the sediment surface also highlights the danger of considering whole groups such as the protobranch pelecypods as deposit feeders.

With these difficulties in mind, the living invertebrates of the Southern California Shelf were categorized as indicated in Table 1. Both published works and specialists have been consulted to establish the feeding characteristics of the organisms in this study. Literature sources that have been most useful are Yonge (1954), Bousefield (1973), Day (1967), and Jorgensen (1966).

Table 1
Invertebrates of the Southern California Shelf.

Source of Food		Feeding Method		Code
In the sediment	(I)	Swallower (nonselective)	(S)	IS
		Collector (selective)	(C)	IC
On the sediment	(O)	Swallower	(S)	OS
		Collector	(C)	OC
From the water	(W)			
High above the substrate	(H)	Active filtering	(A)	WHA
		Passive filtering	(P)	WHP
Low above the substrate	(L)	Active filtering	(A)	WLA
		Passive filtering	(P)	WLP
Carnivore, including parasitic				K
Unknown				U

ANALYSIS OF TROPHIC STRUCTURE

The most abundant species of each community of the total fauna have been categorized by feeding type, ranked by relative abundance within the community, and listed in Appendixes 4 to 7 in four ways: (I) the top 25 species; (II) the noncarnivorous species within list I; (III) the shelled, potentially fossilizable species in list I plus additional shelled taxa in order of dominance to include either all those present or at least 25 of them; (IV) the noncarnivorous species in list III. The rank of each species within a community is based upon the percentage of localities within the biofacies at which it occurs. The frequency for each species is given in the lists of Appendixes 4 to 7. This estimate of dominance does not differ greatly from an estimate based on numerical abundance of individuals, and corresponds closely to the measures available for macrofossils in paleontologic analysis.

The few shelled species present among the dominants of each community are indicated by asterisks in list I of Appendixes 4 to 7. Their relative rarity is also evident from the low rank of shelled species included in lists III and IV of Appendixes 4 to 7.

Polychaetes and crustaceans are the major groups in each total community, and brittle stars are locally a major component. Nemerteans and nematodes rank high, but this may be because they have not been subdivided into individual species as have the other major taxonomic groups.

The trophic structures at the primary consumer level are portrayed for each community in list II of Appendixes 4 to 7. The works of Turpaeva (1957) for modern communities and of Walker (1972) for Paleozoic communities suggest that successive species should utilize different food resources. However, some of the Mesozoic communities mentioned earlier were homogeneous, with only a single feeding type present. Each community in this study falls in a different position between these two extremes. For example, the top three species in community A are all suspension feeders, whereas the top three species in community C or D each utilize a different resource. The extent to which each community is dominated by a single trophic group also differs. In community A, the most abundant species and the community in general utilize waterborne food primarily, and in-sediment and on-sediment feeders are about equally abundant. In community D, on the other hand, the most abundant species is a suspension feeder, but in-sediment feeders are less abundant. In general, resources in suspension and within the sediment are most important and relatively few organisms utilize food on the sediment surface. Each community differs also in the relative abundance and dominance of carnivores.

The number of species used to define the trophic structure of a community has a significant effect on the results. For this reason, the proportions of species utilizing resources in the sediment, on the sediment, or in the water are plotted for the top-ranked 25 species in Fig. 6a and for the top-ranked five species in Fig. 6b. The proportion of the living or fossil assemblage being used for trophic-group analysis needs to be specified and standardized in future work in order for the trophic structure of different communities from different studies to be compared.

Lists III and IV portray the characteristics of the fossil communities likely to be derived from the original biocoenoses approximated by lists I and II. It is apparent that the dominants of the fossil communities would be only minor components scattered randomly down the species lists of the original biocoenoses. The trophic

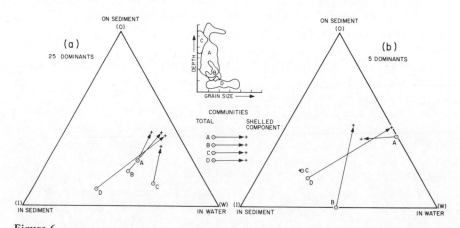

Figure 6
Proportions of dominant 25 (a) or 5 (b) species utilizing trophic resources in suspension, in the sediment, or on the sediment. Values circled are for total communities; values indicated by cross are for shelled communities derived from total communities. Measures derived from data in Appendixes 4 to 7. Depth–texture characteristics of communities generalized from Fig. 4.

composition of each shelled community differs significantly from that of the total community from which it was derived (Fig. 6). Because of the shift in trophic proportions from the total to shelled (original to fossil) community, the amount of in-sediment food utilized in the original environment is underestimated. In these modern communities this shift corresponds to the taxonomic change from abundant polychaetes, crustaceans, and brittle stars in the total community to abundant mollusks in the shelled community.

If one attempts to describe community structure in terms of the proportions of species or individuals at different levels of the food pyramid, similar problems are encountered. The proportion of carnivores to primary consumers in each shelled community bears little relation to that in the total community. For example, the proportion is relatively low in the shelled component of C and high in A (lists III), whereas in the total communities (lists I) it is relatively high in C and low in A.

The trophic structure of a fossil community is not a good estimate of that of the original biocoenosis.

DISCUSSION

The significance of these results must be considered for each type of application of trophic structure in paleontology. The interpretation from trophic structure of ancient water energy, and thus of associated environmental parameters such as water depth and degree of exposure, is based on the observations that proportions of deposit- and suspension-feeding organisms are correlated with water energy and sediment texture in modern environments (Driscoll, 1969). The results of this study indicate that environmental interpretation of trophic structure must be made with caution, for the structure of the fossil community is not a good estimate of that of the biocoenosis. In addition, the expected correlation of structure with environment is absent. The proportion of suspension- to deposit-feeders in the total communities of the Southern California Shelf should presumably increase in the sequence C-A-B-D (Fig. 6) on the basis of their relative depths and substrate texture. In fact, however, the sequence is D-B-A-C (based on 25 most dominant species) or C-D-B-A (based on five most dominant species). The results are opposite or nearly so to what would be expected from the theory.

The application of trophic structure to describe community structure and evolution through geologic time has been based primarily upon the work of Russian fisheries workers on the faunas of the Arctic and Pacific seas adjacent to Russia, and of the Black, Azov, and Caspian seas. These are all high-stress environments characterized by faunas of low diversity and high dominance. Consequently, they are probably poor modern analogues for geologic examples, which are typically viewed in terms of moderate climate and normal marine salinity. They are certainly very different from the Southern California Shelf environment of this study.

In studies of Paleozoic faunas where trophic-structural relations are similar to those of Turpaeva (e.g., Walker, 1972), two explanations seem possible. Either the Paleozoic faunas were diverse like that of the modern California Shelf and the species of the small samples preserved in the fossil record only fortuitously alternated in feeding type, or the Paleozoic communities were, like the modern Russian communities, low in diversity, high in dominance, and composed primarily of the shelled taxa preserved in the fossil record. If this latter hypothesis is correct, either the ancient environment was much more rigorous than we have normally considered it to be, or community structure has undergone significant evolution during geologic time. This evolution, however, would have to have consisted primarily of the addition of the many soft-bodied taxa so abundant and dominant in the modern fauna. This last possibility seems unlikely in view of the relative antiquity of annelids and arthropods in geologic history.

CONCLUSIONS

The community is a valuable analytical entity in paleontology because characteristics of communities of living organisms potentially may be extrapolated into the geologic record and be interpreted from fossil communities. These characteristics include relationships of the community to the physical environment and trophic-structural features of the community.

Much of the usefulness of the community in paleontology requires that two assumptions be valid: (1) each fossil community is a sample drawn from an original biocoenosis, and the boundaries and distribution of the fossil community coincide with those of the biocoenosis, and (2) the trophic structure of the biocoenosis is preserved in the fossil community.

To test these assumptions, communities of the living total benthic macroinvertebrate fauna and of the shelled component of the fauna were determined on the Southern California Shelf. The former communities approximate biocoenoses, whereas the latter correspond to the communities that would be preserved in the fossil record.

High correlations in geographic, depth, and substrate distributions of the total and shelled communities indicate that patterns in the physical environment interpreted from fossil communities coincide with those of the original biocoenoses. The first assumption is valid, and the community provides a valuable source of information for paleoenvironmental analysis.

Poor correlation in trophic structure of corresponding total and shelled modern communities indicates that the trophic structure of the original biocoenosis cannot be estimated satisfactorily from that of the fossil community. The second assumption is not valid, and interpretations from the trophic structure of a fossil community about topics correlated with the trophic structure of the biocoenosis (such as community evolution and aspects of the physical environment) must be made with caution.

ACKNOWLEDGMENTS

Gilbert F. Jones generously provided a card deck of the data upon which this study is based. Financial support has been provided by the American Petroleum Institute and the National Science Foundation.

Larry D. McKinney and Donald Harper generously provided information about the behavior and feeding characteristics of amphipods and polychaetes, respectively. Ian Evans, Ernest Mancini, and Penelope Nelson have provided assistance and stimulating discussions. J. Robert Dodd has read and criticized the manuscript.

REFERENCES

Allan Hancock Foundation, University of Southern California. 1965. An oceanographic and biological survey of the Southern California mainland shelf. California State Water Quality Control Board Publ. 27:1-232; Appendix-Data: 1-445.

Aller, R. C., and R. E. Dodge. 1974. Animal-sediment relations in a tropical lagoon, Discovery Bay, Jamaica. Jour. Marine Res., 32:209-232.

Bousefield, E. L. 1973. Shallow-water gammaridean Amphipoda of New England. Cornell University Press, Ithaca, N.Y. 312p.

Bretsky, P. W. 1968. Evolution of Paleozoic marine invertebrate communities. Science, 159:1231-1233.

Day, J. H. 1967. A monograph of the Polychaeta of Southern Africa. Brit. Museum Nat. Hist., Publ. 652, 2 pt., 878p.

Dodd, J. R., and R. J. Stanton, Jr. 1975. Paleosalinities within a Pliocene bay, Kettleman Hills, California: a study of the resolving power of isotopic and faunal techniques. Geol. Soc. Amer. Bull., 86:51-64.

Driscoll, E. G. 1969. Animal-sediment relationships of the Coldwater and Marshall Formations of Michigan. *In* K. S. W. Campbell (ed.), Stratigraphy and paleontology—essays in honour of Dorothy Hill. National University Press, Canberra, Australia, 337-352.

Fager, E. W. 1963. Communities of organisms. *In* M. N. Hill (ed.), The sea, Vol. 2. John Wiley & Sons, Inc. (Interscience Division), New York, p. 415-437.

Johnson, R. G. 1960. Models and methods for analysis of the mode of formation of fossil assemblages. Geol. Soc. Amer. Bull., 71:1075-1086.

Jones, G. F. 1969. The benthic macrofauna of the mainland shelf of Southern California. Allan Hancock Monog. Marine Biol. 4, 219p.

Jorgensen, C. B. 1966. Biology of suspension feeding. Pergamon Press, Inc., Elmsford, N.Y., 357p.

Lane, N. G. 1964. Paleoecology of the Council Grove Group (Lower Permian) in Kansas, based upon microfossil assemblages. Kansas Geol. Surv. Bull. 170, Pt. 5, 23p.

Lawrence, D. R. 1968. Taphonomy and information losses in fossil communities. Geol. Soc. Amer. Bull., 79:1315-1330.

Newell, N. D., John Imbrie, E. G. Purdy, and D. L. Thurber. 1959. Organism communities and bottom facies, Great Bahama Bank. Amer. Museum Nat. Hist. Bull., 117:177-228.

Parker, R. H. 1960. Ecology and distributional patterns of marine macroinvertebrates, northern Gulf of Mexico. *In* F. P. Shepard, F. B. Phleger, and T. H. van Andel (eds.), Recent sediments, northwest Gulf of Mexico. American Association of Petroleum Geologists, Tulsa, Okla., p. 302-337.

———. 1964. Zoogeography and ecology of some macro-invertebrates, particularly mollusks, in the Gulf of California and the continental slope off Mexico. Vidensk. Medd. Dansk Naturh. Foren., 126:178p.

Pohlo, Ross. 1969. Confusion concerning deposit feeding in Tellinacea. Proc. Malac. Soc. London, 38:361-364.

Rhoads, D. C., and D. K. Young. 1970. The influence of deposit-feeding organisms on bottom-sediment stability and community trophic structure. Jour. Marine Res., 28:150-178.

———, I. G. Speden, and K. M. Waagé. 1972. Trophic group analysis of Upper Cretaceous (Maestrichtian) bivalve assemblages from South Dakota. Amer. Assoc. Petrol. Geol. Bull., 56:1100-1113.

Sanders, H. L. 1956. Oceanography of Long Island Sound, 1952-1954. X. The biology of marine bottom communities. Bingham Oceanog. Coll. Bull., 15:345-414.

Savilov, A. I. 1957. Biological aspect of the bottom fauna groupings of the North Okhotsk Sea. In B. N. Nikitin (ed.), Transa. Inst. Oceanol. Marine Biol. USSR Acad. Sci. Press, 20:67-136. (Published in U.S. by American Institute of Biological Science, Washington, D.C.)

Stanton, R. J., Jr., and J. R. Dodd. 1970. Paleoecologic techniques—comparison of faunal and geochemical analyses of Pliocene paleoenvironments, Kettleman Hills, California. Jour. Paleont., 44:1092-1121.

———, and Ian Evans. 1971. Environmental controls of benthic macrofaunal patterns in the Gulf of Mexico adjacent to the Mississippi Delta. Gulf Coast Assoc. Geol. Soc. Trans., 21:371-378.

Stasek, C. R. 1965. Feeding and particle-sorting in *Yoldia ensifera* (Bivalvia: Protobranchia), with notes on other nuculanids. Malacologia, 2:349-366.

Turpaeva, E. P. 1957. Food relationships of dominant species in marine benthic biocoenoses. In B. N. Nikitin (ed.), Trans. Inst. Oceanol. Marine Biol. USSR Acad. Sci. Press, 20:137-148. (Published in U.S. by American Institute of Biological Science, Washington, D.C.)

Valentine, J. W. 1973. Comments. Geology, 1:59-60.

Walker, K. R. 1972. Trophic analysis: a method for studying the function of ancient communities. Jour. Paleont., 46:82-93.

———, and L. F. Laporte. 1970. Congruent fossil communities from Ordovician and Devonian carbonates of New York. Jour. Paleont., 44:928-944.

Watkins, Rodney, W. B. N. Berry, and A. J. Boucot. 1973. Why "communities"? Geology, 1:55-58.

Wright, R. P. 1974. Jurassic bivalves from Wyoming and South Dakota: a study of feeding relationships. Jour. Paleont., 48:425-433.

Yonge, C. M. 1949. On the structure and adaptations of the Tellinacea, deposit-feeding Eulamellibranchia. Phil. Trans. Roy. Soc. Lond., Ser. B, 234:29-76.

———. 1954. Feeding mechanisms in the Invertebrata. Tabulae Biologicae, 21:46-68.

Ziegler, A. M. 1965. Silurian marine communities and their environmental significance. Nature, 207:270-272.

Appendix 1

Samples used in study (numbers correspond to numbers in Allan Hancock Foundation, 1965) and total and shelled biofacies memberships of each sample. (− indicates that sample did not cluster into one of the biofacies.)

Sample No.	Total	Shelled	Sample No.	Total	Shelled
4718	A	1	4863	A	1
4719	−	1	4871	−	1
4720	B	−	4885	A	4
4721	−	−	4886	A	1
4722	A	1	4917	B	2
4743	B	1	4922	−	4
4745	−	−	4927	A	2
4746	−	−	4928	−	−
4758	B	2	4931	−	−
4759	A	2	4938	−	−
4762	A	1	4984	A	3
4764	A	1	5003	−	3
4768	A	1	5043	B	2
4774	B	2	5045	A	1
4775	A	1	5087	A	1
4777	−	4	5098	A	3
4779	−	2	5109	A	1
4781	A	1	5161	A	1
4782	B	2	5166	A	1
4785	A	1	5167	−	3
4787	−	2	5180	B	2
4806	−	−	5185	−	2
4810	−	−	5187	−	4
4819	−	−	5189	−	2
4823	A	1	5190	−	−
4826	C	3	5202	A	3
4827	A	1	5205	A	1
4829	B	2	5256	A	2
4836	−	4	5260	C	−
4839	A	1	5261	C	3
4840	B	2	5262	C	2
4841	B	2	5263	−	4
4843	A	1	5330	A	1
4844	A	2	5331	A	3
4850	A	−	5354	A	−
4856	−	−	5372	C	3
4862	A	1	5400	C	3

(*continued*)

Appendix I (*continued*)

Sample No.	Total	Shelled	Sample No.	Total	Shelled
5402	C	3	5772	A	1
5404	C	—	5808	—	1
5406	A	1	5810	—	3
5410	C	3	5820	—	3
5413	C	3	5831	A	1
5414	C	3	5832	A	1
5419	C	3	5833	B	2
5500	A	1	5835	B	2
5501	—	4	5836	A	1
5508	B	2	5844	—	—
5533	A	3	5845	A	1
5535	—	3	5897	B	2
5538	A	1	5964	A	1
5539	A	4	5965	B	2
5557	—	—	5966	B	2
5583	—	3	5967	B	2
5607	B	2	5968	A	1
5610	—	4	6000	A	1
5614	—	—	6059	A	1
5617	—	—	6100	—	2
5629	B	2	6101	—	1
5630	B	2	6102	A	1
5729	A	1	6155	A	1
5730	A	2	6157	A	1
5731	A	1	6158	A	1
5732	A	1	6297	D	4
5735	A	1	6298	D	4
5739	B	2	6299	D	4
5740	B	2	6300	D	4
5741	A	2	6301	D	2
5742	—	2	6302	D	4
5743	A	1	6303	D	4
5744	—	2	6304	D	4
5750	—	—	6362	D	4
5751	—	—	6364	D	4
5752	A	2	6365	D	4
5753	B	2	6366	D	4
5754	—	—	6368	D	4
5756	B	2	6369	D	4
5757	A	1	6370	D	4
5758	B	2	6371	D	4
5765	B	—	6372	D	4

Appendix 1 (*continued*)

Sample No.	Total	Shelled	Sample No.	Total	Shelled
6373	D	4	6416	D	4
6381	—	—	6418	D	4
6383	D	4	6421	D	4
6384	D	4	6422	D	4
6385	D	4	6686	D	4
6386	D	4	6687	D	4
6387	B	2	6688	D	4
6388	D	4	6694	—	2
6392	D	4	6721	D	4
6396	D	4	6731	D	4
6397	D	4	6736	D	4
6398	D	4	6738	D	—
6399	D	4	6744	—	1
6408	—	4	6749	D	—
6410	D	4	6750	D	4
6411	—	4	6761	D	—
6412	D	4	6762	D	4
6413	D	—	6763	D	4
6414	D	—			

Appendix 2
Comparison of the taxonomic composition of total and shelled faunas in all samples.

	Total		Shelled	
	Genera	Species	Genera	Species
Phylum Coelenterata	8	11	—	—
Class Anthozoa				
Order Actinaria	2	2		
Order Pennatulacea	2	2		
Order Gorgonacea	1	2		
Order Ceriantharia	2	4		
Order Unknown	1	1		
Phylum Bryozoa	1	1	—	—
Order Cheilostomata	1	1		
Phylum Brachiopoda	2	2	2	2
Class Articulata				
Order Terebratulida	1	1	1	1

(*continued*)

Appendix 2 (*continued*)

	Total		Shelled	
	Genera	*Species*	*Genera*	*Species*
Class Inarticulata				
Order Lingulida	1	1	1	1
Phylum Platyhelminthes	4	9	–	–
Class Turbellaria	4	9		
Phylum Rynchocoela (Nemerteans)	6	12	–	–
Phylum Nematoda	1	1	–	–
Phylum Annelida	212	402	–	–
Class Polychaeta	210	400		
Class Oligochaeta	1	1		
Class Hirudinea	1	1		
Phylum Sipunculida	4	6	–	–
Phylum Echiurida	3	4	–	–
Phylum Phoronida	3	5	–	–
Phylum Arthropods	167	266	–	–
Class Pycnogonada	1	1		
Class Crustacea	166	265		
Phylum Mollusca	139	277	130	271
Class Gastropoda	67	132	60	218
Class Amphineura	1	1	1	1
Class Pelecypoda	67	138	67	138
Class Scaphopoda	3	5	2	4
Class Cephalopoda	1	1	–	–
Phylum Echinodermata	29	40	1	1
Class Asteroidea	2	4	–	–
Class Ophiuroidea	12	17	–	–
Class Echinoidea	8	8	1	1
Class Holothuroidea	6	10	–	–
Class Crinoidea	1	1	–	–
Total	579	1036	133	274

Appendix 3
Comparison of the number of species in each taxonomic group in total and shelled communities.

	Total					Shelled				
	All	A	B	C	D	All	1	2	3	4
Coelenterates	11	7	4	5	4	–	–	–	–	–
Bryozoans	1	1	–	–	–	–	–	–	–	–
Brachiopods	2	1	1	1	1	2	1	1	1	1
Polychaetes, etc.[a]	439	323	189	124	115	–	–	–	–	–
Arthropods	266	182	112	50	66	–	–	–	–	–
Gastropods	132	78	55	15	26	128	57	69	32	30
Pelecypods	138	95	56	31	18	138	88	69	49	31
Other mollusks[b]	7	7	3	3	–	5	4	3	2	–
Echinoderms	40	24	18	11	10	1	1	1	–	1
Total	1036	718	438	240	240	274	151	143	84	66

[a] Primarily Polychaete annelids, but includes other annelids, platyhelminthes, nemerteans, nematodes, sipunculids, echiurids, and phoronids.
[b] Amphineura, scaphopods, and cephalopods.

APPENDIX 4

List of dominants in total communities A, B, C, and D. Rank in each community based on percentage of samples of that community in which species occurs. Feeding resource and method described in text. Taxonomic group: C – arthropods; E – echinoderms; M – mollusks; O – all groups in Appendix 2 other than polychaetes; P – polychaetes. List I – dominant species in total community. List II – dominant species in total community but with carnivores excluded. List III – dominant shelled species in total community. List IV – dominant shelled species in total community but with carnivores excluded.

Rank	Frequency (%)	Shelled	Taxonomic Group	Name	A – Total, I — Feeding Method										Rank	A – Total Noncarnivore, II — Feeding Method							
					IC	IS	OC	OS	WLA	WLP	WHA	K	U		IC	IS	OC	OS	WLA	WLP	WHA	U
1	98		P	Prionospio pinnata						X				1						X		
2	97		O	Nemertean unknown								X										
3	97		P	Prionospio malmgreni						X				2						X		
4	95	*	O	Glottidia albida					X					4					X			
5	93		E	Amphiodia urtica			X							5			X					
6	88		E	Amphipholis squamata			X							6			X					
7	87		P	Spiophanes missionensis						X				7						X		
8	85		P	Haploscoloplos elongatus		X								8		X						
9	83		C	Ampelisca cristata						X				9						X		
10	82		C	Ampelisca brevisimulata						X				10						X		
	82		C	Ostracode other					X					10					X			
	82	*	M	Nuculana taphria				X						10				X				
13	80		P	Pholoe glabra								X										
	80		P	Nereis procera								X										
	80		C	Amphideutopus oculatus			X							13			X					

			Species			
16	75	* M	*Cadulus* sp.	X		
17	73	C	Tanaid unknown		X	X
	73	P	*Aricidea lopezi*		X	
	73	P	*Nephtys* sp.	X		
	73	P	*Tharyx tesselata*		X	X
21	72	P	*Sternaspis fossor*	X		X
	72	P	*Goniada brunnea*			X
23	70	P	*Sthenelanella uniformis*			X X
	70	P	*Cossura candida*	X		
25	68	* M	*Tellina buttoni*	X		X
	68	O	Nematode unknown			X
	68	P	*Pectinaria californiensis*	X		X

A – Shelled, III / A – Shelled Noncarnivores, IV

		Species					
4	95	*Glottidia albida*	X			X	
10	82	*Nuculana taphria*	X			X	
16	75	*Cadulus* sp.	X	X		X	
25	68	*Tellina buttoni*	X			X	
28	66	*Compsomyax subdiaphana*		X			X
32	65	*Macoma yoldiformis*	X			X	
35	62	*Cylichna attonosa*	X			X	
	62	*Turbonilla* sp.			X		
40	58	*Axinopsis serricatus*	X			X X	
49	53	*Rochefortia aleutica*	X			X X	
52	52	Pelecypod unknown			X		X
59	50	*Periploma discus*	X			X	
64	48	Chaetodermatida unknown	X			X	
83	40	*Acteon punctocoelata*		X		X	
91	37	Gastropod unknown			X		X

(continued)

131

Appendix 4 *(continued)*

Rank	Frequency (%)	Name	A – Shelled, III									Rank	A – Shelled Noncarnivores, IV							
			\multicolumn{9}{c}{Feeding Method}		\multicolumn{8}{c}{Feeding Method}
			IC	IS	OC	OS	WLA	WLP	WHA	K	U		IC	IS	OC	OS	WLA	WLP	WHA	U
95	35	*Volvulella tenuissima*								X										
99	33	*Nassarius perpinguis*			X							99			X					
104	32	*Modiolus sp.*							X			104							X	
109	30	*Epitonium bellastriatum*								X										
	30	*Polinices lewisii*								X										
118	28	*Balcis sp.*								X										
	28	*Odostomia sp.*								X										
	28	*Solen sicarius*					X					118					X			
123	27	*Solamen columbianum*							X			123							X	
	27	*Rochefortia sp.*					X					123					X			
132	25	*Mangelia sp.*								X										
	25	*Nucula tenuis*				X						132				X				
	25	*Olivella baetica*								X										
	25	*Tellina idae*				X						132				X				

APPENDIX 5

List of dominants in total communities A, B, C, and D. Rank in each community based on percentage of samples of that community in which species occurs. Feeding resource and method described in text. Taxonomic group: C – arthropods; E – echinoderms; M – mollusks; O – all groups in Appendix 2 other than polychaetes; P – polychaetes. List I – dominant species in total community. List II – dominant species in total community but with carnivores excluded. List III – dominant shelled species in total community. List IV – dominant shelled species in total community but with carnivores excluded.

Rank	Frequency (%)	Shelled	Taxonomic Group	Name	B – Total, I Feeding Method									B – Total Noncarnivore, II Feeding Method							
					IC	IS	OC	OS	WLA	WLP	WHA	K	U	Rank	IC	IS	OC	OS	WLA	WLP	WHA
1	100		P	*Haploscoloplos elongatus*		X								1		X					
2	93		O	Nemertean unknown								X									
3	89		P	*Prionospio malmgreni*						X				3						X	
3	89		P	*Prionospio pinnata*						X				3						X	
5	86		P	*Goniada littorea*								X									
5	86		C	*Paraphoxus epistomus*	X									5	X						
5	86		C	*Diastylopsis tenuis*	X									5	X						
5	86		C	*Ampelisca cristata*						X				5							
9	82	*	M	*Tellina buttoni*				X						9				X			
11	82		P	*Nothria elegans*								X									
11	79		P	*Thalenessa spinosa*								X									
	79		E	*Amphiodia urtica*			X							11			X				

(continued)

Appendix 5 *(continued)*

B – Total, I / B – Total Noncarnivore, II

Rank	Frequency (%)		Name	\multicolumn{9}{c}{Feeding Method (I)}	Rank	\multicolumn{8}{c}{Feeding Method (II)}
				IC	IS	OC	OS	WLA	WLP	WHA	K	U		IC	IS	OC	OS	WLA	WLP	WHA	U
	79	C	*Argissa hamotipes*			X							11			X					
14	75	P	*Goniada brunnea*								X										
14	75*	M	*Olivella baetica*								X										
16	71	P	*Chaetozone corona*							X			16						X		
16	71	E	*Amphipholis squamata*			X							16			X					
16	71	P	*Nephtys caecoides*								X										
19	68	P	*Magelona sacculata*							X			19						X		
19	68	P	*Amaeana occidentales*							X			19						X		
21	64*	M	*Turbonilla sp.*								X		21					X			
21	64	C	Ostracode unknown										21	X							
24	64	C	*Synchelidium shoemakeri*	X																	
24	61	P	*Nephtys sp.*								X										
24	61	P	*Spiophanes bombyx*					X					24					X			
24	61*	M	*Odostomia sp.*								X		24						X		

B – Shelled, III / B – Shelled Noncarnivore, IV

Rank	Frequency (%)	Name	\multicolumn{9}{c}{Feeding Method (III)}	Rank	\multicolumn{8}{c}{Feeding Method (IV)}
			IC	IS	OC	OS	WLA	WLP	WHA	K	U		IC	IS	OC	OS	WLA	WLP	WHA	U
9	82	*Tellina buttoni*				X						9			X					
14	75	*Olivella baetica*								X										
21	64	*Turbonilla sp.*								X										
24	61	*Odostomia sp.*								X										

	27	31	31	41	45	51	51	51	60	60	60	70	70	79	79	88	98	98	112	112		
Chione undatella																			X		27	57
Cadulus sp.																			X		31	54
Nuculana taphria															X						31	54
Pelecypod unknown				X																	41	46
Siliqua lucida																			X		45	43
Lucinisca nuttalli																			X		51	39
Mangelia sp.							X														51	39
Rochefortia sp.								X											X		51	39
Gastropod unknown						X															51	39
Volvulella sp.								X													51	39
Acteon punctocoelata									X	X											60	36
Nassarius perpinguis									X	X											60	36
Glottidia albida												X									60	36
Rochefortia aleutica												X							X		70	32
Macoma yoldiformis												X	X	X	X						70	32
Macoma sp.														X	X						79	29
Cylichna attonosa														X	X						79	29
Poromya sp.																X					88	25
Dendraster excentricus																	X			X	98	21
Modiolus sp.																	X				98	21
Terebra pedroana																		X			21	
Aglaja sp.																			X	X	112	18
Asthenothaerus villosior																			X	X	112	18
Balcis rutila																				X		18
Polinices lewisii																				X		18

| 27 | 31 | | 41 | 45 | 51 | | | | 60 | | | 70 | | 79 | | 88 | 98 | | 112 | | | |

APPENDIX 6

List of dominants in total communities A, B, C, and D. Rank in each community based on percentage of samples of that community in which species occurs. Feeding resource and method described in text. Taxonomic group: C – arthropods; E – echinoderms; M – mollusks; O – all groups in Appendix 2 other than polychaetes; P – polychaetes. List I – dominant species in total community. List II – dominant species in total community but with carnivores excluded. List III – dominant shelled species in total community. List IV – dominant shelled species in total community but with carnivores excluded.

Rank	Frequency (%)	Shelled	Taxonomic Group	Name	C – Total, I — Feeding Method									Rank	C – Total Noncarnivores, II — Feeding Method						
					IC	IS	OC	OS	WLA	WLP	WHA	K	U		IC	IS	OC	OS	WLA	WLP	WHA
1	100		P	*Terebellides stroemi*						X				1							X
1	100		E	*Amphiodia urtica*			X							1			X				
1	100		C	*Heterophoxus oculatus*	X									1	X						
4	92		P	*Sternaspis fossor*		X								4		X					
4	92		O	Nemertean red-banded								X									
4	92		P	*Pectinaria californiensis*	X									4	X						
7	83		O	*Listriolobus pelodes*					X					7					X		
7	83		P	*Cossura candida*						X				7						X	
7	83		P	*Hesperone laevis*									X								
7	83	*	M	*Compsomyax subdiaphana*					X					7					X		
7	83		P	*Prionospio pinnata*						X				7						X	
12	75		C	*Listriella goleta*					X					12					X		
12	75		C	*Haliophasma geminata*				X						12				X			
12	75	*	M	*Saxicavella pacifica*					X					12					X		

		Species								
	O	Phoronid unknown					X		12	
75	P	Ceratocephala americana						X X		
75	P	Glycera capitata						X X		
75	P	Poecilochaetus johnsoni				X			12	X
75	P	Paraonis gracilis	X						12	
19 67	P	Nephtys ferruginea						X X X	19	X
67	P	Lumbrineris sp.						X X X		
67	P	Glycera robusta						X X X		
67	C	Callianassa sp.	X						19	X
67	C	Ampelisca brevisimulata				X X			19	X X
67	C	Byblis veleronis				X X			25	X X
58	C	Gnathia crenulatiformis		X					25	X
25 58	M *	Axinopsis serricatus			X				25	X
58	P	Boccardia basilana				X			25	
58	C	Listriella eriopisa			X				25	X
58	P	Marphysa disjuncta	X						25	
58	P	Spiophanes missionensis				X			25	X
58	P	Pholoe glabra					X		25	

C — Shelled, III

	Species							
7 83	Compsomyax subdiaphana			X X			7	X
12 75	Saxicavella pacifica			X X			12	X
25 58	Axinopsis serricatus			X X			25	X
33 50	Glottidia albida			X X			33	X
47 41	Macoma yoldiformis				X X		47	X X
41	Nucula tenuis				X X		47	X X
41	Nuculana taphria				X X		47	X X
59 33	Bittium sp.		X				59	X
33	Cadulus sp.	X X					59	X X
33	Dentalium neohexagonum	X X					59	X X
71 25	Cardita ventricosa				X		71	X

C — Shelled Noncarnivore, IV

(continued)

137

Appendix 6 *(continued)*

Rank	Frequency (%)	Name	C – Shelled, III — Feeding Method									Rank	C – Shelled Noncarnivore, IV — Feeding Method							
			IC	IS	OC	SO	WLA	WLP	WHA	K	U		IC	IS	OC	SO	WLA	WLP	WHA	U
	25	*Periploma discus*					X					71					X			
	25	*Rochefortia* sp.					X					71					X			
92	17	*Asthenothaerus villisior*					X					92					X			
	17	*Bittium subplanatum*			X							92			X					
	17	Chaetodermatida unknown	X									92	X							
	17	Pelecypod unknown									X	92								X
	17	*Cylichna diegensis*				X						92				X				
	17	*Mangelia* sp.								X										
	17	*Nucula* sp.				X						92				X				
	17	*Nuculana* sp.				X	X					92				X				
	17	*Rochefortia tumida*					X					92					X			
	17	*Tellina idae*				X						92				X				
	17	*Thracia* sp.																		
	17	*Thyasira* sp.																		

138

APPENDIX 7

List of dominants in total communities A, B, C, and D. Rank in each community based on percentage of samples of that community in which species occurs. Feeding resource and method described in text. Taxonomic group: C – arthropods; E – echinoderms; M – mollusks; O – all groups in Appendix 2 other than polychaetes; P – polychaetes. List I – dominant species in total community. List II – dominant species in total community but with carnivores excluded. List III – dominant shelled species in total community. List IV – dominant shelled species in total community but with carnivores excluded.

				D – Total, I										D – Total Noncarnivore, II							
					Feeding Method										Feeding Method						
Rank	Frequency (%)	Shelled	Taxonomic Group	Name	IC	IS	OC	OS	WLA	WLP	WHA	K	U	Rank	IC	IS	OC	OS	WLA	WLP	WHA	U
1	98		P	*Prionospio malmgreni*						X				1						X		
2	92		O	Nemertean unknown								X										
3	88		C	*Paraphoxus epistomus*	X									3	X							
4	85		O	Nematode unknown									X	4								X
5	81		P	*Nephtys* sp.								X										
6	77	*	M	*Tellina buttoni*				X						6				X				
	77		C	*Diastylopsis tenuis*	X									6	X							
8	75		P	*Goniada littorea*								X										
9	65		C	*Synchellidium* sp.	X									9	X							
10	60	*	M	*Olivella baetica*								X										
11	48		P	*Spiophanes bombyx*						X				11						X		
11	48		C	*Paraphoxus bicuspidatus*	X									11	X							

(continued)

Appendix 7 (continued)

					C – Shelled, III									C – Shelled Noncarnivore, IV							
					Feeding Method										Feeding Method						
Rank	Frequency (%)			Name	IC	IS	OC	OS	WLA	WLP	WHA	K	U	Rank	IC	IS	OC	OS	WLA	WLP	WHA
13	46		P	*Haploscoloplos elongatus*	X									13		X					
14	42		P	*Diopatra ornatus*		X								14	X						
15	38		P	*Lumbrineris* sp.								X									
17	38		.P	*Chaetozone spinosa*						X				15						X	
17	35	*	M	*Mangelia* sp.								X									
18	33	*	E	*Dendraster excentricus*						X				18						X	X
18	33		P	*Prionospio pinnata*						X				18						X	X
	33		C	*Monoculodes hartmanae*	X									18	X						
	33		P	*Thalenessa spinosa*									X								
22	29		P	*Glycera convoluta*									X								
23	27		P	*Scoloplos armiger*		X								23		X					
	27		O	Polyciad false segmentation		X							X								
25	25		P	*Aricidea lopezi*										25	X	X					
25	25		P	*Pectinaria californiensis*	X									25	X						
25	25		P	*Tharyx* sp.						X				25						X	

				D – Shelled, III										D – Shelled Noncarnivore, IV							
6	77			*Tellina buttoni*				X						6			X				
10	60			*Olivella baetica*								X									
17	35			*Mangelia* sp.								X									

		Species									
18	33	*Dendraster excentricus*							X		18
28	21	*Rochefortia* sp.		X	X				X	X	28
32	19	*Poromya* sp.		X	X				X	X	32
	19	*Crepidula* sp.		X	X				X	X	32
34	17	*Siliqua lucida*				X					34
	17	*Turbonilla* sp.				X					
42	15	*Balcis* sp.									
53	10	*Modiolus* sp.					X				53
62	8	*Macoma* sp.	X	X							62
	8	*Macoma yoldiformis*		X			X				62
	8	*Nassarius fossatus*	X	X	X						62
78	6	*Nassarius perpinguis*								X	78
	6	*Tricolia* sp.						X	X		78
	6	*Nuculana* sp.	X	X							78
	6	*Lacuna* sp.									78
	6	*Chlamys* sp.		X	X				X		78
	6	*Aglaja* sp.		X	X				X	X	78
	6	*Solen* sp.		X	X				X	X	78
99	4	*Ophiodermella incisa*				X			X		99
	4	*Lucinisca nuttalli*		X							
	4	*Retusa* sp.				X					
	4	*Glottidia albida*		X					X		99
	4	*Mitrella carinata*				X					
123	2	*Chione* sp.		X	X				X	X	123
	2	*Chione undatella*		X	X				X	X	123
	2	*Acteocina intermedia*				X	X				123
	2	*Acteocina* sp.		X	X						123
	2	*Cerithiopsis* sp.					X				
	2	*Crepipatella lingulata*		X					X		123

(continued)

Appendix 7 (continued)

Rank	Frequency (%)	Name	D – Shelled, III										Rank	D – Shelled Noncarnivore, IV						
			Feeding Method											Feeding Method						
			IC	IS	OC	OS	WLA	WLP	WHA	K	U		IC	IS	OC	OS	WLA	WLP	WHA
	2	Ensis myrae					×					123							
	2	Jaton festivus								×									
	2	Aligena sp.				×						123					×		
	2	Asthenothaerus villosior				×						123					×		
	2	Pitar sp.				×						123					×		
	2	Volvulella sp.								×									
	2	Odostomia sp.								×									
	2	Olivella pedroana								×									
	2	Polinices lewisii								×									
	2	Polinices sp.								×									

Raw Material
of the Fossil Record

J.E. Warme
A.A. Ekdale
S.F. Ekdale
C.H. Peterson

Rice University
University of Utah
Salt Lake City
University of Maryland,
Baltimore County

ABSTRACT

The paleoecological potential of two skeletonized taxa (mollusks and benthic foraminifera) was investigated in a temperate terrigenous locality (coastal southern California) and a tropical carbonate locality (coastal eastern Yucatan). Bulk sediment samples were collected, and the distributions of mollusk and foraminifera species were analyzed numerically.

Q-mode and R-mode cluster analyses identified molluscan and foraminiferal communities, which we define as recurrent groups of numerically co-occurring species. Q-mode clusters based on information from dead mollusk and dead foraminifera occurrences correlated directly with obvious habitats. In fact, for the mollusks, clusters obtained from dead mollusk data more accurately reflected the major physical environments than did those from live mollusk data.

Live–dead comparisons of mollusks in sample couplets and in single localities sampled periodically over 3 years indicate that habitat migration and postmortem transport of mollusk shells are minimal. Moreover, species diversities and relative species abundances in living communities appear to be preserved in the dead assemblages of empty mollusk shells.

INTRODUCTION

Purpose and Scope

In four separate studies, we have investigated both the ecological and paleoecological significance of the abundances and distributions of shallow-water marine organisms. We concentrated on living species and their mortality, represented by the accumulating empty skeletons that provide the raw material of the fossil record. We intend to draw attention to what we believe are ecological realities in modern marine habitats and to discuss the potential of the accumulating skeletal remains in providing accurate reflections of natural communities.

Our purposes in this paper are (1) to judge the fidelity with which live species distributions and abundances are preserved in the dead assemblages, (2) to show that species diversity and relative abundances of living communities are potentially recoverable from the fossil record, (3) to emphasize that two shallow marine areas (coastal Yucatan and California), each with very different physical regimes, yield comparable results, (4) to compare the correlation of two well-studied skeletonized groups (mollusks and benthic foraminifera) with particular habitat indicators, and (5) to demonstrate that a standard ecological census can benefit from consideration of both live species and their associated empty skeletons.

Individuals within a species live in populations intermingled with other species populations. All suffer mortality, and some portion is buried and preserved. These events can be reconstructed by collecting live and dead animals and comparing various aspects of the two in different ways. Specifically, our approach is to conduct an ecological census by sampling benthic invertebrates, separating the live and dead skeletonized taxa, and comparing them on a species-by-species basis, sample-by-sample basis, and total "community" basis.

Many workers have examined both modern habitats and the fossil record in order to judge their adequacy for community reconstruction (see papers of Macdonald and Stanton, this volume). Fossil localities have been inspected for evidence of shells in "living position" and for in situ communities (e.g., Johnson, 1960). Studies in modern habitats or in flumes have documented the movement and sorting of shells and show important biasing effects (see review by Warme, 1971a).

In contrast, our method of investigation has been to collect bulk samples and to analyze species distributions statistically and numerically. Postmortem perturbations that may have influenced some specimens or even whole assemblages are purposefully ignored. Our data base is large, involving thousands of specimens. Most information on the precise physical relationship of any single skeleton to bedding or to other potential fossils is lost during collection, but for our purposes the immense volume of useful data generated by bulk collections compensates for any disadvantages. Moreover, these data seem adequate for the examination of many basic ecological and paleoecological problems, such as the preservability of diversity and relative species abundances in fossil communities.

Analysis of the living biota clearly has both ecological and paleoecological significance. However, analysis of the dead remains can also have importance for both fields, and this is not always realized by workers in modern environments. Comparisons of live and dead assemblages and studies of their areal distributions have been fruitful in judging both the ecological nature of modern communities and the paleoecological fidelity of their ancient counterparts.

Postmortem events, such as the transportation, breakage, and dissolution of shells and the lithification and diagenesis of sediment involve geological processes. Although important in the environments we studied, these processes involve haphazard future events whose influence on any single skeletal accumulation is somewhat unpredictable. The detailed investigation of particular postmortem processes is largely beyond the scope of this report.

The Community Problem

Most ecologists and paleoecologists believe that natural populations of several species co-occur in such a way that in theory one can define an ecological community which, despite its temporal changes and interrelationships with other communities, possesses some structural and functional integrity. From a vast literature on community ecology, however, there appears to be little agreement on just how to define a real community in nature (Macfadyen, 1963). Community ecology, especially for marine communities, largely exists as community theory.

Because there is little consensus on the definition of a community, the identification of communities and interpretation of community data are rendered difficult. It is imperative that a clear definition be established before proceeding with such analyses. We use the term "community" as a recurrent group of numerically co-occurring species, and we identify communities by means of cluster analysis (Fager, 1957; Sokal and Sneath, 1963). In this way we can profitably study communities of live and dead organisms independent of presumed habitat boundaries, niche dimensions, and so on. Furthermore, communities defined and identified in such a manner can subsequently be studied for structural properties with some success (Warme, 1971a).

A fresh look at ancient communities, with the time perspective that they provide, will hopefully lead to new testable hypotheses about communities, which have not arisen from studies of living organisms alone. In this light, however, we believe it is valuable to investigate modern environments for their resident species and to judge what aspects of the ecology of these species are transmitted to the sedimentary record.

How different is paleoecology from ecology, given the present state of community ecology? How important is it that soft-bodied organisms usually are not preserved? What is the real effect of shell transportation or mixing of community components in one habitat or between habitats? The results of the investigations reported in this paper apply to these questions.

COLLECTING LOCALITIES

The results and discussions presented herein stem from four separate studies. A. A. Ekdale (1972, 1974a, b) and S. F. Ekdale (1974) worked together in coastal Yucatan, sampling separately for mollusks and benthic foraminifera, respectively, but utilizing the same sample stations. Warme (1969, 1971b) and Peterson (1972, 1975a, b) completed separate studies at Mugu Lagoon, coastal southern California. These two geographic regions, the first tropical and the second warm temperate, were sampled for live and dead individuals of these two skeletonized taxa.

The tropical sampling sites are located along the northeastern coast of the Yucatan Peninsula, Territory of Quintana Roo, Mexico, in the vicinity of the islands of Contoy, Mujeres, and Cancun. Habitats sampled there included a coral reef and back reef, current-swept strait, open bays and restricted lagoon-mangrove complexes. The sediments are almost pure calcium carbonate sand and mud. Coral reef rock and cemented Pleistocene beach ridges provide hard substrates for attached epifauna and boring infauna. Data on the environments, sediments, fauna, climate, hydrography, and geologic history of the region are given in Weidie (1974).

The temperate locality sampled for this report is Mugu Lagoon in southern California. Warme (1969, 1971b) sampled several habitats of the lagoon, including sand channels, eelgrass, tidal flats, and marsh. Peterson (1972, 1957a, b) restricted his sampling to the sand channel habitats. Only the molluscan fauna of Mugu Lagoon is discussed herein, although the foraminifera were studied briefly by Warme (1971b). Marine faunas from this part of the Pacific coast are treated by MacGinitie and MacGinitie (1968), Ricketts and Calvin (1968), Warme (1971b), Peterson (1972, 1975a, b), and many others.

COLLECTING METHODS

Sampling for mollusks was accomplished in intertidal and shallow subtidal areas by shoveling and screening sediment from samples that were isolated by a rigid frame or cylinder pushed into the sediment to prevent caving. Quadrat sizes varied from 0.06 to 0.25 m^2, and depth of sampling varied from 30 to 75 cm. Subtidal samples from deeper water were collected by means of a diver-operated waterlift suction dredge in Yucatan (Ekdale and Warme, 1973) and Mugu Lagoon (Warme, 1971b). Screen sizes for final sieving were similar in all of our studies and limited collections to shells retained by 2- or 3-mm-mesh screens (about 4-mm-maximum diagonal openings).

Some samples from Yucatan, where much of the sediment itself is skeletal debris, were estimated to contain in excess of 2 million shells per m^3. These samples contained too many shells to be counted practically, so they were subsampled and the resulting counts multiplied by an appropriate factor. In Mugu Lagoon, a maximum of only a few tens of thousands of shells per cubic meter were present, partly owing to dilution by terrigenous sediments. Rare pockets of pure shell debris can be accumulated by currents in California lagoons, but no such situations were recognized in our sampling.

Samples for foraminifera were taken in Yucatan by skin diving to the bottom and pushing plexiglass tubes 2.5 cm in diameter into the sediment to a depth of approximately 1 cm. All but three localities sampled were the same as those for the mollusks; 50 molluscan and 30 foraminiferal samples from Yucatan were processed.

SUMMARY OF SAMPLING RESULTS

We believe that the results summarized below have ecological and paleoecological significance and are applicable to the interpretation of extant benthic associations as well as to the interpretation of fossil occurrences. Data and details on the areas studied, the species encountered, and our ecological and paleoecological analyses appear elsewhere (Warme, 1969, 1971b; A. A. Ekdale, 1972, 1974a, b; S. F. Ekdale, 1974; Peterson, 1972, 1975a, b).

Yucatan Mollusks

Fifty samples were collected in June 1971 with the suction dredge at random points along predetermined transects that were selected to intersect a variety of marine habitats (A. A. Ekdale, 1972, Figs. 2 and 3). The most numerous macroinvertebrates were mollusks; 290 different species included 180 that were gastropods, 105 bivalves, 4 scaphopods, and 1 amphineuran. Of these, 82 species were collected alive, and the remainder were present as empty shells only.

Any species can occur in any sample alive, dead, or both; 288 live occurrences of species in samples were realized, 96.5 percent of which were in samples also containing dead specimens of their species. The 10 "live only" occurrences, involving only 22 specimens, perhaps represent migrating or newly recruited populations where they were found. In contrast, 2,401 "dead only" occurrences, involving more than 150,000 specimens, were counted in the 50 samples.

Table 1 describes the diversities (290 species total) and abundances per sample expressed as quartiles of the cumulative number of samples collected (fractions rounded to the next highest whole number).

If the 50 samples are ranked in order of diversity (i.e., number of species contained), particular samples appear in approximately the same position relative to the other samples whether live occurrences only or dead occurrences only are considered. The same relationships can be observed when ranking the samples in terms of their total abundances (i.e., number of individual specimens contained). These observations suggest that diversity and abundance are preserved in the death assemblages that ultimately enter the fossil record. Within the time period represented by our samples, habitat migration and postmortem transportation of mol-

Table 1

Category	% Samples (N = 50)				
	Lowest	25	50	75	Highest
Live species	0	3	5	8	17
Live individuals	0	6	18	77	3,352
Dead species	5	43	52	69	93
Dead individuals	5	358	1,262	3,264	37,423

lusks were minimal. Areas of current large standing crop presumably have had higher densities through some period of history.

Of the total 290 molluscan species, 66 occurred alive in the first 25 samples processed and were included in a series of computerized cluster analyses. The use of cluster analysis (see Fager, 1957; Sokal and Sneath, 1963) for constructing recurrent groups of organisms is discussed by Valentine and Peddicord (1967) and many others. This numerical classification technique constructs clusters of similar samples based on the number of species that they have in common (Q-mode) or clusters of species based on the number of samples in which they co-occur (R-mode). Cluster analyses of the mollusks (from both Yucatan and California) were developed first on the basis of the live animals collected and then independently on the basis of the empty shells in the same samples. Presence–absence as well as relative-abundance data were used.

The Yucatan molluscan data from the first 25 samples processed were compared using several different coefficients of mutual occurrence: Jaccard, Dice, simple match. Fager, taxonomic distance, and correlation coefficients (Cheetham and Hazel, 1969). The correlation coefficient gave the most readily interpretable Q-mode results when using relative-abundance data; that is, the sample clusters matched obvious environments (Fig. 1). In general, the clusters produced by data on live and dead shells for any given coefficient resulted in roughly the same environmental associations, although a greater amount of spread (i.e., lower level of association) usually was shown in the samples when using live-animal data only. Other interpretations of clusters, such as trophic analyses, are difficult. Many species share the same food resources, and details of feeding habits of each species are not known.

Using all 50 Yucatan mollusk samples and relative-abundance data from the 180 species that occurred in three or more samples, the resulting clusters showed an excellent correspondence with obvious habitats. The 50-sample dendrogram (Fig. 2) contains clusters representing the open Caribbean, the current-swept strait, the coral reef–back reef area, open and restricted lagoons, and mangrove tidal channel complexes. Two clusters of samples from one of the islands separates lagoonal mouth samples from others collected in the more restricted portions of the lagoons.

R-mode analyses of the 66 common species (dead shells only) in the first 25 samples yielded two large communities and a third small but very distinctive community (Fig. 3). The first two communities contained 29 and 32 molluscan species and were interpreted as representing conditions of restricted circulation and open circulation, respectively. The two species in the third community were the only ones that represented the open sandy strait habitat. Only three species did not join any of the clusters in the R-mode analysis.

Yucatan Foraminifera

Twenty-seven of the 30 foraminiferal samples from Yucatan were collected in June 1971 at the same sample sites as the mollusks; they thus encompass the same range of habitats as the molluscan samples and comparison of these two fossilizable

Raw Material of the Fossil Record   149

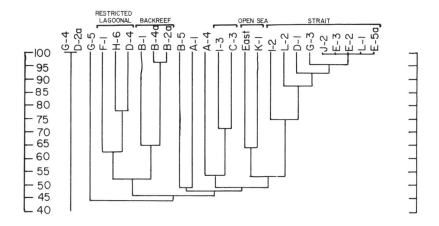

Q-mode analysis of 25 samples (live species) with correlation coefficient based on relative abundance

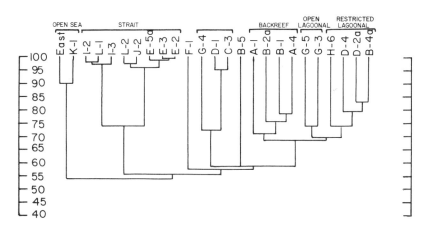

Q-mode analysis of 25 samples (dead species) with correlation coefficient based on relative abundance

Figure 1
Q-mode cluster analysis dendrograms utilizing a product-moment correlation coefficient and species data based on the relative abundances of 66 species of live mollusks (top) and dead mollusks (bottom) in 25 samples from coastal Yucatan. (Ekdale, A. A., 1972, Fig. 12.)

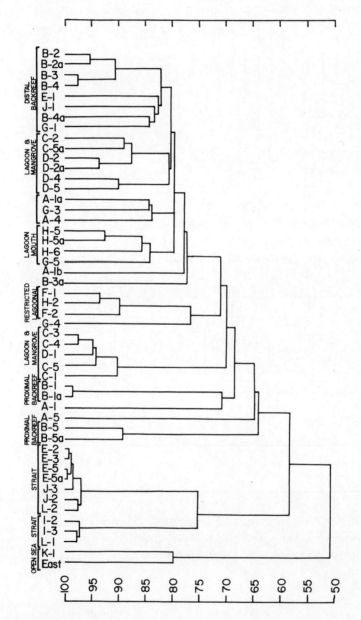

Figure 2

Q-mode cluster analysis dendrograms utilizing a product-moment correlation coefficient and species data based on the relative abundance of 180 species of dead mollusks in 50 samples from coastal Yucatan. (Ekdale, A. A., 1972, Fig. 13.)

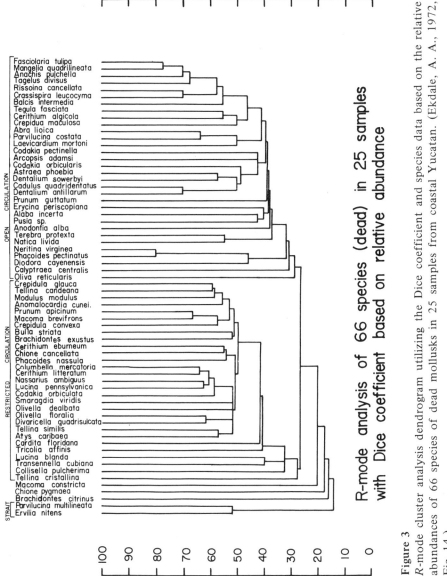

Figure 3
R-mode cluster analysis dendrogram utilizing the Dice coefficient and species data based on the relative abundances of 66 species of dead mollusks in 25 samples from coastal Yucatan. (Ekdale, A. A., 1972, Fig. 14.)

Table 2

	Least Diverse Sample	% Samples			Most Diverse Sample
		25	50	75	
Number of species dead or alive	14	44	61	77	130

groups is facilitated. Owing to difficulties in separating live foraminifera from empty tests and to the very low number of live individuals present in any one sample (a common circumstance in tropical environments), an effort to differentiate between live and dead forms was fruitless.

The 30 samples yielded 222 species of foraminifera (219 benthic species and 3 planktic species). Table 2 presents the foraminiferal diversities in the 30 samples, expressed as quartiles of cumulative samples (fractions rounded to next higher whole number). As was the case for mollusks (Table 1), some of the foraminifera were environmentally restricted, whereas others existed under a variety of conditions. No single foraminiferal species was collected in all 30 samples. Table 3 shows the quartile distributions of the 222 species collected, giving a measure of the ubiquity of the foraminifera in the environments sampled (fractions rounded to next highest whole number).

The Dice, Jaccard, and simple match coefficients were used in presence-absence cluster analyses of the 219 benthic foraminiferal species. Q-mode clusters appear to be most meaningful when using the simple match coefficient (Fig. 4), which yielded eight clusters, seven of which correlated with readily discernible habitats. The eighth cluster appeared as an artifact of this coefficient owing to a large number of mutual absences in the samples. The habitats identified are current-swept strait, submerged Pleistocene dune ridge, proximal back reef (i.e., adjacent to the coral reef), distal back reef (i.e., farther from the reef), open lagoon, restricted lagoon, and mangrove tidal channel.

Table 3

	Lowest No. of Species in a Sample	% Species			Highest No. of Species in a Sample
		25	50	75	
No. samples	1	2	6	13	29

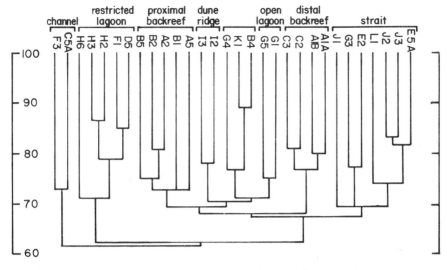

Q-mode analysis of 30 samples with match coefficient based on presence-absence

Figure 4
Q-mode cluster analysis dendrogram utilizing the simple match coefficient and species data based on the presence or absence of 219 foraminiferal species in 30 samples from coastal Yucatan. (Ekdale, S. F., 1974, Fig. 13.)

An R-mode cluster analysis (simple match coefficient) of the 219 benthic species in 30 samples yielded five clusters that also corresponded with particular environmental conditions (S. F. Ekdale, 1974, Fig. 16): three large clusters representing conditions of general open circulation, proximal back reef, and mangrove-rimmed tidal channel, and two tight clusters representing the strait and the lagoons.

R-mode clusters show assemblages that reflect distinct environments. The general open-circulation cluster contains 44 species, 33 of which occurred in 50 percent or more of the open-circulation samples. This is a ubiquitous assemblage. Back-reef samples were highly variable, with species present in from 1 to 24 samples. Within this group a subcluster was identified that came from samples near a submerged Pleistocene beach ridge, which apparently acted as a "pseudo-reef" and attracted the characteristic reef-inhabiting foraminifera. No species was unique to the open bays and lagoons. Instead, they contained vast numbers of a few species that could also live in other habitats. In contrast, mangrove channels, lagoons, and the offshore strait all had tight clusters of species that were unique to these environments.

Inspection of individual samples shows that foraminiferal diversities and densities are related to the cluster environments (Table 4). Diversity here simple refers to the number of species represented in samples, and density refers to the number of indi-

Table 4

Habitat	No. Samples	Diversity	Density
Proximal back reef	5	Moderate to very high (61–130 sp.)	Moderate to high
Distal back reef	4	Moderate to high (61–77 sp.)	Low to moderate
Submerged dune ridge	2	Moderate (59 sp.)	Moderate
Strait	7	Variable (37–55 sp.)	Low
Open lagoon	2	Moderate to high (61–72 sp.)	Moderate to high
Restricted lagoon	5	Low (22–49 sp.)	Very high
Tidal channel	2	High (82–101 sp.)	Moderate

vidual specimens per volume or weight of sediment. In this study the foraminiferal densities were estimated relative to the other samples; no specific value range is intended. Three samples from widely varying habitats did not join any of the Q-mode clusters because of numerous negative matches or mutual absences. All three of these samples were low in both diversity and abundance.

Table 4 presents the seven Q-mode clusters related to distinct habitats, accompanied by foraminiferal diversity and density estimates for those samples. Diversities are highest in the proximal back reef, above average in the tidal channels, variable but lower in the strait, and lowest in the restricted lagoons. In contrast, restricted lagoons exhibit the highest densities. High densities also are present near the reefs, dropping with distance from them. The lowest densities occurred in the strait.

Comparison of Yucatan Molluscan and Foraminiferal Distributions

The main similarity between the distribution patterns of mollusks and foraminifera is their correspondence with obvious physical habitats. R-mode species associations of both groups distinguish between open-circulation and closed-circulation environments, and Q-mode analyses based on data from the two taxa outline the same distinct habitats.

Differences in results of cluster analyses between the mollusks and foraminifera are minor. The molluscan assemblage in the strait had a low diversity and was very distinctive taxonomically. Although the foraminifera of the strait also formed a

Q-mode cluster and an R-mode assemblage, higher diversity and more variability between samples was indicated. Molluscan distributions discriminated between lagoon and lagoon-mouth assemblages, whereas the foraminiferal distributions showed only a generalized lagoon assemblage. Both of the aforementioned contrasts may be due to local mixing of foraminiferal tests, but our data are too scanty to test this suggestion. In contrast, the foraminifera discriminated between the proximal back reef and pseudo-reef dune ridge enivronments, whereas molluscan samples near the submerged dune ridge clustered with the strait samples.

Diversity rankings of the 27 foraminiferal and molluscan sample stations in common show little consistent correspondence between the two groups when based on dead occurrences. Although some of the stations had similar numbers of species and similar rank positions, as a whole they exhibited no clear correlation (S. F. Ekdale, 1974, Fig. 18).

California Mollusks

Mugu Lagoon, a tidally controlled marine lagoon, was sampled for mollusks by Warme during 1964 and 1965 and by Peterson from 1969 to 1972. Warme collected 55 benthic samples in several distinctive habitats: tidal flats, sandy tidal channels, muddy eelgrass ponds, and tidal creeks (Warme, 1971b, Fig. 1). Molluscan species totaled 73. Thirty-three gastropod species were represented by 511 live and 10,313 dead individuals, and 40 bivalve species were represented by 3,237 live and 15,856 dead individuals. The distribution of live and dead molluscan diversities and densities among the samples is shown in Table 5, which describes cumulative quartile distributions of mollusks in 55 samples (fractions rounded to the next highest whole number).

Based on both Q-mode and R-mode cluster analyses of the live animals, a distinctive sand-channel community and diffuse mud-tolerant community are identified (Warme, 1971b, Fig. 9, Figs. C, D in pocket). The mud-tolerant group contains subgroups of species that prefer eelgrass or tidal flats or that flourish in clean sand as well as in muddy substrates. Absolute species abundances correlate with tidal circulation. Once inside the unstable sand environment at the lagoon inlet, samples with 100 or more live bivalves were situated where water exchange was vigorous. In contrast, few or no live animals were found in samples at the end of tidal creeks

Table 5

	Lowest Sample Value	% Samples			Highest Sample Value
		25	50	75	
Live species	0	3	6	8	14
Live individuals	0	26	46	117	264
Dead species	7	17	24	32	44
Dead individuals	21	259	455	1,142	2,589

or in isolated ponds where water exchange appeared to be negligible and the substrate commonly very fluid.

Comparison of both Q-mode and R-mode cluster analyses of live and dead animals strongly suggests that skeletal remains accumulating in the lagoon adequately mirror distributions of the live populations from which they came. Little between-habitat transport of dead remains can be inferred, and the dead assemblages apparently are satisfactory for identifying both the living assemblages and the habitats where they existed (Warme, 1969, 1971b).

In March 1971, Peterson collected 64 samples of both live and dead mollusks from randomly selected sites in the subtidal sand-channel habitat (Peterson, 1972, 1975b). A pair of adjacent samples was taken at each sampling site. A total of 47 live and dead mollusk species were identified: 17 gastropod species represented by 9 live and 836 dead individuals, and 30 bivalve species represented by 1,636 live and 3,764 dead individuals. Table 6 describes the sample molluscan densities and diversities by presenting sample extremes and quartile points, analogus to Tables 2, 3, and 5. The table describes cumulative quartile distributions of mollusks in 64 samples of the sand-channel habitat.

A species-by-species comparison of the live and dead mollusks found together in the same samples revealed that even within this high-energy habitat, postmortem transport of molluscan remains is insignificant (Peterson, 1975b). Spatial patterns in the distribution of dead mollusks apparently also mirror similar patterns still exhibited by the living species.

A comparison between the molluscan densities and diversities of this sand-channel habitat and those found by Warme for all the habitats in Mugu Lagoon reveals trends of additional ecological and paleoecological significance. Warme's sampling revealed that bivalves are better represented than gastropods in terms of diversity and abundance in Mugu Lagoon. Bivalves comprise the majority of the live resident molluscan species (55 percent) and the majority of living individuals (86 percent). This pattern also is reflected in the dead assemblages, where 62 percent of the species and 66 percent of the individuals are bivalves. In Peterson's data for the sand-channel habitat alone, the bivalves are even more dominant: 77 percent of the live species and 99 percent of the live individuals are bivalves. Furthermore, this greater dominance of bivalves in the sand-channel habitat is mirrored in the dead

Table 6

	Lowest Sample Value	% Samples			Highest Sample Value
		> 25	> 50	> 75	
Live species	1	3	3	4	7
Live individuals	1	12	18	37	83
Dead species	1	9	11	13	21
Dead individuals	18	47	72	91	150

remains, where 64 percent of the dead species and 82 percent of the dead individuals are bivalves. Clearly, the correspondence between live and dead is not precise in any of these comparisons. However, it is significant that even without adjusting for differing average life spans of species (Van Valen, 1964) and differing rates of postmortem dissolution (Peterson, 1975b), within-habitat and between-habitat patterns of faunal representation are recognizable in the dead assemblages. Adjustment for differing life spans might improve the correspondence in faunal patterns between live and dead, because the bivalves of Mugu Lagoon tend to live longer than the gastropods (Peterson, 1972).

COMPARISONS OF LIVE AND DEAD MOLLUSCAN ASSEMBLAGES

With few exceptions, each sample contained more species and more individuals of dead mollusks than live. In fact, most of the collections from California and Yucatan alike contained 10 to 100 times more dead than live specimens. Live-dead ratios may not have much meaning, however, because at a long-occupied site the relative proportion of dead shells versus live will depend on the depth to which the sample was collected, the rate of sedimentation, and the number of empty shells recovered at successively lower sedimentary layers. In newly deposited sediment, settling larvae or migrating postlarval forms (see Peterson, 1972) will provide the collector with an initial living assemblage only, unless dead shells were deposited along with the sediment. None of our samples was composed entirely of live specimens; moreover, very few species occurred alive in any sample without dead representatives. With time and ensuing mortality, an accumulating dead assemblage quickly surpasses the standing crop in numbers. Virtually all our samples were collected where dead assemblages had accumulated to this point.

Table 7 shows by quartiles the species as they occurred alive and dead in Warme's 55 samples from all habitats of Mugu Lagoon. Half the species collected alive were found in only three or fewer samples; this points out the difficulty of collecting the rarer and patchily distributed animals in the lagoon. Half the species collected dead were in 15 to 54 samples, reflecting the cumulative or historical character of the dead assemblages.

Table 7

	Lowest No. of Samples	% Species			Highest No. of Samples
		25	50	75	
No. samples with live species (N = 39 species)	1	1	3	10	45
No. samples with dead species (N = 71 species)	1	4	15	30	54

Table 7 describes the cumulative quartile distributions, showing the number of Warme's samples in which live and dead species occurred (fractions rounded to next higher whole number).

The number of dead *individuals* in a sample should be greater than the number of live ones because the dead shells usually have accumulated over a long time. Why should the number of dead *species* be greater than the number of live ones sampled at any given time? Both ecological and historical factors account for this generalization.

Ecological Factors

Our data show that many live species are very rare in the environment, never being present in great numbers. This situation exists in most natural communities (Hairston, 1964; Williams, 1964). In Yucatan, for example, one fourth of all the molluscan species and about one sixth of all the foraminiferal species are known from single sample occurrences only (A. A. Ekdale, 1972; S. F. Ekdale, 1974). Thus, some species are difficult to obtain in samples simply because they are very sparsely distributed.

Almost all species populations also exhibit spatial heterogeneity in their distribution; that is, they occur in patches or clumps (Peterson, 1972, 1975a; Warme, 1971b) and easily can be missed during sampling. This phenomenon of patchiness accounts for the very uneven counts of live specimens of most species in samples collected from the same habitat (Warme, 1971a; A. A. Ekdale, 1972).

Opportunistic species are those that flourish in a habitat in great numbers for a brief time, perhaps owing to a disturbance or some transistory change in the physical enivronment (Levinton, 1970). Under normal conditions, these species are absent or present only in small numbers, and chances are that most sampling programs would encounter none or only a few opportunists at any one time. Although it is possible for samples to document opportunistic outbreaks, usually as vast numbers of empty shells, none of our samples contained obvious examples of opportunism.

Historical Factors

The dead assemblage represents more time and, thus, is a cumulative record of occupation, showing all the species that have lived at the sample site during the period of sedimentation. Disturbances in the environment occasionally destroy or severely reduce the numbers of shells of some species, especially those that are fragile; but our data indicate that the dead assemblage precisely documents all the species that live and die in a particular habitat over time.

The historical aspects of habitat changes or habitat shifts can be extremely important. At any benthic site, conditions that dictate the distribution of plants and animals can and usually do change with time. For example, the substrate may change in composition, texture, surface structure, and plant cover. Hydrographic conditions may alter owing to freshwater flooding by storms or hurricanes, or by very local events, such as channel migration. Any number of other natural and perhaps unpredictable environmental changes can alter the historical species composition of assemblages accumulating at a particular site.

Species that are sensitive to environmental fluctuations may be over-represented in the dead assemblage as young individuals, because they could not survive a sudden change in physical environment and died prematurely. Subsequently, they may be recruited anew in the same habitat, perhaps to experience again the same fate. In contrast, a more permanent environmental change can produce a whole new suite of taxa. Theoretically, a succession of rapidly changing environmental conditions could cause large numbers of organisms with different ecological requirements to accumulate in the same dead assemblage. Such a turn of events would promote a very low correlation between the accumulating dead assemblage and the current live assemblage. Obviously, without precise stratigraphic data, it would be difficult if not impossible to unravel the effects of such a sequence of events merely by inspecting the accumulation of dead shells, because the spectrum of possible organism responses to physical environment through time is immense.

Fortunately, however, rapid environmental fluctuations sufficient to produce anomalous composite dead assemblages appear to be rare in marine habitats. Instead, sudden and extreme changes in the physical environment usually cause mass mortalities of certain species, which may be recognized in the fossil record more easily. In all four of our studies in Yucatan and California the live assemblages did indeed correlate well with the dead assemblages, and the dead assemblages appeared to correspond to obvious existing habitats. Our evidence from modern marine environments, therefore, suggests that composite faunas of ecologically unrelated organisms produced by sudden habitat shifts may not be common in the marine fossil record. This generalization deserves further testing.

FAUNAL COMMUNITIES AND PHYSICAL HABITATS

As mentioned earlier, in both Yucatan and Mugu Lagoon we used Q-mode cluster analysis to define groups of similar samples and thus provide us with generalized biofacies in the two regions. Communities of co-occurring species were defined using R-mode cluster analysis. For both geographic areas the Q-mode analyses yielded similar clusters based on live-only and dead-only occurrences, and these clusters corresponded directly with obvious physical habitats. This numerical correspondence was developed solely on the basis of species occurrence independent of any additional information pertaining to habitat description or areal distribution of samples. R-mode clusters developed from data on the distribution of dead mollusks and foraminifera showed a similar correlation of species groups to obvious habitats. We believe that these results are good evidence for the fidelity of dead assemblages as both recurrent groups (i.e., communities) and as habitat indicators (i.e., biofacies).

The correspondence between live and dead species and communities in the habitats sampled appears to hold regardless of the level of physical energy in these habitats. The Yucatan locality is in the Caribbean hurricane tract, and several major storms have crossed the sampling area within the last two decades. Even so, mixing of mollusk shells and foraminifera tests between habitats appears to be numerically insignificant and does not hinder our efforts to reconstruct living recurrent groups. Moreover, few examples of physically abraded mollusk shells were encountered in either the Yucatan or California locality, and counts of disarticulated bivalve shells in the Yucatan samples revealed nearly equal numbers of left and right valves for the most abundant species.

In both the Yucatan and California localities, habitats with the most vigorous waves and currents (the current-washed strait in Yucatan and the main sand channel in Mugu Lagoon) also exhibited the most distinct molluscan communities and the best correlation between live and dead specimens. The molluscan assemblages of these habitats are low in diversity, appearing to contain only a few species specialized for high-energy conditions; thus they are easily identified by numerical analysis. That the same assemblages also are easily identified through their empty shells indicates that the shells remain after death in the habitat in which they lived. Furthermore, they do not become seriously diluted by other species that wash in from adjoining habitats.

It is interesting that the most environmentally rigorous habitats, those characterized by unstable substrates of shifting sand, exhibit the most distinctive clusters of species. We suspect that there are ecological reasons for this, because only a few species have adapted for survival under such high-energy conditions. Indeed, our data indicate markedly low molluscan diversities in the sand channel habitat of Mugu Lagoon and the current-swept strait of northeastern Yucatan. Because there are few competing species and because food-bearing currents are strong, these few species are uncommonly abundant (in terms of millions per cubic meter in the winnowed sand of the Yucatan strait habitat). Yet these same species apparently are poorly adapted for lower-energy conditions, because rarely are they found in quieter lagoonal or back-reef environments.

The foraminifera in Yucatan showed both higher diversity and more variability in the strait than did the mollusks. Because live foraminifera were not separated from dead, these elements were not compared. Foraminiferal tests are very likely to be transported as sedimentary clasts, and some mixing undoubtedly occurs in the strait. However, because those foraminiferal samples taken in the strait near coral reefs and submerged Pleistocene dune ridges reflected the proximity of these features, large-scale, postmortem mixing of different faunas apparently is insignificant.

WITHIN-HABITAT STUDIES

The spatial and temporal variability of molluscan species *within* habitats in the study areas was investigated by intensively collecting in one distinctive habitat over a period of 3 years (Peterson, 1972) and by comparing closely spaced couplet samples (Warme, 1971b; A. A. Ekdale, 1972; Peterson, 1972). Both methods demonstrated that the empty shells represented more species and more even distributions between samples than shown by the live animals.

Temporal Accumulation

Peterson (1975b) sampled the sand habitat of both Mugu Lagoon and Tijuana Slough (Imperial Beach, California) on ten occasions over a 3-year period. Data on dead specimens from each lagoon were compared to the corresponding data on live specimens tabulated over longer periods of time. The cumulative number of live species encountered at any one sample site increased with time, gradually approaching the number of dead species encountered in a single sampling (Fig. 5). Also, using an index of proportionate similarity (Fig. 6), Peterson found that relative species abundances in the living community became more similar to those of the dead community as the live data were tabulated over successively longer time periods.

These results suggest that a fossil assemblage can be viewed as a cumulative ecological history, which reflects the average community ecology in a marine habitat much better than sampling the community at a single moment in time.

Couplet Samples

Data from closely spaced pairs of samples can be used to test significant differences between stations that single samples from a station cannot show. If pairs of adjacent samples show larger differences than do widely separated samples, a large degree of local variability is indicated; such local heterogeneity cannot be discerned in a single bulk sample.

Five pairs of molluscan samples were collected in the back-reef environment of coastal Yucatan (A. A. Ekdale, 1972). Similarities between the two samples of each couplet were computed separately for the live and dead elements using chi-square association tests and four different coefficients of mutual occurrence (the Dice, Jaccard, Fager, and simple match coefficients). Association values obtained for the dead species in the couplets were much higher than those for the corresponding live species. This is due to the fact that the dead assemblages contain more species and more individuals than the live assemblages, and the dead elements better represent the full complement of species that can live within each habitat. The live couplets show more variation between samples because of the vagaries of live species distributions on a local scale.

Similarities between live and dead species found in adjoining samples from given habitats also are illustrated by couplet data from Mugu Lagoon (Warme, 1971b). Four couplets from three habitats (sand channel, tidal creek, and eelgrass bed) were sampled in 1964 or 1965 and then resampled in 1968. Comparison of the two different censuses (Tables 8 to 11) shows that (1) different habitats contain several of the same ubiquitous species (e.g., *Protothaca staminea, Tagelus californicus, Chione undatella*); (2) the relative abundances of ubiquitous live species differ between habitats but remain constant within couplets (and thus within the sampling locality) over the sampling period; (3) the live abundances show a similar ranking of both ubiquitous and habitat-specific species when comparing data from both sampling periods, and this constancy is best displayed by the more abundant species; (4) the absolute value of the dead abundances has little ecological significance, because this value increases with depth of sampling and also is governed by

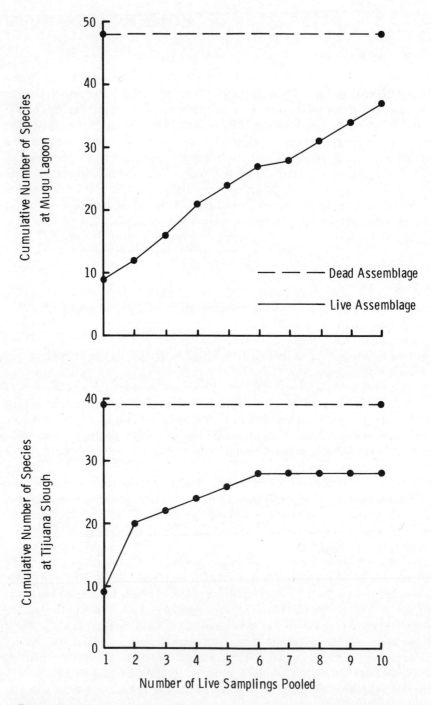

Figure 5
Cumulative numbers of live mollusk species encountered through time (solid line) compared with numbers of dead mollusk species encountered during a single sampling (dashed line) in Mugu Lagoon and Tijuana Slough. (Peterson, 1975b, Fig. 1.)

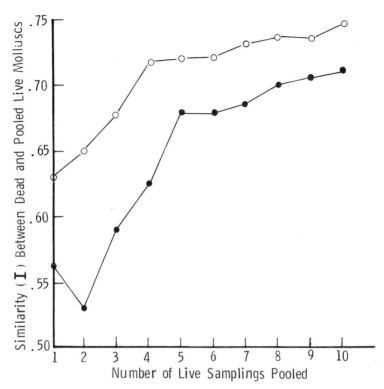

Figure 6
Proportionate similarity (I) between molluscan dead assemblages and average molluscan live assemblages pooled over varying lengths of time for Mugu Lagoon (solid circles) and Tijuana Slough (open circles).

$$I = 1.0 - 0.5 \sum_{i=1}^{S} |a_i - b_i|,$$

where a_i is the proportion of individuals in sample A that belong to species i, b_i is the proportion of individuals in sample B that belong to species i, and S is the total number of species present in the two samples. (Peterson, 1975b, Fig. 2.)

historical circumstances, such as sedimentation rates and past productivity; and (5) with few exceptions, ranking of samples by dead abundances is very similar to ranking by live, thus indicating that community structure in terms of relative abundances is preserved in the dead assemblages.

Although not evident in Tables 8 to 11 it is clear from inspection of the Mugu Lagoon samples that most of the more abundant dead-only species occurrences in each habitat involved individuals that died as juveniles, apparently after encountering unfavorable circumstances after their initial settling.

Table 8
Sand channel couplet samples (¼-m² area) from stations 39 and 40 collected from the same locality in both 1965 and 1968, showing absolute abundances of live and dead bivalves. Species are listed by approximate order of abundances.

	Live Abundances		Dead Abundances	
Species (bivalves only)	1965	1968	1965	1968
Sanguinolaria nuttallii	128–98	50–55	157–184	205–100
Apolymetis biangulata	18–9	1–0	14–14	19–5
Protothaca staminea	1–1	1–2	15–6	11–10
Tagelus californianus	1–4	0–0	5–1	3–3
Cumingia californica	0–5	0–0	0–0	0–0
Tresus nuttallii	0–0	0–0	8–7	0–0
Chione undatella	0–0	0–1	0–0	0–0
C. californiensis	0–0	0–0	0–2	1–0
Macoma secta	0–0	1–0	18–3	8–9
Donax californica	0–0	0–0	10–2	1–0
Leptopecten latiauratus	0–0	1–0	0–2	1–1
Macoma nasuta	0–0	0–0	0–0	0–3
Cryptomya californica	0–0	0–0	17–6	9–2
Macoma irus	0–0	0–0	3–3	0–0
Pecten circularis	0–0	0–0	0–1	0–0
Ostrea lurida	0–0	0–0	0–0	0–2
Laevicardium substriatum	0–0	0–0	0–0	1–0
Diplodonta orbella	0–0	0–0	0–1	0–0
Saxidomus nuttallii	0–0	0–0	1–0	0–0
	148–117	54–58	248–231	259–135
Washed over barrier from beach				
Tivela stultorum	0–0	0–0	1–6	10–5
Donax gouldii	0–0	0–0	10–12	4–3
Total gastropods (6 spp.)	2–3	0–0	16–16	23–20

All these results suggest that ecologists should not ignore the data provided by empty shells in their investigations of benthic communities. The dead assemblages, by smoothing out temporal variations, apparently contain more information about the total community living in a particular habitat than do the live assemblages. Moreover, the dead assemblages seem to be less complicated by random "noise" caused by the spatial and temporal variability of living populations.

CONCLUSIONS

Live-dead comparisons of mollusks in Yucatan and California reveal that species diversities and relative species abundances in these live communities generally are preserved in the dead assemblages of empty mollusk shells. Rarely did live individu-

Table 9

Eelgras couplet samples (¼-m² area) from stations 25 and 26 collected from the same locality in both 1964 and 1968. Note the high abundances of dead species not collected alive in these samples; with few exceptions, those species live in eelgrass, based on their presence in other samples from this habitat, but simply were not collected alive in these couplets. Note the correspondence between the attached bivalves and attached gastropods, the former serving as substrates for the latter. See Table 10 for comparison.

Species	Live Abundances		Dead Abundances	
	1964	1968	1964	1968
Macoma nasuta	85–78	45–64	238–157	81–185
Protothaca staminea	7–18	28–28	195–261	65–85
Tagelus californianus	3–2	0–0	28–18	40–85
Apolymetis biangulata	8–6	2–4	6–14	39–19
Saxidomus nuttallii	1–4	55–44	1–4	2–14
Leptopecten latiauratus	2–0	0–0	36–97	35–22
Tellina modesta	1–0	0–0	12–4	2–1
Cumingia californica	1–0	0–0	14–13	1–6
Chione undatella	1–0	2–4	0–0	0–3
Macoma irus	1–0	0–0	2–2	0–0
Diplodonta orbella	0–3	0–0	0–1	0–0
Tresus nuttallii	0–0	0–0	3–2	0–0
Macoma secta	0–0	0–0	8–7	3–3
Cryptomya californica	0–0	0–0	5–16	0–12
Laevicardium substriatum	0–0	0–0	16–17	14–17
Tellina carpenteri	0–0	0–0	15–4	1–2
Pecten circularis	0–0	0–0	1–10	7–4
Lucinisca nuttallii	0–0	0–0	0–0	0–1
	110–111	132–144	580–627	289–450
Attached bivalves				
Ostrea lurida	0–0	13–25	1–2	8–9
Mytilus edulis	0–0	3–18	2–3	1–0
Attached gastropods				
Crepidula onyx	2–0	60–100	4–5	4–35
C. lingulata	0–0	0–2	0–0	3–8
C. nummaria	0–0	0–0	2–3	0–1
Total other gastropods (9 spp.)	16–1	2–0	233–107	35–68

als occur in samples that did not also contain many more dead individuals of the same species. Molluscan censuses in sample couplets and in single localities sampled periodically over 3 years indicate that habitat migration and postmortem transport of mollusks during the time period represented by our samples were minimal. Even the assemblages of empty foraminiferal tests, which are quite susceptible to agents

Table 10

Eelgrass couplet samples (¼-m² area) from stations 48 and 49 collected from the same locality in both 1965 and 1968. The high abundances of *Leptopecten latiauratus* (also in Table 9) reflect intense settlement and rapid growth. Because this species is calcitic, along with *Pecten circularis* and *Ostrea lurida*, the probability of their dissolution is reduced and may help account for their high abundances as dead shells.

Species	Live Abundances		Dead Abundances	
	1965	1968	1965	1968
Macoma nasuta	23–22	128–119	176–323	137–352
Protothaca staminea	15–12	24–20	330–442	93–72
Apolymetis biangulata	10–16	35–22	57–59	15–11
Saxidomus nuttallii	1–16	11–8	45–59	18–8
Tagelus californianus	1–3	0–0	35–43	23–10
Macoma irus	1–0	14–0	2–5	23–21
Tellina carpenteri	0–0	3–3	5–3	0–0
Leptopecten latiauratus	0–0	1–4	68–251	180–204
Tresus nuttallii	0–1	1–1	4–6	11–3
Cummingia californica	1–0	0–1	3–11	1–2
Lucinisca nuttallii	0–0	1–0	0–1	0–0
Tellina modesta	0–0	0–1	8–13	0–1
Laevicardium substriatum	0–0	0–0	12–38	9–9
Pecten circularis	0–0	0–0	6–14	0–6
Cryptomya californica	0–0	0–0	10–5	0–2
Sanguinolaria nuttallii	0–0	0–0	2–0	1–0
Clinocardium nuttallii	0–0	0–0	1–0	1–0
Macoma secta	0–0	0–0	7–3	0–0
Heterodonax bimaculata	0–0	0–0	1–1	0–0
Diplodonta orbella	0–0	0–0	6–0	0–0
Attached bivalves				
Ostrea lurida	0–0	0–0	7–15	5–5
Mytilus edulis	0–0	0–0	1–0	3–3
Attached gastropods				
Crepidula onyx	0–0	7–2	0–11	8–17
C. lingulata	0–0	0–0	0–1	0–1
C. nummaria	0–0	0–0	0–3	0–0
Total other gastropods				
(12 spp.)	6–3	18–4	92–177	49–67

of physical transport, indicate no large-scale, postmortem mixing of faunas from different major environments.

Q-mode cluster analyses of both the Yucatan and California mollusks resulted in roughly the same environmental associations whether live or dead individuals were considered. However, the level of numerical association between samples obtained from live mollusk data was generally lower than that obtained from dead mollusk

Table 11
Tidal creek couplet samples (¼-m² area) from stations 43 and 44 collected from the same locality in both 1965 and 1968. These samples contained very high densities of three species, both alive and as empty shells. The large numbers of "dead only" *Cryptomya californica*, *Macoma secta*, and *Sanguinolaria nuttallii* possibly represent individuals that settled in the wrong habitat and failed to colonize successfully, or, more likely, individuals that washed into the mouth of this narrow tidal creek from the adjoining sand flats where they live in abundance.

Species	Live Abundances		Dead Abundances	
	1965	1968	1965	1968
Protothaca staminea	154–106	96–45	289–50	125–44
Macoma nasuta	71–80	93–58	414–266	352–59
Tagelus californianus	12–54	5–7	156–124	98–24
Saxidomus nuttallii	10–8	0–0	4–3	2–0
Cumingia californica	2–4	8–1	1–1	0–0
Chione undatella	1–0	0–0	6–0	10–6
Laevicardium substriatum	0–1	0–0	6–9	10–2
Apolymetis biangulata	1–4	1–6	0–0	0–2
Tellina carpenteri	1–1	0–0	18–43	8–5
Mytilus edulis	1–0	0–0	4–1	0–1
Cryptomya californica	0–0	0–0	101–46	74–37
Macoma secta	0–0	0–0	23–15	12–21
M. irus	0–0	0–0	2–3	0–0
Heterodonax bimaculata	0–0	0–0	10–0	0–0
Chione californiensis	0–0	0–0	4–1	8–0
Pecten circularis	0–0	0–0	6–0	6–10
Donax californica	0–0	0–0	4–0	7–5
D. gouldii	0–0	0–0	3–0	6–0
Sanguinolaria nuttallii	0–0	0–0	11–2	2–9
Leptopecten latiauratus	0–0	0–0	2–9	3–0
Zirfaea pilsbryi	0–0	0–0	3–0	0–0
Clinocardium nuttallii	0–0	0–0	0–0	3–0
Tivela stultorum	0–0	0–0	0–0	1–0
Tresus nuttallii	0–0	0–0	1–0	1–0
	253–258	203–117	1068–573	728–225
Total gastropods (14 spp.)	11–6	0–0	928–812	429–136

data. That is, *Q*-mode clusters resulting from information on dead molluscan remains correlated more closely with obvious habitats than did clusters resulting from information on live animals only. Moreover, cluster analysis of dead foraminiferal species from Yucatan also yielded distinct clusters that could be related to particular habitats.

These findings are very significant to paleontologists, who must rely solely on dead organism remains for paleoenvironmental reconstructions. Indeed, the assemblages of dead invertebrates appear to be more useful for environmental interpretations than the actual living assemblages, even when available.

REFERENCES

Cheetham, A. H., and J. E. Hazel. 1969. Binary (presence–absence) similarity coefficients. Jour. Paleont., 43:1130-1136.

Ekdale, A. A. 1972. Ecology and paleoecology of marine invertebrate communities in calcareous substrates, Northeast Quintana Roo, Mexico. Unpublished M.A. thesis, Rice University, Houston, Texas, 159p.

———. 1974. Marine molluscs from shallow-water environments (0 to 60 meters) off the northeast Yucatan Coast. Bull. Marine Sci., 24:638-668.

———. 1974b. Recent Marine molluscs from Northeast Quintana Roo, Mexico. In A. E. Weidie (ed.), Field Seminar on Water and Carbonate Rocks of the Yucatán Peninsula, Mexico. Geological Society of America Guidebook. No. 2, Field Trip to Yucatan Peninsula. p. 199-218.

———, and J. E. Warme, 1973. A diver-operated dredge for collecting quantitative benthic samples in soft sediments. Jour. Paleont., 47:1119-1121.

Ekdale, S. F. 1974. Recent foraminiferal associations from Northeastern Quintana Roo, Mexico. Unpublished M.A. thesis, Rice University, Houston, Texas, 151p.

Fager, E. W. 1957. Determination and analysis of recurrent groups. Ecology, 38:586-595.

Hairston, N. G. 1964. Studies on the organization of animal communities. Jour. Animal Ecol., supp. to v. 33 (Brit. Ecol. Soc. Jubilee Symp.):227-239.

Johnson, R. G. 1960. Models and methods for analysis of the mode of formation of fossil assemblages. Geol. Soc. Amer. Bull. 71:1075-1086.

Levinton, J. S. 1970. The paleoecological significance of opportunistic species. Lethaia, 3:69-78.

Macfadyen, A. 1963. Animal Ecology, Aims and Methods. Sir Issac Pitman & Sons Ltd., London 344p.

MacGinitie, G. E., and N. MacGinitie. 1968. Natural History of Marine Animals (2nd ed.). McGraw-Hill Book Company, New York. 523p.

Peterson, C. H. 1972. Species diversity, disturbance and time in the bivalve communities of some coastal lagoons. Unpublished Ph.D. dissertation, University of California, Santa Barbara, 229p.

———. 1975a. Stability of species and of communities for the benthos of two lagoons. Ecology, 56:958-965.

———. 1975b. The effects of differential solution rates and temporal variability on the live-dead correspondence of molluscs in two lagoons. Unpublished ms., 20p.

Ricketts, E. F., and J. Calvin. 1968. Between Pacific Tides (4th ed., revised by J. W. Hedgpeth). Stanford University Press, Stanford, Calif., 614p.

Sokal, R. R., and P. H. A. Sneath. 1963. Principles of Numerical Taxonomy. W. H. Freeman and Company, San Francisco, 359p.

Valentine, J. W., and R. G. Peddicord. 1967. Evaluation of fossil assemblages by cluster analysis. Jour. Paleont. 41:502-507.

Van Valen, L. 1964. Relative abundance of species in some fossil mammal faunas. Amer. Naturalist, 98:109-116.

Warme, J. E. 1969. Live and dead molluscs in a coastal lagoon. Jour. Paleont., 43:141-150.

———. 1971a. Fidelity of the fossil record and paleoecology. *In* J. D. Howard, J. W. Valentine, and J. E. Warme (eds.), Recent Advances in Paleoecology and Ichnology. American Geological Institute Short Course Notes. Washington, D.C., p. 11–28.

———. 1971b. Paleoecological aspects of a modern coastal lagoon. Univ. Calif. Publ. Geol. Sci., 87:131p.

Weidie, A. E. (ed.). 1974. Field Seminar on Water and Carbonate Rocks of the Yucatan Peninsula, Mexico. Geological Society of America Guidebook. No. 2, Field Trip to Yucatan Peninsula, 274p.

Williams, C. B. 1964. Patterns in the Balance of Nature. Academic Press, Inc., New York. 624p.

Comparison of Seven Lingulid Communities

R. R. West   Kansas State University

ABSTRACT

Community properties based on preservable or potentially preservable components will not reflect properties of the "total" community because the nonpreservable components are lacking.

Comparisons between fossil and Holocene communities can be made using potentially preservable structural properties. Lingulid communities, reconstructed from preservable or potentially preservable components, are compared spatially and temporally using (1) Brillouin diversity, (2) equitability, (3) ratio of infaunal to epifaunal species, (4) trophic structure, and (5) substrate grain size. Trophic structure is subdivided into (1) rank of trophic categories based on number of species in each and (2) species distribution within each trophic category based on species density.

Determination of these properties and their comparison with geologic age and substrate grain size for seven lingulid communities (four Paleozoic, two Mesozoic, and one Holocene) suggest several generalizations for further testing. (1) More communities must be studied to determine the relationship, if any, between Brillouin diversity, equitability, and substrate grain size. (2) In general, epifaunal species dominate the lingulid communities irrespective of grain size and are more prominent in the Paleozoic than in the post-Paleozoic. (3) The predominance of epifaunal species on Paleozoic clay substrates (a) suggests less substrate mobility than post-Paleozoic clay substrates, (b) illustrates the post-Paleozoic adaptive radiation of infaunal siphonate bivalves or (c) both. (4) Low-level suspension feeding species dominate five of the seven communities and increase through time at the expense of the other three trophic groups. (5) The total number of species is more equally distributed in the four trophic groups on a silt substrate, but is this because of the substrate grain size or because it is the oldest community studied? (6) Two of seven possible niche groups, epifaunal high-level and infaunal low-level suspension feeding species, are present in all seven communities but fewer niche groups (three or four) are represented in post-Paleozoic communities, with five or six represented in Paleozoic communities.

INTRODUCTION

Most paleobiologists are interested in communities because organisms evolve within a framework created by biotic and abiotic factors, a community framework. By recognizing communities we should be able to define trends or patterns in the structure and/or other properties of communities that would more fully explain morphological and other biological changes in the organisms comprising the communities. Studies by Olson (1961, 1966) on the evolution of Permian vertebrate communities are a good example of this approach.

In community paleoecology, it is essential to differentiate between "total" communities and the "preserved" (shelled) or "potentially preservable" communities. There is no logical reason to believe that the preserved or potentially preservable components of a community will reflect accurately the composition, structure, or dynamics of the "total" community unless the percentages of preservable and nonpreservable components have been constant for all communities through time; of course, this is not true. These preserved or potentially preservable components only reflect a part of the community. Driscoll and Swanson (1973) have shown that diversity of preservable epifaunas on mollusk valves does not reflect the diversity of the living faunas from which they are derived. Stanton (this volume) shows that the trophic structure of the "total" community and that of the "shelled" (potentially preservable) community are different, and validly questions whether the trophic structure of an ancient biocoenosis is preserved in a paleocommunity. Stanton's conclusions are based on samples taken over a large geographic area and over a long period of time and, therefore, are more representative of the "time-averaged assemblages" encountered in the fossil record. Concerning the potentially preservable components of a community, Warme et al. (this volume) show that the death assemblages of mollusks and foraminiferids of some biocoenoses off the coasts of Yucatan and California are a true reflection of the life assemblage of these "shelled" components; they appear to be more useful for environmental interpretations than the living assemblages because they represent a "time-averaged" assemblage rather than an assemblage from a temporal "instant."

As community paleoecologists, we must realize that we are dealing with only the preservable part of the community in our reconstructions and comparisons and are in no way reconstructing and comparing "total" communities. In the remainder of this paper usage of the term "community" should be understood to refer to preserved or potentially preservable community.

Recognition of community trends or patterns requires that we (1) define and recognize the same or similar communities spatially and temporally, a problem we have not yet solved to everyone's satisfaction, and (2) can compare these communities using properties common to all communities. I have chosen lingulid communities for comparison because (1) the geologic range of lingulid brachiopods (? Ordovician, Silurian-Holocene) provides a long temporal common denominator for community comparisons, (2) I had available quantitative data on a Holocene lingulid community and similar data on six fossil (Ordovician to Cretaceous) lingulid communities, and (3) lingulids are evolutionarily rather conservative (at least in hard-part morphology); therefore, it might be reasonable to assume, as a first approximation, that properties of these communities would be similar.

In this study, all organisms preserved with the lingulids are assumed to be members of the lingulid community. However, because of their infaunal habit, it is possible that the lingulids belong to the community in the bed above them, rather than the bed in which they are preserved. In all the communities considered, lingulid brachiopods (*Lingula* or *Glottidia*) rank as one of the six most abundant taxa on the basis of percentages of individuals. Usage of the term "community" in this paper is the same as that of plant ecologists of the dominance-constancy school and as that of Petersen in his work on bottom faunas. Problems involved in the usage of this concept in marine biology are discussed by Stephenson (1973) and also apply to its use in paleoecology. For fossil lingulid occurrences there is another problem, as pointed out by Ager (oral comm., 1974): in many fossil assemblages where lingulid brachiopods rank first (based on number or percentage of individuals) other fossils are rarely found. My usage of the dominance-constancy concept does not mean that I think it is necessarily the one to be accepted; rather it is a way for the reader to know what criteria were employed in defining a lingulid community in this paper. This study does not define statistical parameters for lingulid communities, nor does it intend to define *the* typical lingulid community.

As pointed out previously (West, 1972), paleoecological studies of communities are most significant when undertaken within a realistic environmental framework. For the most meaningful community comparisons, both spatially and temporally, we should compare communities from the same or similar environments to minimize what Thayer (1973) has called environmental "noise," which includes preservational "noise" (effects of differential preservation). In this paper, environments of all seven communities can be characterized as nearshore, high stress, and hence unstable; although more precisely defined depositional environments would be desirable, I think the main objectives of this paper can be obtained within this general environmental setting. These objectives are to (1) suggest via readily testable generalizations some common community properties useful for comparative purposes, both spatially and temporally, (2) emphasize the need for more and better quantivative data on fossil assemblages for community studies, and (3) point out the need for more functional morphological studies of invertebrates.

PROPERTIES OF SOME LINGULID COMMUNITIES

General

Properties common to all communities and potentially useful for comparison are (1) species rank based on the percentages of individuals of each species, (2) Brillouin diversity index (Lloyd et al., 1968), (3) equitability (Lloyd and Ghelardi, 1964), (4) ratio of epifaunal to infaunal species, (5) trophic structure in terms of high- and low-level suspension-feeding species, collecting and/or swallowing species (Turpaeva, 1957; Walker, 1972; Walker and Bambach, 1974), and (6) grain size of substrate. Item 5, trophic structure, can be subdivided into (1) rank of trophic groups based on the number of species in each group and (2) species distribution based on species density within each trophic group. Calculations of these properties require a list of taxa and their density (i.e., number of individuals of each taxa) for each fossil community. Density recorded in terms of rare, common, abundant, and so on, is inadequate for determining these properties. Other criteria used in reconstructing these communities are given by Johnson (1960).

Holocene Lingulid Community

The sample of the Holocene lingulid community was collected off Mangrove Point in Charlotte Harbor on the west coast of Florida by M. A. True of Bio-Oceanic Research on December 17, 1971. The area sampled was nearshore in 1.8 to 2.1 m of water at the confluence of the Peace and Myakka rivers (a tropical estuary). Using a benthic suction sampler (True et al., 1968) with a sieve mesh size not larger than 2.0 mm or smaller than 0.5 mm, an 0.3-m^2 area was sampled. Composition of this community is given in the Appendix, part A, and the dominant organism is the bivalve *Mulinia lateralis* with *Glottidia pyramidata* ranking third. Brillouin diversity is 0.5235 (Table 1), and the substrate is a well-sorted, mature, subrounded, fine-grained quartz sand with a few skeletal (bivalve) grains (West and Twiss, 1973). Numbers of epifaunal and infaunal species are equal; however, in terms of the number of individuals there are half again as many infaunal as epifaunal individuals (Table 1). Low-level suspension-feeding species and individuals dominate the trophic structure, and swallowers are absent. Number of species in each life mode and trophic group are evolutionarily significant because they indicate spatial and temporal diversity changes within these categories of the preserved or potentially preservable community.

Cretaceous Lingulid Communities

Two Cretaceous lingulid communities were recognized, one from mudstone–shale, the other from sandstone, by analyzing data recorded by Scott (1970) in his study

Table 1
Structure of the seven lingulid communities.

Age	Substrate	Brillouin Diversity Index	Equitability	Mode of Life			Trophic Structure[a]				
				Infaunal	Epifaunal	HLS	LLS	C	S	S[b]	
Holocene	Fine sand (2∅)	0.5235	0.2057	4 (50) 90 (60)	4 (50) 61 (40)	2 (25) 56 (37)	4 (50) 90 (60)	2 (25) 5 (3)	0 (0) 0 (0)	S I
Cretaceous	Very fine to fine sand	0.5662	0.0772	11 (50) 746 (86)	11 (50) 122 (14)	8 (36) 116 (13)	10 (45) 744 (86)	3 (14) 6 (0.7)	1 (4) 2 (0.2)	S I
	Clay	0.7527	0.1606	11 (92) 40 (85)	1 (8) 7 (15)	1 (8) 7 (15)	7 (58) 31 (66)	0 (0) 0 (0)	4 (33) 9 (19)	S I
Permian	Clay	0.4379	0.1184	4 (31) 88 (28)	9 (69) 299 (72)	6 (46) 18 (6)	5 (38) 295 (93)	1 (8) 1 (0.3)	1 (8) 3 (0.9)	S I
Lower Carboniferous	Clay	0.5776	0.1007	5 (29) 199 (31)	12 (71) 448 (69)	4 (24) 400 (62)	6 (35) 192 (30)	5 (29) 38 (6)	2 (12) 17 (3)	S I
Silurian	Sand	0.5887	0.1232	5 (36) 101 (27)	9 (64) 267 (72)	4 (29) 35 (10)	5 (36) 284 (77)	3 (21) 8 (2)	2 (14) 41 (11)	S I
Ordovician	Silt	0.8605	0.1106	6 (32) 90 (50)	13 (68) 90 (50)	7 (37) 45 (25)	2 (10) 30 (17)	5 (26) 43 (24)	5 (26) 62 (34)	S I

[a]HLS, high-level suspension feeders; LLS, low-level suspension feeders; C, collectors; S, swallowers.
[b]S, species; I, individuals. First figure is actual number of species or individuals. Figure in parentheses is percentage of species or individuals.

of the Kiowa Formation of Kansas. The Kiowa in central Kansas is 100 to 125 ft thick, dominated by dark-gray shale and brown sandstone (Scott, 1970, p. 5). Localities from which data were used represent a total area of several thousand square miles in Clark, Kiowa, Rice, McPherson, and Saline counties of Kansas.

Sand substrate. Part B of the Appendix shows the composition of the lingulid community on this substrate and was compiled by combining data from localities C1:7, R2:5, R6, and S4 (Scott, 1970, Table 3). The dominant species is the infaunal bivalve, *Corbula? fenti* with *Lingula subspatulata?* ranking fifth. Diversity is about the same as for the Holocene lingulid community (Table 1). The very fine to fine quartz sand substrate is generally similar to that of the Holocene sample, and the number of epi- and infaunal species is equal (Table 1). Ranking of trophic groups reveals that low-level suspension-feeding species are dominant, with high-level suspension-feeding species, collectors, and swallowers second, third, and fourth, respectively (Table 1). This same trophic structure is also reflected by the number of individuals in each feeding group.

Clay substrate. Data from Scott (1970), localities C1:5, C3:3, C3:4, C3:7, K1:3, and M4:8-9, were combined and are tabulated in part C of the Appendix. The total number of taxa in this community decreases to 12 from 22 in the sand community; the infaunal bivalve *Trachycardium kansasense*, which ranks second in the sand substrate community, is now dominant, and the lingulid, *Lingula subspatulata?*, has moved from fifth to second place. Three other taxa are common to both Cretaceous communities (Appendix, parts B and C); *Scabrotrigonia emoryi, Crassinella semicostata,* and *Drepanochilus kiowanus;* in terms of percentages of individuals, they are more abundant in the clay substrate community than in the sand substrate community. The greater percentages of individuals of *Trachycardium kansasense, Lingula subspatulata?, Scabrotrigonia emoryi, Crassinella semicostata,* and *Drepanochilus kiowanus* on the clay substrate suggests (1) preference for a clay substrate (i.e., selected against in the sand substrate community), (2) more favorable preservational conditions of the clay substrate, (3) collecting bias, or (4) a combination of two or more of the first three. It is not possible with available data to speculate which of the four might be the correct interpretation. Diversity and equitability have both increased, and infaunal species and individuals are clearly dominant (Table 1). Dominance of low-level suspension-feeding species might suggest a firm clay substrate with little turbidity near the sediment-water interface, but the absence of collectors (epifaunal gastropods) and presence of infaunal swallowers would imply a loose clay bottom. Perhaps the low-level suspension feeders were selective in particle choice, and a turbid zone near the bottom, if present, was not detrimental.

Permian Lingulid Community

Yarrow (1974) collected data on a Permian lingulid community from part of the Hughes Creek Shale Member (Foraker Limestone, Gearyan Stage, Lower Permian Series) from north-central Kansas. The tabulation (Appendix, part D) represents 10 horizontal mapped surfaces (15 cm × 15 cm) of a mudstone interval (approximately 1 dm thick) from five localities (representing a geographic area of several hundred square miles). *Crurithyris expansa,* an epifaunal articulate brachiopod, is the dominant species, with *Lingula carbonaria* ranking second. Substrate characteristics are (1) grain size of 9.2ϕ ($\bar{\bar{x}}$), (2) coarsest 1 percent grain size of 4.5ϕ ($\bar{\bar{x}}$), (3) weight percent of insolubles, 42.3 ($\bar{x}$), and (4) percentage of organic carbon, 0.044 g ($\bar{x}$) (weight loss per gram). Compared to the Cretaceous lingulid (clay) community, diversity and equitability have decreased and epifaunal species are over twice as abundant as infaunal species (Table 1). For the first time, high-level suspension-feeding species dominate the trophic structure; however, the collector and swallower are tentative because the former is an ostracode, which could probably be assigned to all four of the trophic groups, and the latter is indicated by a horizontal burrow.

Lower Carboniferous Lingulid Community

A marine transgression was described by Ferguson (1962) at a single locality of Visean shale in Scotland. He recognized what he called "topozone 1," on the basis of definitely indigenous *Lingula squamiformis* and *Streblopteria ornata,* in the lower 105 cm of this 2.8-m-thick interval. Tabulating data for topozone 1 from Ferguson (1962, Table 1) in the same way as for previous lingulid communities shows that *Streblopteria ornata* is dominant, with *Lingula squamiformis* ranking second (Appendix, part E). Diversity and equitability values are similar to those for the Permian lingulid community, as is the ratio of epi- to infaunal species (Table 1). Trophic structure is dominated by low-level suspension-feeding species, followed by collectors, high-level suspension-feeding species, and swallowers, but there are more high-level suspension-feeding individuals (Table 1). This trophic structure could be interpreted as reflecting a firm clay substrate with low turbidity at the sediment-water interface. Careful examination of the exposure would be required to test this interpretation.

Silurian Lingulid Community

The *Lingula* community of Ziegler et al. (1968) occurs in sandstone and is characterized in Table 1 and Figure 1 in their paper. I have reanalyzed these data and tabulated them in part F of the Appendix. As with the Scottish Visean lingulid community, the data for this Welsh Silurian community are from one locality (U.S.N.M. locality No. 10234) selected as representative by Ziegler et al. (1968). An unidentified species of *Lingula* and *Lingula pseudoparallela* occur in this community, the former ranking second and the latter sixth (Appendix, part F). The epifaunal low-level suspension-feeding brachiopod *"Camarotoechia" decemplicata* is most abundant. Compared to the Holocene and Cretaceous lingulid communities with sand substrates, the diversity is about the same, while equitability is closer to that of the Cretaceous (sand) community (Table 1). With coarser substrate grain size one would expect the epifaunal species to dominate the community, as indicated by Craig and Jones (1966), and epifaunal species are almost twice as abundant as infaunal species (Table 1). In terms of food source, the low-level suspension-feeding species again predominate, with four, three, and two species in the high-level suspension, collecting, and swallowing categories, respectively.

Ordovician Lingulid Community

The oldest lingulid community for which I have usable data comes from a study by Bretsky (1970) on the upper Ordovician of the Appalachians. After studying information provided by Bretsky (pers. comm., 1973) on all collections of his linguloid population, I selected his locality 203 and tabulated the data in the same form as for the other six lingulid communities. Locality 203 consists of approximately 50 ft of the Reedsville Formation at one geographic point. *Lingula?* sp. ranks third in this community, and again an infaunal bivalve, *Tancrediopsis cuneata,* is the dominate invertebrate (Appendix, part G). This community has the highest diversity, and equitability is close to that of the Lower Carboniferous and Permian lingulid communities (Table 1). The substrate is a poorly sorted, phosphatic, "clayey," fine sand- to silt-sized quartz, and epifaunal species are over twice as abundant as infaunal species, with equal numbers of individuals in these two categories. Species are distributed more "evenly" among the four trophic groups in this lingulid community than in any of the other lingulid communities analyzed. Collecting and swallowing species tie for second rank, with high-level suspension-feeding species first and low-level suspension-feeding species third and last (Table 1).

COMPARISON OF LINGULID COMMUNITIES

General

One should first compare contemporaneous lingulid and other communities from the same and different depositional environments, substrates, and so on, to establish the amount of variability in the community at or during a particular time (age, epoch, or period). This approach requires the accurate and careful accumulation of numerical and other data for all fossils obtained by prudent dissection of blocks of the entombing rock. Additionally, more functional morphological studies of fossils are needed so that their ecological role in the community and the community structure will be better understood. Within this framework temporal comparisons would be more meaningful.

Unfortunately such studies are time consuming, and to my knowledge adequate data for lingulid communities are available only for the seven considered in this paper. It is pertinent to compare these seven at this stage in our study of communities because we must analyze fossil assemblages with an idea toward not only defining and recognizing a community, but also toward determining properties that will be useful and meaningful in comparing communities in the study of community evolution. With this in mind, I propose some comparisons for consideration and evaluation as well as some testable generalizations.

Brillouin Diversity and Equitability

Small sample size (seven communities) and lack of replication in the analysis of any of the seven individual lingulid communities prevents making meaningful comments about diversity or equitability of lingulid communities. Additionally, Driscoll (pers. comm., 1975) has pointed out that diversity cannot be expected to stabilize until about 400 individuals per community have been examined. Only two of the seven lingulid communities contain over 400 individuals.

Based on work by Sanders (1968) and Driscoll (1973), diversity in recent marine benthic environments is greater on stable sand substrates than on mud substrates. Of the seven lingulid communities, three occupied a sand substrate and three a clay substrate. There is no statistically significant difference between the average sand and clay substrate diversities and equitabilities at the 95 percent level (Table 2). The same is true when F tests are performed on the variances of these two variables for the sand and clay substrate (Table 2). However, the differences observed between sand and mud substrates in recent marine benthic environments do not mean that the same or similar differences will also exist in fossil marine benthic communities. Unfortunately, data necessary to test this for lingulid communities, or for most other fossil marine benthic communities that have been documented, are still lacking.

Ratio of Infaunal to Epifaunal Species

From Table 1, it is clear that during the Paleozoic epifaunal species dominate lingulid communities (averaging 68 percent for four Paleozoic communities), whereas infaunal species are dominant in post-Paleozoic lingulid communities (averaging 64 percent for three post-Paleozoic communities). The high percentage of infaunal species (92 percent) in the Cretaceous (clay) community is responsible for this apparent infaunal domination. Epifaunal and infaunal species are equal in the two post-Paleozoic communities occurring on a sand substrate. In general, epifaunal

Table 2
Comparison of diversity and equitability of lingulid communities.

Sub-strate	N	Diversity		Equitability	
		Mean	Variance	Mean	Variance
Sand	3	0.5595 ± 0.0191	0.001	0.1354 ± 0.0376	0.004
Clay	3	0.5844 ± 0.0910	0.024	0.1266 ± 0.0178	0.0009
		$t_s = 0.274$	$F_s = 0.042$	$t_s = 0.212$	$F_s = 0.224$
		$t_{.05} = 2.57$ (calc)	$F_{.05[2.2]} = 19.0$	$t_{.05} = 2.57$ (calc)	$F_{.05[2.2]} = 19.0$

species dominate the lingulid communities irrespective of substrate grain size, and the similarity of values for all Paleozoic communities suggests that substrate was of little importance during the Paleozoic, but again more data are needed.

The high percentage of infaunal species in the Cretaceous (clay) lingulid community can be explained, in part, by the post-Paleozoic adaptive radiation of infaunal siphonate bivalves (Stanley, 1968) and the inability of epifaunal brachiopods to compete successfully in post-Paleozoic communities. Six of the ten (60 percent) bivalve species in the Cretaceous (clay) community belong to superfamilies classed as infaunal siphon feeders by Stanley (1968). In the Cretaceous (sand) lingulid community, only 40 percent (6 of 15) of the bivalve species are infaunal siphon feeders, and in all four Paleozoic lingulid communities there is only one bivalve (*Edmondia* sp.) that might have been an infaunal siphon feeder.

Trophic Structure

Three ways of comparing communities involving this community property are as follows:

General. There is, in general, a gradual increase in low-level suspension-feeding species (10 percent in Ordovician to 38 percent in Permian) at the expense of the other three trophic groups through the Paleozoic, with little change from Cretaceous (45 and 58 percent) to Holocene (50 percent). This same general pattern is observed for lingulid communities from sand- and clay-sized substrates through geologic time (Table 1).

Through time, in lingulid communities inhabiting a sand substrate, swallowers (Holocene) disappear, and in communities inhabiting a clay substrate, collectors disappear (Cretaceous clay). In the case of the sand substrate this is because the swallowers present in the "total" Holocene lingulid community (West and Twiss, 1973) have no preservable hard parts. Absence of collectors in the community on the clay substrate may reflect an inability of collectors to utilize available nutrients and/or mobility of the substrate. In the Ordovician lingulid community there is more equal distribution of species in the four trophic categories (Table 1). It is then reasonable to ask whether this distribution is because of the silt-sized substrate or because it is the oldest lingulid community analyzed. Because this is the only Ordovician lingulid community studied to date, and also the only one from a silt substrate, we cannot yet answer this question. More studies of lingulid communities from substrates of all grain sizes and from the same and different geological periods are necessary.

Ranking of trophic groups within communities. The feeding category with the greatest number of species is ranked first, and the category with the least number of species is ranked last. If the number of species is equal, the two categories share the same rank. In five of the seven lingulid commnities (71.4 percent) analyzed, low-level suspension-feeding species dominate irrespective of age or substrate (Table 3).

Silurian and Cretaceous lingulid communities on a sand substrate are structured the same trophically (Table 3), which indicates (1) a "stable" trophic relationship

Table 3
Comparison of lingulid communities by ranking of trophic group.[a]

Age[b]	Subtrate	Rank of Trophic Group[c]			
		1	2	3	4
H	Sand	LLS	HLS/C		
K	Sand	LLS	HLS	C	S
	Clay	LLS	S	HLS	
P	Clay	HLS	LLS	S/C	
LC	Clay	LLS	C	HLS	S
S	Sand	LLS	HLS	C	S
O	Silt	HLS	S/C	LLS	

[a]Trophic group abbreviations same as those in Table 1.
[b]H, Holocene; K, Cretaceous; P, Permian; LC, Lower Carboniferous; S, Silurian; O, Ordovician.
[c]By number of species.

and/or similar environments for these two communities or (2) a chance event. There appears to have been more diversification or partitioning of the suspension-feeding niche in the Cretaceous because the number of species in the two suspension-feeding categories doubles from Silurian to Cretaceous [LLS ratio = 5(S) : 10(K), HLS ratio = 4(S) : 8(K)]. However, this increase in suspension-feeding organisms could reflect the greater availability of suspended nutrients. Investigations of the planktic content of the rocks containing these communities would be helpful.

Whether or not this ranking technique is useful in studies of community evolution and which, if any, of the relationships suggested here are accurate must await analysis of other communities.

Species distribution within each trophic group based on species density. Turpaeva (1957) indicated that for Arctic and Boreal marine benthic communities each trophic group will have a dominant species in terms of number of individuals, with the most abundant species in one group, the second most abundant species in a different group, and so on. Such a distribution eliminates intense competition for the same or a limited resource. Walker (1972) applied this scheme successfully to some Paleozoic communities. The Ordovician lingulid community best fits this pattern, with the most abundant species a swallower (deposit-feeding bivalve), the second most abundant a collector (snail), the third a *Lingula* (low-level suspension feeder), and the fourth a high-level suspension-feeding bivalve (Table 4). The predicted pattern also holds, in general, for the Lower Carboniferous and Holocene lingulid communities; however, it does not apply well to the remaining four lingulid communities (Table 4).

Table 4
Comparison of lingulid communities by species distribution within each trophic group.

Age[a]	Sub-strate	Trophic Group[b]						
		HLS		LLS		C		S
		Taxa	% Indiv.	Taxa	% Indiv.	Taxa	% Indiv.	Taxa	% Indiv.
H	Sand	Amygdalum papyria	36.4	Mulinia lateralis	39.7	Mitrella lunata	2.6		
		Barnacles	0.6	Glottida pyramidata	18.5	Nassarius vibex	0.6		
				Anodontia alba	0.6				
				Mysella planulata	0.6				
K	Sand	Ostrea rugosa	4.3	Corbula? fenti	68.1	Euspira smolanensis	0.5	Drepanochilus kiowanus	0.2
		Ostrea arcuata	4.1	Trachycardium kansasense	9.1	Otostoma sp.	0.1		
		Ostreidae sp. indet.	3.5	Lingula subspatulata?	3.6	Pirsila? sp.	0.1		
		Crassostrea kiowana	1.0	Protocardia texana	1.7				
		Lopha quadriplicata	0.1	Corbulids	1.0				
		Lopha subovata	0.1	Turritella belviderei	1.0				
		Lopha? kansasensis	0.1	Flaventia belviderensis	0.5				
		Modiolus sp. A	0.1	Scabrotrigonia emoryi	0.3				
				Crassinella semicostata	0.2				

		Species	%	Species	%	Species	%
K	Clay	Breviarca habita	14.9	Turritella kansasense	0.1		
				Trachycardium kansasense	27.6	Drepanochilus kiowanus	8.5
				Lingula sub-spatulata?	23.4	Nuculana mutata	6.4
				Corbula smolanensis	6.4	Yoldia microdonta	2.1
				Scabrotrigonia emoryi	2.1	Tellina? sp. indet.	2.1
				Crassinella semicostata	2.1		
				Pholadomya? belviderensis	2.1		
				Corbulids	2.1		
P	Clay	Volsellina? sp.	4.1	Crurithyris expansa	66.0	Ostracode	0.3
		Pinna? sp.	0.3	Lingula carbonaria	22.4	Burrows (horizontal)	0.9
		Septimyalina sp.	0.3	Orbiculoidea missouriensis	4.1		
		Aviculopecten cf. A. artisulcata	0.3	Acanthocrania sp.	0.3		
		Rhipidomella carbonaria	0.3	Petrocrania sp.	0.3		
		Crinoid	0.3				
LC	Clay	Streblopteria ornate Actinopteria fluctuosa	56.4	Lingula squamiformis	26.0	Bucaniopsis decussatus	2.9
			4.8	Crurithyris urei	1.2	Donaldina pulchra	1.4
		Dunbarella papyraceus	0.5	Eomarginifera longispinus	1.1	Soleniscus (S.) typicus	0.8
						Nuculopsis gibbosa	1.8
						Nuculana laevistriata	0.8

(continued)

Table 4 (continued)

Age[a]	Sub-strate	Trophic Group[b]							
		HLS		LLS		C		S	
		Taxa	% Indiv.	Taxa	% Indiv.	Taxa	% Indiv.	Taxa	% Indiv.
LC	Clay	Naiadites crassa	0.1	Edmondia sp.	0.9	Soleniscus (M.) brevis	0.6		
				Camarotoechia pleurodon	0.1	Natacopsis sp.	0.1		
				Orbiculoidea nitida	0.1				
S	Sand	Pteronitella cf. P. retroflexa	4.9	"Camarotoechia" decemplicata	60.0	Gastropods	1.1	Palaeoneilo rhomboidea	10.0
		Cornulites sp.	3.3	Lingula sp.	13.0	Tropidodiscus sp.	0.5	Lyrodesma sp.	1.1
		Bryozoans	1.1	Lingula pseudo-parallela	3.0	Plectonotus trilobatus	0.5		
		Modiolopsis sp.	0.3	Burrow	0.3				
				Leptostrophia sp.	0.3				
O	Silt	Ischyrodonta truncata	8.3	Lingula? sp.	15.5	Plectonotus sp.	20.5	Tancrediopsis cuneata	30.0
		Orthorhynchula linneyi	6.1	Zygospira modesta?	1.1	Loxoplocus (L.) abbreviata	1.6	Lyrodesma post-striatum	1.6
		Ambonychia praecursa	4.4			Liospira? sp.	0.5	Palaeoneilo? sp.	1.1
		Pterinea (C.) demissa	3.3			Flexicalymene? sp.	0.5	Nuculites? sp.	1.1
		Modiolopsis modiolaris	1.6			Bucania? sp.	0.5	Praenucula? sp.	0.5
		Cymatonota sp.	0.5						
		Conularia sp.	0.5						

[a] As in Table 3.
[b] See note c in the Appendix.

Assuming that Turpaeva's scheme is correct, there are some possible explanations for the anomalies encountered. (1) Classifications are artificial and no one individual or species will necessarily belong exclusively to one trophic group. (2) Accurate assignment of modern and fossil organisms to a trophic group most often occupied requires careful analysis of morphology in terms of function; more functional morphological studies of all invertebrates are needed. (3) Absence of organisms with nonpreservable hard parts affects the outcome of trophic analysis, and studies incorporating the recognition and interpretation of trace fossils in community analysis (Walker, 1972) are needed. (4) Turpaeva's method is based on the assumption that food is the limiting resource; suppose that it is not? Can similar patterns be expected if space, or some other requirement, is limiting? It should also be pointed out that Turpaeva studied communities from Arctic and Boreal regions; perhaps communities in tropical regions are different (see Stanton, this volume).

Niche Group

A more complete idea of the ecological niche occupied in these communities can be obtained by combining mode of life with trophic group. This combination produces eight niche groups: epi- and infaunal high-level suspension feeders, epi- and infaunal low-level suspension feeders, epi- and infaunal swallowers, and epi- and infaunal collectors. Because collectors (browsers of Walker and Bambach, 1974) are, by definition, epifaunal, there are only seven niche groups. The percentages of species in each niche group for each of the seven lingulid communities are given in Table 5, epifaunal high-level and infaunal low-level suspension-feeding species are present in all seven. Epifaunal high-level suspension-feeding species dominate the two (Ordovician and Silurian) Paleozoic lingulid communities with a sand substrate. Epifaunal collectors dominate the Lower Carboniferous (clay) community, and equal numbers of epifaunal high- and low-level suspension-feeding species dominate the Permian (clay) community. Infaunal low-level suspension-feeding species dominate the three post-Paleozoic lingulid communities irrespective of substrate.

In post-Paleozoic communities there are fewer niche groups represented (three or four), whereas in the Paleozoic communities five or six of the seven possible niche groups are represented. Infaunal high-level suspension feeders (two bivalves, *Volsellina* and *Pinna;* the latter is semi-infaunal) are represented in the Permian but lacking in the other three Paleozoic communities. Bivalve diversification is suggested because no other Paleozoic lingulid community contains infaunal high-level suspension-feeding bivalves. Is this diversification because of limited space, keen competition for, or limited amounts of suspended or other nutrients near or at the sediment–water interface, or a combination? These questions and others can only be answered with more study.

Table 5
Comparison of lingulid communities by niche groups.[a]

Substrate	Age[b]	Niche Group[c]						
		EHLS	ELLS	EC	ES	IHLS	ILLS	IS
Sand	H	25	–	25	–	–	50	–
	K	36.4	–	13.6	–	–	45.4	4.5
	S	28.6	14.3	–	21.4	–	21.4	14.3
Silt	O	33.3	5.5	27.8	–	–	5.5	27.8
Clay	K	8.3	–	–	–	–	58.3	33.3
	P	30.8	30.8	7.7	–	15.4	7.7	7.7
	LC	23.5	17.6	29.4	–	–	17.6	11.8

[a] Values in percent species.
[b] Abbreviations same as Table 3.
[c] EHLS, epifaunal high-level suspension feeders; ELLS, epifaunal low-level suspension feeders; EC, epifaunal collectors; ES, epifaunal swallowers; IHLS, infaunal high-level suspension feeders; ILLS, infaunal low-level suspension feeders; and IS, infaunal swallowers.

CONCLUSIONS

Based on these comparisons it is possible to suggest the following testable generalizations:

1. More communities must be studied to determine the relationship, if any, between Brillouin diversity, equitability, and substrate grain size.
2. In general, epifaunal species dominate lingulid communities irrespective of grain size and are more prominent in Paleozoic than in post-Paleozoic communities.
3. Clay substrates support more epifaunal species during the Paleozoic than during the post-Paleozoic, (a) suggesting that possibly Paleozoic clay substrates of lingulid communities were less mobile (i.e., firmer), (b) illustrating the post-Paleozoic adaptive radiation of infaunal siphonate bivalves, or (c) both.
4. Low-level suspension-feeding species dominate five of the seven communities and increase through time at the expense of the other three trophic groups.
5. Silt substrate supports a more equal number of species in the four trophic groups; but is this because of substrate grain size or because it is the oldest community studied?
6. Two of seven possible niche groups, epifaunal high-level and infaunal low-level suspension-feeding species, are present in all seven communities, but fewer niche groups (three or four) are represented in post-Paleozoic communities, with five to six represented in Paleozoic communities.

There is no question that before these or any other community properties, comparisons, or generalizations are meaningful we need (1) more quantitative data on fossil assemblages, including possible food sources (i.e., planktic components),

(2) petrology of the rocks containing the fossil assemblages, including organic and insoluble residue content, and (3) more functional morphological studies of invertebrates. These data are needed for contemporaneous and noncontemporaneous fossil assemblages from similar and different depositional environments. It may be considered premature to make such comparisons and to suggest generalizations with no clearly defined concept of a community; however, a possible approach to community comparisons and ultimately to community evolution is indicated. This paper attempts, in part, to answer the question posed by Scott in a letter to all symposium participants (July 2, 1974): "Whither goest paleoecology?"

ACKNOWLEDGMENTS

I thank R. W. Scott (Amoco Research, Tulsa, Oklahoma) for inviting me to participate in this symposium and for many helpful discussions and comments. Page C. Twiss (Kansas State University) willingly discussed petrologic and other aspects of the study with me. Merrill and Renate True (Bio-Oceanic Research, New Orleans), L. Ferguson (Mount Allison University, Sackville, New Brunswick), and P. W. Bretsky (State University of New York, Stony Brook), respectively, provided access to and discussions on the Holocene lingulid assemblage, provided samples of topozone 1 for petrologic study and loaned thin sections of rocks containing his Appalachian linguloid population. D. V. Ager, E. G. Driscoll, L. E. Lindemuth, and A. J. Rowell critically reviewed the manuscript and provided many helpful suggestions that improved it, but I accept full responsibility for failings that remain. I acknowledge the donors of the Petroleum Research Fund, administered by the American Chemical Society (P.R.F. Grant 2077-G3) and the Bureau of General Research of Kansas State University for financial support.

REFERENCES

Bretsky, P. W. 1970. Upper Ordovician ecology of the central Appalachians. Peabody Museum, Nat. Hist. Bull., 34, 150p.

Craig, G. Y., and N. S. Jones. Marine benthos, substrate and palaeoecology. Palaeontology, 9:30-38.

Driscoll, E. G. 1973. Mollusc-sediment relationships in northwestern Buzzards Bay, Massachusetts, U.S.A. Malacologia, 12:13-46.

——, and R. A. Swanson. 1973. Diversity and structure of epifaunal communities on mollusc valves. Buzzards Bay, Massachusetts. Palaeogeog. Palaeoclimat. Palaeoecol., 14:229-247.

Ferguson, L. 1962. The paleoecology of a lower Carboniferous marine transgression. Jour. Paleont., 36:1090-1107.

Johnson, R. G. 1960. Models and methods for analysis of the modes of formation of fossil assemblages. Geol. Soc. Amer. Bull., 71:1075-1086.

Lloyd, M., and R. J. Ghelardi. 1964. A table for calculating the "equitability" component of species diversity. Jour. Animal Ecol., 33:217-225.

——, and others. 1968. On the calculation of information theoretical measures of diversity. Amer. Midland Naturalist, 79:257-272.

Olson, E. C. 1961. The food chain and the origin of mammals. Intern. Colloq. on Evol. Mammals. Kon. Vlaamse Acad. Wetensch. Lett. Sch. Kunsten Belgie., I:97–116.

——. 1966. Community evolution and the origin of mammals. Ecology, 47:291–302.

Sanders, H. L. 1968. Marine benthic diversity: a comparative study. Amer. Naturalist, 102:243–282.

Scott, R. W. 1970. Paleoecology and paleontology of the Lower Cretaceous Kiowa Formation, Kansas. Univ. Kansas Paleont. Contrib., Art. 52 (Cretaceous I), 94p.

Stanley, S. M. 1968. Post-Paleozoic adaptive radiation of infaunal bivalve molluscs – a consequence of mantle fusion and siphon formation. Jour. Paleont., 42:214–229.

Stephenson, W. 1973. The validity of the community concept in marine biology. Roy. Soc. Queensland Proc., 84:73–86.

Thayer, C. W. 1973. Taxonomic and environmental stability in the Paleozoic. Science, 182:1242–1243.

True, M. A., and others. 1968. Progress in sampling the benthos: the benthic suction sampler. Deep-Sea Res., 15:239–242.

Turpaeva, E. P. 1957. Food interrelationships of dominant species in marine benthic biocoenoses. *In* B. N. Nikitin (ed.), Trans. Inst. Oceanol. Marine Biol. USSR Acad. Sci. Press, 20:137–148. (Published in U.S. by American Institute of Biological Science, Washington, D.C.).

Walker, K. R. 1972. Trophic analysis: a method for studying the function of ancient communities. Jour. Paleont., 46:82–93.

——, and R. K. Bambach. 1974. Feeding by benthic invertebrates: classification and terminology for paleoecological analysis. Lethaia, 7:67–78.

West, R. R. 1972. Relationship between community analysis and depositional environments: an example from the North American Carboniferous. 24th Intern. Geol. Congr. Proc., Sec. 7:130–146.

——, and P. C. Twiss. 1973. Modern lingulid community. Geol. Soc. Amer. Abst., 5:285–286.

Yarrow, G. R. 1974. Paleoecologic study of part of the Hughes Creek Shale (Lower Permian) in north-central Kansas. Unpublished M.S. thesis, Kansas State University, 247p.

Ziegler, A. M., R. N. Cocks, and R. K. Bambach. 1968. The composition and structure of Lower Silurian marine communities. Lethaia, 1:1–27.

APPENDIX: COMPOSITION OF THE SEVEN LINGULID COMMUNITIES

Rank[a]	Taxa	Mode of Life[b]	Trophic Group[c]		No. of Indiv.	% Indiv.
A — Holocene						
1	*Mulinia lateralis*	I	LLS	(1)	60	39.7
2	*Amygdalum papyria*	E	HLS	(1)	55	36.4
3	*Glottidia pyramidata*	I	LLS	(2)	28	18.5
4	*Mitrella lunata*	E	C	(1)	4	2.6
5	*Nassarius vibex*	E	C	(2)	1	0.6
5	*Anodontia alba*	I	LLS	(3)	1	0.6
5	*Mysella planulata*	I	LLS	(3)	1	0.6
5	Barnacles	E	HLS	(2)	1	0.6
8	Totals				151	99.6
B — Cretaceous (sand)						
1	*Corbula? fenti*	I	LLS	(1)	591	68.1
2	+*Trachycardium kansasense*	I	LLS	(2)	79	9.1
3	*Ostrea rugosa*	E	HLS	(1)	37	4.3
4	*Ostrea arcuata*	E	HLS	(2)	36	4.1
5	+*Lingula subspatulata?*	I	LLS	(3)	31	3.6
6	Ostreidae sp. indet.	E	HLS	(3)	30	3.5
7	*Protocardia texana*	I	LLS	(4)	15	1.7
8	Corbulids	I	LLS	(5)	9	1.0
8	*Turritella belviderei*	I	LLS	(5)	9	1.0
8	*Crassostrea kiowana*	E	HLS	(4)	9	1.0
9	*Euspira smolanensis*	E	C	(1)	4	0.5
9	*Flaventia belviderensis*	I	LLS	(6)	4	0.5
10	+*Scabrotrigonia emoryi*	I	LLS	(7)	3	0.3
11	+*Crassinella semicostata*	I	LLS	(8)	2	0.2
11	+*Drepanochilus kiowanus*	I	S	(1)	2	0.2
12	*Lopha quadriplicata*	E	HLS	(5)	1	0.1
12	*Lopha subovata*	E	HLS	(5)	1	0.1
12	*Lopha? kansasensis*	E	HLS	(5)	1	0.1
12	*Turritella kansasensis*	I	LLS	(9)	1	0.1
12	*Modiolus* sp.	E	HLS	(5)	1	0.1
12	*Otostoma* sp.	E	C	(2)	1	0.1
12	*Pirsila?* sp.	E	C	(2)	1	0.1
22	Totals				868	99.8
C — Cretaceous (clay)						
1	+*Trachycardium kansasense*	I	LLS	(1)	13	27.6
2	+*Lingula subspatulata?*	I	LLS	(2)	11	23.4
3	*Breviarca habita*	E	HLS	(1)	7	14.9
4	+*Drepanochilus kiowanus*	I	S	(1)	4	8.5

(*continued*)

Appendix (*continued*)

Rank[a]	Taxa	Mode of Life[b]	Trophic Group[c]		No. of Indiv.	% Indiv.
C — Cretaceous (Clay) (*continued*)						
5	*Corbula? smolanensis*	I	LLS	(3)	3	6.4
5	*Nuculana mutata*	I	S	(2)	3	6.4
6	*Yoldia microdonta*	I	S	(3)	1	2.1
6	+*Scabrotrigonia emoryi*	I	LLS	(4)	1	2.1
6	+*Crassinella semicostata*	I	LLS	(4)	1	2.1
6	*Pholadomya? belviderensis*	I	LLS	(4)	1	2.1
6	*Tellina?* sp. indet.	I	S	(3)	1	2.1
6	Corbulids	I	LLS	(4)	1	2.1
12	Totals				47	99.8
D — Permian						
1	*Crurithyris expansa*	E	LLS	(1)	209	66.0
2	*Lingula carbonaria*	I	LLS	(2)	71	22.4
3	*Orbiculoidea missouriensis*	E	LLS	(3)	13	4.1
3	*Volsellina?* sp.	I	HLS	(1)	13	4.1
4	Burrow (horizontal)	I	S	(1)	3	0.9
5	*Pinna?* sp.	I	HLS	(2)	1	0.3
5	*Septimyalina* sp.	E	HLS	(2)	1	0.3
5	*Aviculopecten* cf. *A. artisulcatus*	E	HLS	(2)	1	0.3
5	*Rhipidomella carbonaria*	E	HLS	(2)	1	0.3
5	*Acanthocrania* sp.	E	LLS	(4)	1	0.3
5	*Petrocrania* sp.	E	LLS	(4)	1	0.3
5	Crinoid	E	HLS	(2)	1	0.3
5	Ostracode	E	C	(1)	1	0.3
13	Totals				317	99.9
E — Lower Carboniferous						
1	*Streblopteria ornata*	E	HLS	(1)	365	56.4
2	*Lingula squamiformis*	I	LLS	(1)	169	26.1
3	*Actinopteria fluctuosa*	E	HLS	(2)	31	4.8
4	*Bucaniopsis decussatus*	E	C	(1)	19	2.9
5	*Nuculopsis gibbosa*	I	S	(1)	12	1.8
6	*Donaldina pulchra*	E	C	(2)	9	1.4
7	*Crurithyris urei*	E	LLS	(2)	8	1.2
8	*Eomarginifera longispinus*	I	LLS	(3)	7	1.1
9	*Edmondia* sp.	I	LLS	(4)	6	0.9
10	*Nuculana laevistriata*	I	S	(2)	5	0.8
10	*Soleniscus (S.) typicus*	E	C	(3)	5	0.8
11	*Soleniscus (M.) brevis*	E	C	(4)	4	0.6
12	*Dunbarella papyraceus*	E	HLS	(3)	3	0.5

Appendix (*continued*)

Rank[a]	Taxa	Mode of Life[b]	Trophic Group[c]		No. of Indiv.	% Indiv.
E – Lower Carboniferous (*continued*)						
13	*Naiadites crassa*	E	HLS	(4)	1	0.1
13	*Camarotoechia pleurodon*	E	LLS	(5)	1	0.1
13	*Natacopsis* sp.	E	C	(5)	1	0.1
13	*Orbiculoidea nitida*	E	LLS	(5)	1	0.1
17	Totals				647	99.7
F – Silurian						
1	"*Camarotoechia*" *decemplicata*	E	LLS	(1)	223	60.0
2	*Lingula* sp.	I	LLS	(2)	48	13.0
3	*Palaeoneilo rhomboidea*	I	S	(1)	37	10.0
4	*Pteronitella* cf. *P. retroflexa*	E	HLS	(1)	18	4.9
5	*Cornulites* sp.	E	HLS	(2)	12	3.3
6	*Lingula pseudoparallela*	I	LLS	(3)	11	3.0
7	*Lyrodesma* sp.	I	S	(2)	4	1.1
7	Bryozoans	E	HLS	(3)	4	1.1
7	Gastropods	E	C	(1)	4	1.1
8	*Plectonotus trilobatus*	E	C	(2)	2	0.5
8	?*Tropidodiscus* sp.	E	C	(2)	2	0.5
9	?Burrow	I	LLS	(4)	1	0.3
9	*Leptostrophia* sp.	E	LLS	(4)	1	0.3
9	?*Modiolopsis* sp.	E	HLS	(4)	1	0.3
14	Totals				368	100
G – Ordovician						
1	*Tancrediopsis cuneata*	I	S	(1)	54	30.0
2	*Plectonotus?* sp.	E	C	(1)	37	20.5
3	*Lingula?* sp.	I	LLS	(1)	28	15.5
4	*Ischyrodonta? truncata*	E	HLS	(1)	15	8.3
5	*Orthorhynchula linneyi*	E	HLS	(2)	11	6.1
6	*Ambonychia praecursa*	E	HLS	(3)	8	4.4
7	*Pterinea (Caritodens) demissa*	E	HLS	(4)	6	3.3
8	*Modiolopsis modiolaris*	E	HLS	(5)	3	1.6
8	*Loxoplocus (Lophospira) abbreviata*	E	C	(2)	3	1.6
8	*Lyrodesma poststriatum*	I	S	(2)	3	1.6
9	*Palaeoneilo?* sp.	I	S	(3)	2	1.1
9	*Nuculites?* sp.	I	S	(3)	2	1.1
9	*Zygospira modesta?*	E	LLS	(2)	2	1.1
10	*Liospira?* sp.	E	C	(3)	1	0.5
10	*Praenucula?* sp.	I	S	(4)	1	0.5

(*continued*)

Appendix (*continued*)

Rank[a]	Taxa	Mode of Life[b]	Trophic Group[c]		No. of Indiv.	% Indiv.
G — Ordovician (*continued*)						
10	*Flexicalymene?* sp.	E	C	(3)	1	0.5
10	*Bucania?* sp.	E	C	(3)	1	0.5
10	*Conularia?* sp.	E	HLS	(6)	1	0.5
10	*Cymatonota* sp.	E	HLS	(6)	1	0.5
19	Totals				180	99.2

[a]The higher the number, the fewer number of individuals.
[b]I, infaunal; E, epifaunal.
[c]HLS, high-level suspension feeders; LLS, low-level suspension feeders; C, collectors; S, swallowers. Number in parentheses indicates rank (based on number of individuals) of taxa within trophic group.
+Species common to both Cretaceous communities.

Temporal Changes in a Pleistocene Lacustrine Ostracode Association; Salt Lake Basin, Utah

Kenneth R. Lister

54 Tuscaloosa Avenue
Atherton, California

ABSTRACT

Stratigraphic collections from a 200-m-long core of Quaternary sediments in the Great Salt Lake Basin, Utah, revealed structural changes in the lacustrine ostracode populations. Sediments of the Great Salt Lake Basin were deposited during a number of successive lacustrine episodes that were terminated by shrinkage and drying of the lake. Each episode can be treated as a separate period of ecosystem development. Data on faunal composition and relative abundance of ostracode species were collected from samples taken at intervals of 10 to 20 cm. These data were compared to each other and to data on the physical stratigraphy of the core. Information-theory diversity, equitability, and number of species were calculated; moving-average curves of each of these parameters were generated for the core. The data obtained from the core were correlated with Pleistocene paleoclimatic data. Four pronounced drops in diversity delineate four faunal episodes that probably result from the effects of climatic change. These episodes correlate within the core with climatic changes inferred from other studies. Change in diversity within each episode is probably due to gradual climatic changes, immigration of species new to the community, or chance events.

INTRODUCTION

In this study the hypothesis which states that with time a community should increase in biotic complexity was examined. Investigation of the hypothesis proceeded in three stages. (1) Complexity and specific composition of stratigraphically arranged lacustrine ostracode samples were analyzed to find whether changes in the proportion of each species in the population occurred in a regular and predictable manner. (2) When the composition of the assemblage was found to change through time, it was noted that, in relatively stable environments, change was toward a more diverse assemblage. (3) It was noted that disturbance of the physical environment, as interpreted from stratigraphical evidence, correlated with decreased diversity of the assemblage. Superficially, the results could be regarded as a test of the stability-time hypothesis (Sanders, 1968), which is concerned with increases in community complexity with time. Circumstances surrounding development of the fauna studied here precluded its use as a test of this hypothesis because the stratigraphic span of the ostracode species studied exceeds that of lacustrine episodes. Because it was found that the increase in complexity could not be attributed to evolutionary or successional mechanisms, it was postulated that changes in the Great Salt Lake Basin ostracode fauna could be explained by a model similar to MacArthur and Wilson's (1967) island ecosystem model. Changes in faunal diversity with time were found to be similar to changes experienced by island faunas during colonization. The mechanism for faunal change appears to have been chance migration into the basin, interrupted by periods of extermination of species due to environmental change.

The term "community structure" represents the organization intrinsic to a group of species living together. Communities can be compared in structure as well as in specific membership, and two communities differing radically in species may be structurally similar in other ways. Certain aspects of community structure can be represented quantitatively as numbers of species and relative proportions of species abundances. Diversity measures the number of species in a community or collection, and dominance (or the converse, equitability) measures the relative proportion of individuals per species. Dominance-diversity indexes measure both properties. MacArthur (1955) stated that these diversity measures are related to the complexity of community organization. According to Margalef (1963), diversity indexes that measure the information content of a biotic community, as dominance-diversity does, are proportional to the complexity of community structure. Perhaps diversity measures of taxonomically dissimilar communities when compared will indicate relative levels of biological organization. It has been found that the diversity of one taxonomic group of a community can, in many cases, be used to represent the relative diversity of the entire community. In the present study, dominance-diversities of ostracode assemblages sampled from different stratigraphic levels in a sedimentary sequence were calculated. Comparison of these diversities was assumed to give an indication of the differences in diversity of the entire lacustrine community existing at different times, corresponding to each stratigraphic level.

Some environmental influences affecting diversity are outlined by Pianka (1966). These can be broken down into two factors, which probably explain most diversity differences: (1) physical heterogeneity, in general, lessens competition. It may be the result, in some cases, of increased predation, or may result from the history of an environment (during a long history many areas may "accumulate" habitats). Biotic heterogeneity brought about by physical heterogeneity or otherwise can itself increase diversity by increasing the number of habitats available. (2) Long periods of environmental stability permit greater productivity on the average and thus accumulation of immigrating or evolving species. High diversity may be the result of the long history of a community in a stable, nonvigorous environment.

According to Margalef (1963), a community living in a stable environment is thought to become more diverse with time. This tendency, if it exists, may be due to (1) successional modification of the physical environment by organic activity leading to greater heterogeneity and thus greater diversity, (2) evolutionary development of new species due to specialized adaptation to exploit one factor of the environment, thus leading to finer partitioning of available resources, and (3) immigration leading to accumulation of species in an environment. Extrinsic physical change in the environment may come into play, but is as likely to lead to lower as to higher diversity. The result would be an environment capable of supporting more or less organisms. Sudden changes in the environment are likely to lead immediately to lower diversity because some species are not able to accommodate to new conditions and are exterminated (Levinton, 1970).

An important factor in the development of a community is probably the ratio between the rate of evolution of organisms in the community and the rate at which environmental changes occur. If this ratio is high, the community may have time to evolve more species with more specialized habits until a point is reached when niche width reaches an optimum value (dependent on type and amount of resources). If the ratio of evolutionary rate to environmental variability is low, organisms do not have time to evolve narrow specializations, nor will these have been advantageous in rapidly changing environments. If the ratio is near 1, and evolution was rapid, there might result a situation in which species evolved and became extinct in a relatively short span of time. Past environments that yield good biostratigraphic index fossils may have been of this sort.

If one examines communities that develop over much shorter periods of time, during which evolution is not a factor in increasing the numbers of species present, an increase in complexity is often noted. In the case of terrestrial plants, this sort of community development is termed "succession." Progressively more complex communities made up of longer-lived individuals with more total biomass succeed each other. In most cases this sort of community development can be explained by gradual change in the physical environment. This change may be brought about by the organisms themselves, as in the case of a classical prairie-to-forest succession, or by extraneous factors, as in the case of a seasonal succession of lacustrine phytoplankton. Although successional change does not depend directly on evolution, it may lead to more diverse and stable communities with time.

Few data are available to support or refute the positive correlation between biotic complexity, age of the ecosystem, and physical stability that is suggested by theory. Goulden (1969) presented evidence to indicate that undisturbed ecosystems tend toward higher dominance-diversity through time, but in his study the change was correlated with gradual amelioration of physical conditions affecting the community. Some published evidence (Tsukata and Deevey, 1967; Levinton, 1970; Goulden, 1971), as well as many casual observations, indicates that an ecosystem in a physically unstable environment will not increase in diversity with time as rapidly as a community in a stable environment. Yet, more evidence is needed before one can predict that physical stability leads to increased diversity as the general case. The fact remains that not all short-term experiments indicate a positive correlation between diversity and biotic stability (Hairston et al., 1968) or between diversity and physical stability (Hurd et al., 1971).

The purpose of the study described herein was to determine whether systematic changes in community structure can be expected to occur over spans of time longer than those which can be examined by observation of extant biotas and longer than those studied in lake cores to date. Ostracodes from lacustrine deposits representing about 700,000 years were studied by the author. Although this represents a span of time shorter than that during which evolutionary changes of specific level were expected to occur, it does represent a span of time during which the physical environment changed considerably.

Although paleontological investigation of community structure allows a greater time span of observation, there are problems associated with the measurement of diversity for fossil samples. Paleontological evidence tends to be biased by many factors (Fagerstrom, 1964; Lawrence, 1968). Ostracode assemblages in Pleistocene Lake Bonneville sediments lend themselves to a study in which biases in the fossil record are reduced. Ostracodes possess hard parts of the same general size and composition and with the same number of molts per lifespan. Fine-grained sediments in which the ostracodes were enclosed indicate that postmortem transport of valves was negligible. Fossils show little effect of crushing and no effect of solution. Each sample was taken from a stratigraphically and spatially restricted area. Within one lake, variations in conditions due to patchiness of the habitat are minimized because of similarities in water chemistry, temperature, and available nutrients over much of the lake. Depth, temperature, and substrate composition are probably the most important physical variables within the basin at any one time. Analysis of sediments along with the organisms permits investigation of the effects of substrate control on faunal composition. Influence of depth is more difficult to estimate, but examination of ostracodes from nearshore beds of Little Valley, Promontory Point, Utah, has shown that the same species occur there as in the core samples of generally deeper water beds of the same age.

Calculated diversities were compared only to other diversities calculated in the same manner for the same area and using equal volumes of sediment to avoid comparing different types of collections. The diversity calculated for each collection of ostracodes is not necessarily equal to the diversity of the organisms that were living in the basin. However, if the reservations expressed above are kept in mind, the diversity calculated can be taken as an indication of the overall diversity of the ostracodes which were living in that environment.

GREAT SALT LAKE BASIN

The Great Basin area of the western United States is a region of generally north-south trending fault block mountains and closed intermontane basins containing thick sequences of Cenozoic alluvial and lacustrine strata. Today the climate in this region is arid and few intermontane basins contain intermittent lakes. During times of Pleistocene glacial advance, cooler and moister climates prevailed over much of this region, and many of the basins contained permanent bodies of water. The Great Salt Lake Basin and several adjoining valleys contained a lake that at times was the largest of the pluvial lakes, Lake Bonneville. Wave-cut benches found high up on the Wasatch Range, which formed the eastern boundary of the lake, indicate how much larger this lake was than the present saline remnant, the Great Salt Lake (Fig. 1).

The latest and probably the most extensive of the Pleistocene lakes to occupy the Great Salt Lake Basin has been named Lake Bonneville (Gilbert, 1890), and the sediments deposited in this lake are the Lake Bonneville Group. Morrison (1965) has included two major lacustrine episodes within the Lake Bonneville Group and has named two corresponding formations: the Little Cottonwood and the Draper. The Little Cottonwood Formation, which includes the Alpine, Bonneville, and Provo members, is comprised of gravel, sand, silt, and clay, and is up to 100 m thick. It is bounded stratigraphically by soils: the Dimple Dell below and the Graniteville above; the latter separates the Little Cottonwood Formation from the Draper Formation. The Draper Formation consists of sand and gravel tongues interbedded with lacustrine silt and clay, and is up to 50 m thick. The Draper Formation is overlain by the Midvale Soil, which is overlain by Holocene alluvium and lake deposits. Formations older than the little Cottonwood have not been named. These older formations make up the bulk of the strata sampled in this study.

The lake level has been high several times in the past; each time high level has been followed by drying up or by drastic shrinkage in the dimensions of the lake. During the Pleistocene, lacustrine deposits accumulated in the subsiding Great Salt Lake Basin and left a rather thick (perhaps up to 300 m) sedimentary record of the Quaternary of the Great Basin. According to Eardley and Gvosdetsky (1960), these sediments indicate that the basin was occupied by a lake during much of the late Pleistocene. Occasionally the lake dried, allowing formation of soil, and occasionally lake level was low enough to be marked by indicators of low lake level: (1) plant root tubes in sediment, (2) coarse clastics, (3) reddish sediment coloration, (4) ooids, or (5) brine shrimp fecal pellets.

In five places lacustrine sediments of the Great Salt Lake Basin have been cored under the direction of the late Armand J. Eardley. Three of the cores are Saltair (depth of 200 m, taken in 1956), Section 28 (depth of 236 m, taken in 1960), and Burmester (depth of 307 m, taken in 1970). The Saltair and Section 28 cores were extensively sampled (Fig. 1). This paper will be concerned only with the Saltair

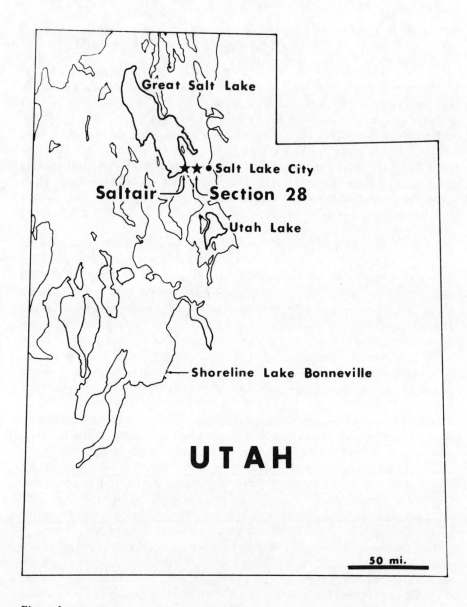

Figure 1
Index map of Utah, showing location of cores used in this study (stars) and the shoreline of Lake Bonneville at its maximum extent.

core, which is the most complete and longest of the cores available for study at the present time.

METHODS

Samples 2.5 cm long were sawed from the 2-cm-diameter Saltair core. Samples were spaced 10 to 20 cm apart where recovery was complete. The sediment was soaked in a cold, dilute solution of sodium carbonate and most samples had disaggregated after a day. The disaggregated sediment was washed through 20-, 40-, 60-, and 80-mesh screens. The fraction retained on each screen was dried and ostracodes removed.

Numbers of adult and immature individuals of each species in each sample were counted. For each collection, dominance–diversity was calculated using Brillouin's (1962) formula for information content:

$$H = \frac{1}{N} \log_e \left(\frac{N!}{N_1! N_2! \cdots N_s!} \right)$$

where N = number of individuals in the collection and N_i = number of individuals in the i-th species, and s = number of species per collection. Equitability was calculated by a method suggested by Pielou (1966):

$$E = \frac{H}{H_{max}}$$

where H_{max} is the maximum possible diversity in a collection of s species, and

$$H_{max} = \frac{1}{N} \log [N! (m!)^{s-r} (m+1)!^r]$$

In this equation, m is the largest interger less than N/x and $r = N - sm$.

Moving averages of dominance–diversity, equitability, and number of species calculated over intervals of 5 through 30 samples were plotted versus depth in the core. The most satisfactory curve of dominance–diversity versus sample depth was based on averages of 10 samples.

Species were clustered using transformed ($\log y + 1$) numerical data on species abundances. The NY-SYS programs of Rohlf and others were used; the 1969 version of NY-SYS is in the library of the University of Kansas Computation Center. R-mode clusters were formed by the unweighted pair-group method using arithmetic averages (Sokal and Sneath, 1963) applied to a matrix of correlation coefficients. Cophenetic correlation values were also computed. Clusters obtained were used to define the dominant Pleistocene ostracode assemblages of the Great Salt

Lake Basin and several minor assemblages. Interpretations of paleoenvironments based on the rather limited knowledge of the biology of Holocene freshwater ostracode species was possible for some stratigraphic intervals.

Correlations were made between the Saltair ostracode data and Ruddiman's (1971) data on oceanic paleotemperatures by using Eardley and Gvosdetsky's (1960) age correlation of Saltair core stratigraphy. Interpretations of lacustrine conditions were also based on the faunal data gathered in this study and on the sedimentologic and stratigraphic data as interpreted by Eardley and Gvosdetsky.

RESULTS

Results of data analysis can be summarized in moving-average curves of dominance-diversity, equitability, and number of species versus depth-in-core in meters (Fig. 2). Of interest is the existence of a pattern in the change in dominance-diversity with time. If the curve is broken into four segments, the diversity is seen to increase during the bulk of each segment. The segment boundaries are (1) 178 to 145 m, (2) 145 to 119 m, (3) 119 to 77 m, and (4) 77 to 5 m. Within each segment diversity gradually increases (although the curve is broken by fluctuations) up to a point near the top of each segment, where the curve begins to decline steeply. A linear regression was calculated for diversity within each segment. The regression shows significant increase in diversity at the 0.001 level for episodes 1, 2, and 4 of the Saltair core and at the 0.05 level for episode 3.

If the measurement of dominance-diversity is divided into two component parts, number of species and equitability, one can see from the curves that equitability increased during the early part of each episode of community development. Average equitability remained high throughout the later part of each episode. Most of the increase in dominance-diversity within episodes was due to an increase in the number of species. The number of species fluctuated increasingly during each episode as greater numbers of less common species became established within the basin. These species were most likely to experience extermination if the lacustrine environment was disturbed.

The boundaries between segments based on diversity minima probably correspond to times of environmental change in the ancient lake. According to Eardley and Gvosdetsky's (1960) interpretation of lacustrine conditions, based on the Saltair core, these faunal boundaries correspond to times of low lake level, some of which are indicated by soil. Ostracode assemblages occur within crude intervals interpreted as low lake level or even lake drying (indicated by soils) (Fig. 3).

The most common ostracode species found in the Saltair core include *Candona acutula* Delorme, *Candona adunca* Lister, *Candona rawsoni* Tressler, *Cyprinotus glaucus* Furtos, *Cypridopsis vidua* (Müller), *Ilyocypris gibba* (Ramdohr), *Cyprideis salebrosa* van den Bold, *Limnocythere itasca* Cole, *Limnocythere friabilis* Benson and MacDonald, and *Limnocythere staplini* Gutentag and Benson. These species occur in samples from all parts of the core. About 20 less common species also occur. All species are listed in Table 1 and described by Lister (in press).

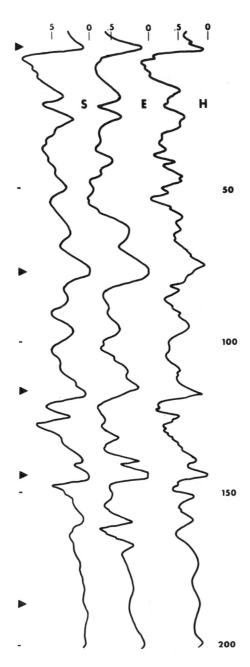

Figure 2
Moving-average curves of dominance-diversity (H), equitability (E), and number of species (S) versus depth in core (meters) for the Saltair core. Arrows on left side indicate important diversity minima.

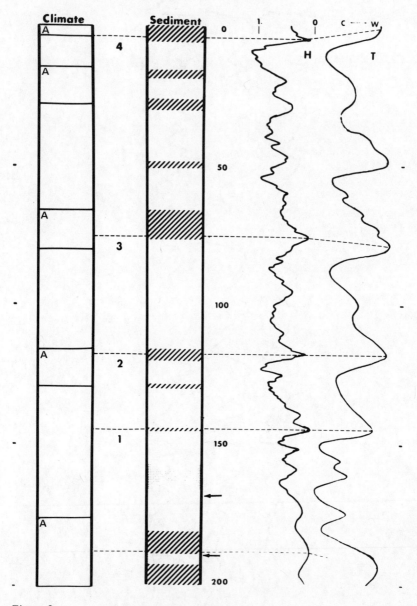

Figure 3

Stratigraphy of the Saltair core, indicating sediment type (diagonal pattern: shallow lacustrine and soil; no pattern: lacustrine; tuffs at 170 m and 190 m), inferred climate (A: arid; unmarked: glaciopluvial), division of the core into four faunal episodes based on the moving average curve of diversity (H), and the curve of marine paleotemperatures (T) for this interval of the Pleistocene (c: cool; w: warm); depth in core given in meters. [Modified in part from Eardley and Gvosdetsky (1960) and Ruddiman (1971).]

Table 1
Summary of environmental preferences of living representatives of ostracode species found in the Great Salt Lake Basin sediments. Taken from various sources (Lister, 1974).

Species	Habitats
Candona actula	Shallow water, lakes, nearctic
C. camuramarginis	Temporary ponds, nearctic
C. caudata	Streams and deeper parts of lakes, cosmopolitan
C. crogmaniana	Permanent and temporary ponds, nearctic
C. lactea	Lakes and ponds, holarctic
C. adunca	–
C. pronopa	–
C. rawsoni	Ponds and lakes, tolerant of moderately high amounts of dissolved minerals, nearctic
C. wanlessi	–
Cyclopcypris ampla	Wide tolerance, ponds and lakes, nearctic
C. serena	Wide tolerance, ponds and lakes, nearctic
Physocypria pustulosa	Lakes and ponds, not found in very cold water, holarctic
Cyprinotus glaucus	Lakes, not found in very cold water, holarctic
Eucypris serrata	Lakes and ponds, not found in very cold water, holarctic
Cypridopsis vidua	Lakes and ponds, in weeds, cosmopolitan
Potamocypris granulosa	Lakes and ponds, holarctic
Potamocypris pallida	Springs, holarctic
P. smaragdina	Lakes and ponds, not found in arid regions, holarctic
P. unicaudata	Lakes and ponds, holarctic
Ilyocypris biplicata	Streams and lakes, holarctic
I. gibba	Streams and lakes, holarctic
Pelocypris tuberculata	Playa lakes, nearctic, possibly neotropical
Cypredeis salebrosa	Tolerant of saline water, lakes and ponds, neotropical and nearctic
Cytherissa lacustris	Cold, deep lakes, holarctic
Limnocythere ceriotuberosa	Lakes, nearctic
L. friabilis	Lakes, nearctic
L. illinoisensis	Cool lakes, nearctic
L. itasca	Cool lakes, nearctic
L. paraornata	Streams, nearctic
L. parascutariense	Streams and lakes, nearctic
L. pseudoreticulata	Lakes and permanent ponds, nearctic
L. staplini	Lakes, tolerant of moderate to high salinities, nearctic
Darwinula stevensoni	Large, cool lakes, cosmopolitan

On the basis of co-occurrence in collections, the ostracode species may be clustered into several groups (Fig. 4). Although general knowledge of freshwater ostracode environmental preference is meager (Table 1), it is possible to define several ecologically meaningful groups based on cluster analysis: (1) weedy pond or shallow lake assemblage: *Candona acutula* Delorme, *Cycloypris ampla*, *Cypridopsis vidua* (Müller), *Limnocythere itasca* Cole; (2) lake assemblage indicating moderate amounts of dissolved salts: *Candona rawsoni* Tressler, *Cyprinotus glaucus* Furtos, *Limnocythere friabilis* Benson and MacDonald; (3) saline lake assemblage: *Cypri-*

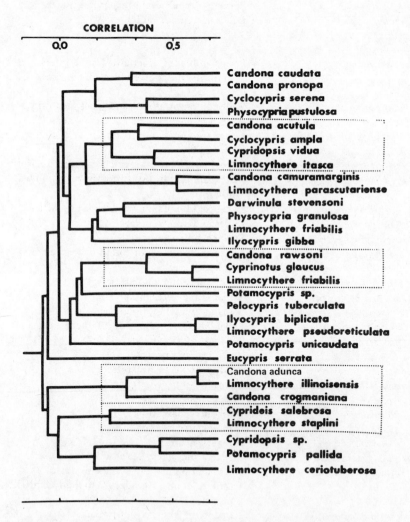

Figure 4
Phenogram of R-mode cluster analysis showing correlations among species for Saltair fauna. Cophenetic correlation coefficient = 0.732.

deis salebrosa van den Bold, *Limnocythere staplini* Gutentag and Benson; (4) low-diversity, high-abundance assemblage (habitat unknown): *Candona adunca* Lister, *Limnocythere illinoisensis, Candona crogmaniana*. Some species, particularly the rare ones, do not cluster with each other at a high level of correlation. Other clusters apparent in the phenogram consist of species without known modern habitats (such as *Candona pronopa*), with incompletely known habitats, or with very wide environmental tolerances.

Members of assemblage 2 first occur early in each episode and remain the most common species throughout much of each lacustrine period. Members of assemblage 1 occur less commonly in samples scattered throughout each segment. Members of assemblages 3 and 4 are very abundant in some samples (often to the exclusion of other species). Most other species occur in only occasional collections.

Two superimposed patterns of species occurrence are found in each diversity episode: (1) within each episode, species diversity tends to increase because of gradual addition of less abundant ostracode species throughout each episode. At least some of these species occur more commonly in samples above their first occurrence. (2) Superimposed on the first pattern is an alteration of low- to moderate-salinity ostracode assemblage (assemblage 2 plus less common "accessory" species) with high-salinity species (principally assemblage 3).

DISCUSSION

It is not surprising that times of low (and probably saline) water and times of lake drying should also be times of low diversity of ostracodes. What is more noteworthy is that diversity seems to have changed in a systematic way between times of low diversity.

Furthermore, this systematic diversity change has repeated itself during what might be interpreted as lacustrine episodes. At least five processes may have caused the gradual increase in diversity during each lacustrine period: (1) Lacustrine episodes were controlled by climate; perhaps climate itself had a direct effect on the ostracodes in the lake. (2) Successional changes in the flora of the land area surrounding the lake or in the flora and fauna of the lake itself could have had an effect on the ostracodes, although the data do not seem to indicate succession of ostracode communities. (3) Evolutionary change could have occurred in the ostracode species of the lake. (4) The gradual increase in numbers of ostracode species could be the result of the gradual immigration of new species from other areas. (5) Faunal change could be due to chance events that affected the relative population size of the ostracode species in a systematic way.

To test these possibilities, changes in the specific composition of the ostracode fauna within each core segment were examined. Dominance-diversity changes seem to be due principally to additions of ostracode species to the fauna throughout each episode. Species first occurring toward the top of each core segment are apparently less likely to become abundant in samples than those species first appearing toward the bottom of each segment. In general, the same species occur near the bottom of the Saltair core as near the top, and individuals of each species do not seem to undergo evolutionary change. There is, likewise, no evidence of succession within faunal episodes, because there is no orderly sequence of occurrence of ostracode assemblages. It seems most likely that faunal changes within diversity episodes resulted from environmental changes, immigration, or chance events.

Periodic environmental changes are indicated by the recurrence of saline ostracode assemblages. The most probable cause of periodic change in lacustrine conditions during the Pleistocene Epoch was the climatic cycles that occurred in concert with glacial cycles. Glaciopluvial conditions favored development of a permanent lake. During interglacial stages, conditions probably tended to be warm and dry (similar to those at present), favoring shrinkage of the lake. This in turn could have resulted in the concentration of dissolved minerals in the residual lake to a level great enough so that no ostracode species could survive. At present the waters of the Great Salt Lake are close to being saturated in NaCl and no ostracodes live there. Coincidence of boundaries of faunal episodes with sedimentological indicators of lake shrinkage or drying supports the above conclusions.

Evidence from outside the Great Salt Lake Basin confirms the proposed association between faunal episode boundaries and climatic events. Eardley and Gvosdetsky (1960) suggested a correlation between Saltair core stratigraphy and Emiliani's (1958) data on oceanic paleotemperatures. Subsequent oxygen isotopic studies of planktic foraminifera obtained from deep-sea cores (Emiliani, 1966, 1974) have tended to confirm the shape of the original paleotemperature curve (Eardley and Gvosdetsky, 1960). Temperature peaks tend to correlate with episodes of lake drying, unless the correlation proposed by Eardley and Gvosdetsky (1960) is incorrect.

Eardley and Gvosdetsky also presented a time scale for the Saltair core based on several carbon-14 dates near the top of the core, inferred rates of sedimentation, and the presence of a Pearlette-type tuff at 170-m depth. This tuff is probably equivalent to the Pearlette type-0 ash found at Onion Creek in east-central Utah and dated at 0.7 ± 0.2 m.y. by Naeser et al. (1973). Other Pearlette type-0 tuffs were dated at about 0.7 m.y. Another ash bed found near the bottom of the Saltair core (Fig. 3) is mineralogically similar to the 0.7-m.y.-old Bishop Tuff of California (Eardley and Gvosdetsky, 1960). If the total span of the core covers a 0.65- to 0.7-m.y. duration, it can be correlated with dated deep-sea paleotemperatures. Close correlation was found between the Great Salt Lake diversity data and Ruddiman's (1971) curve of oceanic paleotemperature (Fig. 3). Ruddiman's curve was based on environmental analysis of foraminiferal assemblages from long cores. In addition, a close correlation was found between the boundaries of faunal episodes in the upper half of the Saltair core and the warm peaks in Emiliani's paleotemperature curve, although in the lower part of the Saltair core there are fewer faunal boundaries than there are warm peaks in Emiliani's curve. Pleistocene paleotemperature curves from a number of sources can be correlated, and all yield a similar picture of temperature fluctuations (Cooke, 1973). The diversity curve of the ostracode fauna of the Great Salt Lake Basin correlates well with this general picture of climatic change.

If this proposed correlation between global climate and episodes of ostracode community development in the Great Salt Lake Basin is correct, regional climate may have determined the course of faunal development. Climatic cooling is apparently associated with greater rainfall and lower evapotranspiration rates in the Great Basin. Consequently, a permanent body of water may have been established in the Great Salt Lake Basin during glacial times. The important environmental factor for the ostracodes is thought to have been development of a permanent lake and not direct effects of changes in temperature. Development of the lake depended on moisture, which in turn depended greatly on the effect of a continental ice sheet on atmospheric circulation. Development of a permanent lake was a threshold event; once extant, the ostracode populations became well established. Events that occurred during a lacustrine episode had a much less profound effect on the lacustrine community than events which caused destruction of the lake.

Gradual increase in species during each episode may be due to gradual amelioration in the environment, which permitted the existence of more species of ostracodes; alternatively, the lake may have gradually developed more heterogeneous habitats, which resulted in more exploitable subenvironments. Perhaps the ostracode community added species through immigration from outside the basin as long as climatic conditions remained conducive to maintenance of the lake. Minor periods of climatic warming and less hospitable or more unstable lacustrine conditions may have caused temporary fluctuations in ostracode diversity. These conditions may have involved lake shrinkage, greater seasonal temperature fluctuations, or some other factor.

It is a fact that most freshwater ostracode species are adapted to survive interlake migration. Successful immigration of a number of new species into the lake per unit time would tend to build up species numbers until the environment was "full" or environmental change caused extermination (as at the end of each faunal episode). This process seems likely to require less time in most studied examples than the 100,000 or so years that each episode spans (Simberloff and Wilson, 1969).

If the isolated lake is likened to an island, comparison can be made between lake and island faunal history. According to the MacArthur and Wilson (1967) model of the island ecosystem, species diversity is directly proportional to diversity of habitats and to the distance of an island from a source of migrants. Equilibrium diversity is a result of variation in rates of immigration and extinction per unit time with number of species present on an island (Fig. 5). Rate of immigration of *new* species is hypothesized to have fallen as more species became established on the island and fewer immigrants belonged to new species. The rate of immigration may have decelerated with increasing number of species, as only the rarer, less commonly migrating species had not yet established populations on the island. The rate of extinction per unit time was hypothesized to have risen as more species established themselves on the island and were available for extermination. This rate may have accelerated with increasing numbers of species as average species population size declined and chance of extermination increased.

Following the reasoning of MacArthur and Wilson (1967), it may be that species diversity reflects an equilibrium between the rate of immigration of new ostracode species into the basin and the rate of extermination of species within the basin. A gradual change in the equilibrium point, due to increased immigration rates or de-

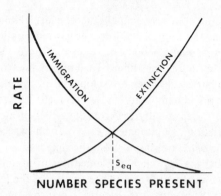

Figure 5
Hypothesized effect of immigration and extinction of species into and within an isolated habitat on the number of species present in the habitat. (After MacArthur and Wilson, 1967.)

creased extermination rates, would cause an increase in equilibrium diversity with time. A change in immigration rates might have been caused by a reduction in distances between western North American lakes when more lakes formed during the wetter and cooler climate. This would have caused an increase in immigration rate. Perhaps as lakes became larger under the influence of climate, extermination rate in any one lake decreased because more environmental niches were available. Perhaps some other environmental factor changed gradually and affected extermination rate in the Great Salt Lake Basin.

The pattern of occurrence of ostracode assemblages outlined above does not seem to be due to purely chance events, but the increase in diversity within core segments may be more apparent than real. It might be due to "random" fluctuations of the diversity curve based on random faunal changes. There is some evidence that the pattern of diversity fluctuation observed in the Saltair core is also found in other cores from the Great Salt Lake Basin (Lister, 1974), but, as yet, stratigraphic correlation between the two cores studied in detail is unclear.

CONCLUSIONS

Study of stratigraphic samples of ostracodes from sediments of the Great Salt Lake Basin indicates that community development in the basin can be traced. Dominance-diversity, equitability, and number of species for the ostracode assemblage were found to have changed in a regular and predictable manner with time. It was found that diversity increased gradually during times which seem to correspond to lacustrine episodes. Change in lacustrine conditions associated with shrinkage and/or drying of the lake corresponds to a decrease in diversity of lacustrine ostracodes. Neither evolution nor succession is implicated in the faunal development of lakes occupying this basin.

The ultimate control of diversity was probably extrinsic to the ostracode community and due to physical environmental factors. The length of a faunal episode apparently was directly dependent on the duration of climatic conditions that allowed the development of a permanent lake in the Great Salt Lake Basin. These episodes of favorable conditions were probably coincident with glaciopuvial stages within the Pleistocene Epoch. During these times, immigration led to the development of freshwater ostracode communities in isolated basins in western North America. Faunal and lacustrine episodes ended after the regional climate became more arid. Such a climate is thought to have been characteristic of interglacial stages. Because the same ostracode species recur throughout the Saltair core and throughout North America in the Pleistocene, it is likely that species immigrated into the Great Salt Lake Basin from other areas after establishment of a lake in the basin.

On the whole, the changes in the ostracode fauna can be explained by using a model similar to that developed by MacArthur and Wilson (1967) for islands. A closed lake basin is probably as isolated for an ostracode population as an island is for a terrestrial animal population; thus, it is not illogical to apply an island ecosystem model to the lacustrine environment. Immigration of new species into the basin may or may not have been in equilibrium with extermination of species in the basin at all times. If equilibrium existed at most times, changing environmental conditions were probably causing the equilibrium point to change. If equilibrium was rarely achieved, perhaps immigration of new species exceeded extermination throughout each diversity episode.

ACKNOWLEDGMENTS

I wish to thank R. L. Kaesler, C. Teichert, W. M. Merrill, E. E. Angino, and W. J. O'Brien of the University of Kansas for their constructive criticism. I am grateful to the faculty and staff of the Department of Geological and Geophysical Sciences of the University of Utah, and in particular R. T. Shuey, J. F. Miller, and the late A. J. Eardley, for their assistance in obtaining access to the Salt Lake cores in 1972 and 1973. Funds for this study were provided in part by Sigma Xi Grant-in-Aid of Research, 1971; Summer Fellowship in Systematics and Evolutionary Biology of the National Science Foundation (principal investigator: R. S. Hoffman, Department of Systematics and Ecology, the University of Kansas), 1972; and the Geological Society of America Grant-in-Aid, 1973. General financial assistance was provided by the Department of Geology, the University of Kansas; Phillips Petroleum Company; and the University of Kansas through a Dissertation Fellowship.

REFERENCES

Brillouin, L. 1962. Science and information theory (2nd ed.), Academic Press, Inc., New York, 347p.

Cooke, H. B. S. 1973. Pleistocene chronology: long or short? Quarternary Res., 3:206-220.

Eardley, A. J., and V. Gvosdetsky. 1960. Analysis of Pleistocene core from Great Salt Lake Basin, Utah. Geol. Soc. Amer. Bull., 71:1323-1344.

Emiliani, C. 1958. Paleotemperature analysis of core 280 and Pleistocene correlations. Jour. Geol. 66:264-275.

———. 1966. Isotopic paleotemperatures. Science, 154:851-857.

———. 1974. The Brunhes epoch: isotopic paleotemperatures and geochronology. Science, 183:511-514.

Fagerstrom, J. A. 1964. Fossil communities in paleoecology: their recognition and significance. Geol. Soc. Amer. Bull., 75:1197-1216.

Gilbert, G. K. 1890. Lake Bonneville. U.S. Geol. Surv. Monog. 438p.

Goulden, C. E. 1969. Temporal changes in diversity. In G. M. Woodwell and H. H. Smith (eds.), Diversity and stability in ecological systems. Brookhaven Symp. Biol., 22:96-102.

———. 1971. Environmental control of the abundance and distribution of the chydorid Cladocera. Limnol. Oceanog., 16:320-331.

Hairston, N. G., R. K. Allen, R. K. Colwe, D. J. Futuyma, J. Howell, M. D. Lubin, J. Mathias, and J. H. Vandermeer. 1968. The relationship between species diversity and stability: an experimental approach. Ecology, 49:1091-1101.

Hurd, L. E., D. V. Melanger, S. J. McNaughton, and L. L. Wolf. 1971. Stability and diversity at three trophic levels. Science, 173:1134-1136.

Lawrence, D. R. 1968. Taphonomy and information losses in fossil communities. Geol. Soc. Amer. Bull., 79:1315-1330.

Levinton, J. L. 1970. The paleontological significance of opportunistic species. Lethaia, 3:69-78.

Lister, K. H. 1974. Paleoecology of Ostracoda from Quaternary sediments of the Great Salt Lake Basin, Utah. Unpublished Ph.D. dissert., University of Kansas, 325p.

———. 1975. Quaternary freshwater Ostracoda fauna from the Great Salt Lake Basin, Utah. Univ. Kansas Paleont. Contr., 34p.

MacArthur, R. H. 1955. Fluctuations of animal populations as a measure of community stability. Ecology, 35:533-536.

———, and E. O. Wilson. 1967. The theory of island biogeography. Princeton University Press, Princeton, N.J., 203p.

Margalef, R. 1963. On certain unifying principles in ecology. Amer. Naturalist, 92:357-374.

Morrison, R. B. 1965. Lake Bonneville: Quaternary stratigraphy of eastern Jordan Valley, south of Salt Lake City, Utah. U.S. Geol. Sur. Profess. Paper, 477, 80p.

Naeser, C. W., G. A. Izett, and R. E. Wilcox. 1973. Zircon fission-track ages of Pearlette family ash beds in Meade County, Kansas. Geology, 1:187-189.

Pianka, E. R. 1966. Latitudinal gradients in species diversity: a review of concepts. Amer. Naturalist, 100:33-46.

Pielou, E. C. 1966. The measurement of diversity of different types of biological collections. Jour. Theoret. Biol., 13:131-144.

Ruddiman, W. F. 1971. Pleistocene sedimentation in the equatorial Atlantic: stratigraphy and faunal paleoclimatology. Geol. Soc. Amer. Bull., 82:283-302.

Sanders, H. L. 1968. Marine benthic diversity: a comparative study. Amer. Naturalist, 102:243-282.

Simberloff, D. S., and E. O. Wilson. 1969. Experimental zoogeography of islands. The colonization of empty islands. Ecology, 50:278-296.

Sokal, R. R., and P. H. A. Sneath. 1963. Principles of numerical taxonomy. W. H. Freeman and Company, San Francisco, 776p.

Tsukata, M., and E. S. Deevey. 1967. Pollen analysis from four lakes in the southern Maya area of Guatemala and El Salvador. Intern. Assoc. Quaternary Res. Proc., VII Congr., 6:303–331.

Diversity of Assemblages of Late Paleozoic Ostracoda

Michael D. Brondos
Roger L. Kaesler

University of Kansas

ABSTRACT

Indexes of diversity and equitability of ostracode assemblages in the Beil Limestone Member (Lecompton Formation) showed that this Upper Pennsylvanian microfauna responded in several ways to changes of depositional environment during regression of the sea. No significant differences were found for diversity or equitability within beds in an outcrop or among the lower three of four beds sampled at five outcrops in northwestern Douglas County, Kansas. However, significant decreases at the 0.05 level were found for both indexes between the two uppermost beds. This decrease in diversity and equitability, and also in the number of species, occurred during a time of shallower water, of periodic increase in the influx of terrigenous sediments, and of higher energy conditions in the last stages of regression.

Q-mode cluster analysis showed that the ostracode assemblages were distributed in a similar way throughout the study area with a higher-diversity assemblage in the lower Beil and a lower-diversity assemblage in the upper Beil. R-mode cluster analysis showed that species present in the lower Beil were replaced gradually during deposition of the upper parts of the Beil, the species composition of the assemblage of the uppermost bed being distinctly different from those of the lowermost bed. From R-mode cluster analysis and principal coordinates analysis, two assemblages were identified: the higher-diversity assemblage characterized by *Cryptobairdia seminalis, Silenites silenus,* and *Orthobairdia texana;* and the lower-diversity assemblage of the upper Beil characterized by *Cavellina nebrascensis, Bairdia beedei, Pseudobythocypris pediformis, Moorites minutus, Healdia simplex,* and *Monoceratina lewisi.* The ostracode assemblages in the Beil, thus, increased in diversity, equitability, and number of species; underwent replacement of species; and finally decreased in diversity, equitability, and number of species with regression of the sea.

INTRODUCTION

Studies of the structure and development of communities of modern organisms have led in recent years to a number of hypotheses relating the diversity of species in a community to the variability of enivronmental parameters (see Woodwell and Smith, 1969). Because the evolution of most communities has occurred over a very long time span, paleontologists can contribute substantially to the development of theories to explain diversity by providing insight into the factors that affected both present and ancient communities. Some investigators have begun to provide such information for variations in diversity over relatively large geographic areas or rather long intervals of time (Gibson and Buzas, 1973; Valentine, 1973). However, studies of changes in diversity are also needed for communities that respond to local environmental changes in small geographic areas or over relatively short intervals of geologic time. In this study, variations in diversity and equitability of ostracode assemblages within one Upper Pennsylvanian unit were analyzed in part of northeastern Kansas. The purposes of our research were to study changes in diversity with regression of the Pennsylvanian sea and to identify assemblages of Pennsylvanian ostracodes characteristic of particular environments.

The term "diversity" is used here in the sense of Pielou (1969, p. 222), that is, as a statistic that combines the number of species and the distribution of individuals among the species in a collection. Whereas some authors have used diversity synonymously with number of species, the terms "species abundance" or "number of species" will be used here for that parameter of the community. Use of the term "equitability" follows that of most authors and refers to the relative evenness with which individuals are distributed among species. Highest equitability occurs in a sample where diversity is at a maximum, so that all species have an equal number of individuals.

The Beil Limestone Member of the Lecompton Formation in Douglas County, Kansas, was selected for study because it is richly fossiliferous, and its stratigraphy, paleontology, and petrography have been studied previously. Ostracodes sampled from shaly intervals at five outcrops over a distance of about 10 km were studied for variations in diversity and equitability, both within one bed at a single outcrop and among the same four beds at five outcrops. Variations in diversity and equitability were compared with changes in the sedimentary environment as determined by previous investigators and by our own work. The potential problems of using environmental interpretations based largely on studies of the carbonates while using samples of ostracodes collected from shaly interbeds were judged not to have been serious.

Results showed that diversity and equitability varied little for several samples obtained from within one bed at an outcrop. Among the five outcrops, diversity and equitability generally increased slightly to the third bed sampled, and then decreased significantly in the fourth bed. This decrease occurred with deterioration of the normal marine environment during late stages of regression or filling of the basin in which the Beil was being deposited as the deposition of terrigenous sediments became predominant. Changes of the ostracode assemblages up the section occurred primarily by species replacement, so that many species present in the lower beds were not found in the uppermost bed. The changes in assemblages occurred at approximately the same horizon in all outcrops and indicated similar responses to environment by ostracodes throughout the area studied.

ECOLOGY OF OSTRACODA

The distribution of ostracode species has been emphasized in many previous studies. Apparent species distribution, however, is largely a function of species abundance, which, in turn, may be limited to varying degrees by environmental factors such as salinity, water depth, type and stability of substrate, temperature, pH, oxygen, light, food, water circulation, and parasites (Benson, 1961; Van Morkhoven, 1962). Clearly, many of these ecological parameters are highly intercorrelated, and evaluation of their effects on an assemblage must include consideration of the modes of life of the ostracodes and the effects of the interaction of environmental conditions at a station. However, several factors mentioned above seem to affect living ostracodes in predictable ways, although it is, of course, not certain that Paleozoic ostracodes responded in a similar way. For instance, Elofson (1941) studied Holocene marine ostracodes from Sweden and showed that the number of marine species decreased with lowering of salinity, then dropped abruptly when salinity fell below 10 °/oo. The species inhabiting brackish water, however, are often represented by large numbers of individuals. Elofson also showed the restriction of some species to colder waters, well-defined depth zones, or particular kinds of substrates; he also observed that the carapaces of shelf species inhabiting similar environments were often structured similarly, such as burrowers often having smoother and more strongly calcified carapaces than nonburrowing benthic species.

Swain (1955) attributed the greater abundance of ostracodes in San Antonio Bay than in the nearby Gulf of Mexico to the nutrients available in the bay. He noted also a greater number of species in the more saline water away from the areas of freshwater influx. In some cases, he observed somewhat greater abundances where the bottom was sandier.

Benson (1959) found that the distribution of ostracode species in Todos Santos Bay, Baja California, was markedly affected by lateral and/or vertical changes in salinity, with the greatest abundance of species found in shallow shelf water with relatively stable salinity values. He noted, as had Elofson (1941), that the structure and abundance of plants was a major biological factor in determining the distribution of ostracode species, particularly, in shallower water.

Hulings and Puri (1964) observed that the number of species increased with increase in salinity in waters off the west coast of Florida. They reported seasonal variations in abundance, with some species preferring colder and some warmer temperatures. They found the greatest number of species in sediments of a sand-mud mixture, with fewer species in clean sand.

Puri et al. (1969) studied the distribution of ostracodes in the Mediterranean and concluded that assemblages close to land were favorably influenced by higher salinity and greater abundance of nutrients. Nearshore and offshore species were locally more abundant where the bottom was muddy, somewhat sandy, or vegetated.

Hazel (1975) observed that the diversity of ostracode assemblages in the Atlantic off Cape Hattaras generally increased with depth, finer grain size of the substrate, and stability of temperature. However, he noted that stability of the substrate—a function of wave, swell, and current action—may have exerted the greatest control on diversity, although it could not be readily quantified. Pokorný (1971) found that changes in diversity of assemblages from the Upper Cretaceous of Bohemia generally coincided with transgressive and regressive phases of sedimentation. He interpreted variations from this pattern as the result of differences in rate of sedimentation at different depths within the depositional basin. Krutak (1973, 1975), on the other hand, found that in seven marginal marine environments of coastal Mississippi species abundance increased with decreased salinity or increased fluctuations in salinity. He found that diversity and species abundance were less and equitability was greater in more stable environments.

It is clear that both diversity and abundance of ostracode species are related to environmental conditions, and both generally are greatest for marine species on the offshore shallow shelf where most conditions are relatively stable. Abundance and diversity may also be high in the nearshore area where plants and nutrients are abundant.

STRATIGRAPHY AND PALEONTOLOGY

The stratigraphy and paleontology of the Beil Member (Virgilian) are relatively well known from previous studies (Zeller, 1968), and we have benefited from the paleoenvironmental interpretations made by previous investigators. Brown (1958) discussed the stratigraphy and environmental setting of the Beil in Kansas, and Schrott (1966) included the Beil in his study of depositional environments of the Lecompton Formation. Elements of the Beil fauna have been identified and studied by Purrington (1948), Perkins et al. (1962), Moore (1966), Baird (1971), Koepnick and Kaesler (1971, 1974), Farmer and Rowell (1973), and von Bitter (1972). In addition, R. H. Maerz is now studying the corals of the Beil, and P. Enos is working on the sedimentary petrography of the Lecompton Formation.

The systematics of ostracodes from Upper Paleozoic units stratigraphically above and below the Beil were discussed by Kellett (1933, 1934, 1935, 1936), Cordell (1952), and Sohn (1960a, b). Hattin (1957), Lane (1958, 1964), McCrone (1963). Imbrie (1955), and Bifano et al. (1974) studied ostracodes as part of their work on various Lower Permian rock units of the Midcontinent, and discussed some of the relationships among species as well as environmental factors affecting species abundance. Many of these species found are closely related to species found in the Beil.

The Forest City Basin, in which the Beil sediments were deposited, is asymmetrical with a steep western flank and a gently sloping eastern flank (Jewett, 1951; Merriam, 1963). During Late Pennsylvanian time, deposition on the eastern flank took place under conditions of alternating transgression and regression. Moore (1936, 1950, 1966) has discussed the cyclic sequences that resulted from these alternating conditions. One sequence includes the Beil Limestone as well as the King Hill Shale above and the Queen Hill Shale below the Beil (Fig. 1). The Queen Hill is laterally persistent over a wide area, cropping out from eastern Iowa to southern Kansas. It consists of two facies, a lower black, platy, fissile shale usually less than 1 m thick, and an upper gray, clayey, blocky shale, generally slightly thicker than 1 m. The lower unit has abundant conodonts (von Bitter, 1972) and other fossils, such as bryozoans, *Crurithyris, Orbiculoidea, Aviculopecten*, shark teeth, coprolites, and occasional crinoids and fusulinids, which increase upward in the unit (Schrott, 1966).

The depositional environment of the Queen Hill Shale, and particularly of the black shale facies, is not yet adequately understood. The water was stagnant, and anoxic conditions and a reducing environment near the bottom prevented benthic organisms from inhabiting the sediment. Zangerl and Richardson (1963), on the basis of their work in several quarries in Indiana, argued for a shallow-water origin for black shales. However, Heckel and Baesemann (1975) reviewed the sedimentological and paleontological evidence and suggested that the black shales of eastern Kansas were deposited in the deepest waters of the Midcontinent epeiric sea. The stagnant conditions in which the black shale was deposited gradually gave way to conditions of better circulation, and the rate of influx of detrital sediments probably increased during deposition of the upper Queen Hill. Deposition of the Beil Limestone began with the accumulation of predominantly carbonate sediments.

The Beil may also be divided into two facies. The lower facies, comprising beds from the basal limestone to the limestone above bed 3 (Fig. 2), contains persistent limestones and interbedded highly calcareous fossiliferous shales. The shales increase in thickness up the section and become less calcareous. The limestone beds below bed 2 (Fig. 2) are fusulinid-crinoidal biomicrites of low species richness with 10 to 30 percent fossil fragments (Schrott, 1966). Between beds 2 and 3, the lime-

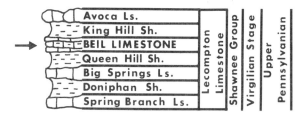

Figure 1
Members and stratigraphic relationships of the Lecompton Formation.

stones are primarily fusulinid biomicrites. Their fauna is much more diverse than that of the lowermost Beil and includes abundant brachiopods, corals, fusulinids, bryozoans, crinoids, mollusks, and other organisms. The abundance of well-preserved fossils and the presence of calcareous shale were interpreted by Brown (1958), Moore (1966), Schrott (1966), and Baird (1971) as indicating a shallow, warm marine environment relatively undisturbed by currents.

The sediments and fauna of the upper Beil, however, show the effects of increasing influx of terrigenous debris and wider fluctuations in current velocities and energy conditions. The limestones become increasingly thinner toward the top of the member; shale beds become much thicker and noncalcareous. The number of species present in the upper Beil is greatly reduced from that of the lower beds, and many organisms, including fusulinids, corals, bryozoans, and crinoids, are rare or absent. Apparently, bottom conditions were unfavorable for the diverse offshore fauna that consequently was replaced by a few dominant species of brachiopods and mollusks.

The limestone beds in the upper Beil have relatively little carbonate mud matrix and have many grains that are uniformly algal coated. R. H. Maerz (1975, pers. comm.)

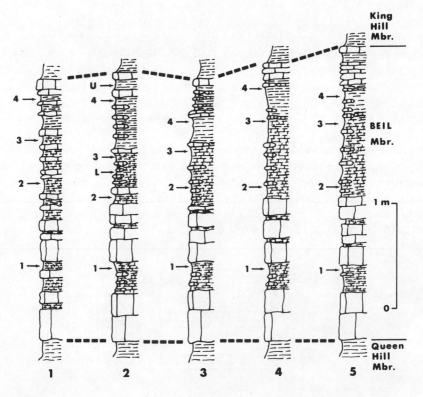

Figure 2
Locations of samples in outcrops.

has found dolomitic algal-laminated microsparite with fenestral fabric in the uppermost bed of the Beil near our study area. The upper part of the Beil, with its alternating beds of shale and thin, nodular limestone and greatly reduced fauna, appears to be the result of an unstable environment with fluctuating energy conditions, perhaps in rather shallow water, and with intermittent influx of terrigenous sediments.

Above the Beil are the unfossiliferous, greenish-gray and yellowish impure limestones of the King Hill. Because of the lack of abundant fossils and sedimentary structures, the environment of deposition of the King Hill is uncertain. However, P. Enos (1973, pers. comm.) found dessication cracks in the limestones in the King Hill about 10 km northwest of locality 1, which suggest a supratidal environment close to the area of study. Other authors, on the basis of fragmentary plant and animal fossils or the color of the rocks, have interpreted the unit as being of intertidal, superatidal, or subaerial origin (Schrott, 1966; Baird, 1971; Yochelson *in* Johnson and Adkison, 1967). The sedimentary sequence from the deep-water Queen Hill Shale to the nonmarine King Hill Shale is a regressive one. To interpret the environments of deposition of the Beil as having changed gradually from deep to shallow water is consistent with this model and with the evidence: decreasing species abundance of the macroinvertebrates, increasing influx of terrigenous sediments, fluctuating energy conditions, and the presence of algal laminated biosparites and fenestral fabric.

METHODS OF STUDY

Five outcrops of the Beil Limestone were selected for this study on the basis of three criteria: (1) availability of a nearly complete section, (2) presence of easily accessible unweathered shales, and (3) previous study by other investigators (e.g., Brown, 1958; Koepnick and Kaesler, 1971, 1974; von Bitter, 1972). Four of the localities were within 1 km of one another, and the fifth (locality 1) was about 10 km northwest of the other four. Detailed descriptions and locations of the localities were given by Brondos (1974, Appendix A). Bed 1 was located in shale immediately below the second persistent massive limestone of the Beil (Fig. 2). Bed 2 was 10 to 15 cm above that same limestone. Bed 3 was the uppermost shale in the calcareous facies; shale beds above 3 are noncalcareous. Bed 4 was just beneath the persistent interbedded thin shales and nodular limestones near the top of the Beil. Even if some of these correlations are not strictly correct, nevertheless the rank order of the beds is the same, and their spacing over the 2.5-m thickness of the Beil is approximately equal at all five localities.

At each outcrop one 3,000-cm^3 sample was collected from each of four shaly intervals and all were studied to determine the effects of changes in environment on the ostracode fauna. The units sampled are identified in Fig. 2. In addition, at locality 2, five 5,000-cm^3 samples were collected 2 m apart laterally from each of two additional beds. These beds are labeled L and U in Fig. 2. The latter samples were used to determine differences in ostracode diversity within strata and among strata within an outcrop.

Samples were processed according to methods described by Collinson (1963,

1965). The shales were soaked in Stoddard solvent and washed through 20- and 100-mesh sieves. Ostracodes were sampled from the residue on the 100-mesh sieve.

Preliminary studies, following the methods of Gibson and Buzas (1973), demonstrated that 300 individuals from a sample were adequate for computing a reliable estimate of diversity of the total sample. The volume of sediment with this number of individuals was determined for each collection by picking first a 2-g sample and then randomly split fractions as necessary. All instars identifiable to the species level and all single valves at least one half complete were counted. When the individuals from each sample had been identified and counted, the samples were reduced randomly to 300 individuals to eliminate effects of differences in sample size.

Values of diversity and equitability were calculated for each unit based on the random sample of 300 individuals. Brillouin's (1962) index from information theory was chosen to measure diversity:

$$H = \frac{1}{N} \log_e \left(\frac{N!}{N_1! N_2! N_3! \ldots N_s!} \right),$$

where N is the total number of individuals in the sample and $N_1, N_2, N_3, \ldots, N_s$ are the numbers of individuals in each respective species. Pielou (1966, 1969) has demonstrated that this equation is appropriate to use with samples being treated as statistical populations and where the true number of species present in the community is not known. Equitability was computed as $E = H/H_{max}$, where H is the diversity and H_{max} is the theoretical maximum diversity possible for a sample with the number of species in the collection being studied (Pielou, 1969, p. 233-234). For the two beds at locality 2 from which multiple samples were drawn, the mean, standard error, and 95 percent confidence limits were calculated for the diversity and equitability of each bed. Diversity and equitability of samples from the five outcrops were analyzed for significant differences among beds using the distribution-free Kruskal-Wallis test and the Mann-Whitney-Wilcoxon U test.

R-mode and Q-mode cluster analyses of matrices of correlation coefficients were computed using the unweighted pair-group method (UPGMA) in order to determine relationships among samples and among species on the basis of the distribution of species among samples (Sokal and Rohlf, 1969). The R-mode techniques grouped those species which were most often associated and indicated which species responded in a similar way to the depositional environment. The Q-mode analysis grouped samples that were similar to each other in the faunas that they contained. For a more thorough discussion of the meaning and application of cluster analysis to paleontological problems, see Kaesler (1966) or Hazel (1970).

Finally, principal coordinates analysis (Gower, 1966) in the R-mode was used to ordinate the species in a sample space for the collections from the latter part of this study. Principal coordinates analysis was selected over principal components analysis because the number of samples was less than the number of species. Rohlf (1972) pointed out that principal coordinates analysis is much more efficient under these circumstances, although the two methods give proportional results when no data are missing. Coordinates of the species were computed by transforming a matrix of distance coefficients between species (R-mode). The first seven eigenvectors were extracted from the transformed matrix; the elements of the eigenvectors were interpreted as coordinates of the species in a space of seven dimensions. The first two coordinates of each species were then plotted, which gave a two-dimensional ordination.

RESULTS AND DISCUSSION

Previous studies generally have not discussed the amount of error present in calculations of diversity and equitability at one locality. Apparently, single values were assumed to give reliable estimates of diversity and equitability for a wide area. This assumption was tested in our study with the multiple samples from beds L and U at locality 2. Results are given in Table 1 and show that both indexes varied little within beds. Standard errors were similar and relatively small for both beds, and the composition of the assemblages changed very little within the beds. Thus, the distribution of ostracodes in each of the two beds was rather uniform, at least over the 10-m lateral intervals sampled at this locality. It was therefore assumed that at each locality in the study area a single sample from a bed provided a reliable estimate of diversity and equitability for that bed.

The Kruskal-Wallis test was used to test the hypothesis that no differences in diversity and equitability existed between the two beds. The test showed that differences in diversity and equitability between the two beds were not significant at the 0.05 level. However, the compositions of the assemblages in the two beds were quite different, and the average number of species in a sample decreased from 15 to 11 from the lower to the upper bed.

Diversity and equitability indexes for the five localities showed slight differences among the four beds (Table 2). The Kruskal-Wallis test partially supported this observation. Equitability was not significantly different among beds at the 0.05 level, but diversity was significantly different ($P = 0.05$). Because the Kruskal-Wallis test does not show where the differences lie, the Mann-Whitney-Wilcoxon test was

Table 1
Diversity (H) and equitability (E) of multiple samples within beds at locality 2.

Bed	Index	Sample					Mean	Std. Error	95 Percent Conf. Limits	
		1	2	3	4	5				
U	H	1.650	1.912	1.846	1.823	1.830	1.812	0.097	1.622	2.002
	E	0.741	0.798	0.797	0.818	0.741	0.734	0.041	0.609	0.859
L	H	1.987	1.891	1.800	1.850	2.004	1.906	0.088	1.734	2.078
	E	0.764	0.711	0.710	0.696	0.790	0.779	0.036	0.708	0.850

Table 2
Diversity (H) and equitability (E) of samples from five localities.

Bed	Index	Locality					Mean	Std. Error	95 Percent Conf. Limits	
		1	2	3	4	5				
4	H	1.904	1.456	1.556	1.605	1.745	1.653	0.175	1.310	1.996
	E	0.771	0.629	0.698	0.754	0.783	0.727	0.064	0.602	0.852
3	H	2.184	2.174	2.391	2.175	2.121	2.210	0.104	2.006	2.414
	E	0.882	0.802	0.882	0.787	0.816	0.825	0.036	0.754	0.896
2	H	2.071	1.891	2.283	2.234	2.324	2.161	0.179	1.810	2.512
	E	0.725	0.684	0.878	0.841	0.802	0.786	0.080	0.629	0.943
1	H	1.168	1.949	1.559	2.399	1.685	1.752	0.458	0.854	2.650
	E	0.487	0.750	0.631	0.840	0.664	0.674	0.132	0.415	0.932

used. It showed that equitability in bed 3 was significantly greater than in bed 4 (Table 3). Diversity was similarly different between these two beds at the 0.05 level (Table 4).

These results show that maximum diversity and equitability of the ostracodes were attained in the presumably normal marine environments in which beds 2 and 3 and adjacent beds were deposited. The depositional environment of the upper Beil differed greatly, and many species abundant in the lower Beil were eliminated or replaced by species tolerant of more turbid water or higher-energy conditions. The loss of species is shown in Fig. 3D. Average abundance of species at the five localities decreased from 18 to 17 to 11 species in beds 2, 3, and 4, respectively. Similar changes were observed for H (Fig. 3A) and H_{max} (Fig. 3C). Changes in equitability were not as dramatic but also followed a similar pattern (Fig. 3B). The replacement of species is illustrated by Figure 4. *Cavellina nebrascensis* (Fig. 4A), for example, increased in abundance from almost zero percent in the first bed to 30 to 50 percent in the fourth bed. *Cryptobairdia seminalis* (Fig. 4B) decreased from between 35 and 72 percent of the assemblage to only a few individuals. Several other species showed trends similar to these two species, although not as striking. Data on the abundances of all the species were given by Brondos (1974, Appendix B).

Conditions were favorable for an abundant and diverse ostracode fauna throughout most of the time the Beil was being deposited. However, the significant decrease in both diversity and equitability at all localities from bed 3 to bed 4 showed that fewer species of ostracodes were adapted to the depositional environment of the upper Beil. Together with the reduction in abundance of species, these indexes showed a situation similar to that reported by Elofson (1941) and Benson (1959), who found only a small number of Holocene species of ostracodes able to flourish in fluctuating marine environments. The drop in abundance of ostracodes was paralleled by the decrease in abundance of species of the macrofauna in the upper Beil. A few species of productid brachiopods and mollusks dominated the fossil assemblages in the upper units.

Table 3
Results of the Mann–Whitney–Wilcoxon U test for comparison of equitability among beds. Numbers of beds increase stratigraphically up the section; * indicates difference at $p = 0.05$; other values not significant.

	Beds			
Beds	4	3	2	1
4	—			
3	25*	—		
2	18	9	—	
1	9	4	5	—

Table 4
Results of the Mann–Whitney–Wilcoxon U test for comparison of diversity among beds. Numbers of beds increase stratigraphically up the section; * indicates significant difference at $p = 0.05$; other values not significant.

	Beds			
Beds	4	3	2	1
4	—			
3	25*	—		
2	24	12	—	
1	15	5	6	—

One interpretation of the Q-mode cluster analysis (Fig. 5) showed three major clusters of samples. Group A consisted of samples from the lowest bed at all localities, plus the second bed from localities 1 and 2. It is apparent that, at the time of deposition of the first bed sampled, the same species were present over the entire study area. The second cluster (B) consisted of samples from the two middle beds and represented a time when conditions of the substrate, depth, and other factors were near the optimum for many species of ostracodes as well as other kinds of organisms. Finally, cluster C consisted solely of samples from bed 4 from all outcrops. The tight clustering of the samples of cluster C (minimum $r = 0.65$) reinforced the concept of a relatively sudden and uniform response of the ostracodes to changes of the environment. Together, the three clusters suggested the following sequence: (1) an assemblage of moderate but highly variable diversity was established when the lowermost Beil sediments were being deposited; (2) the assemblage

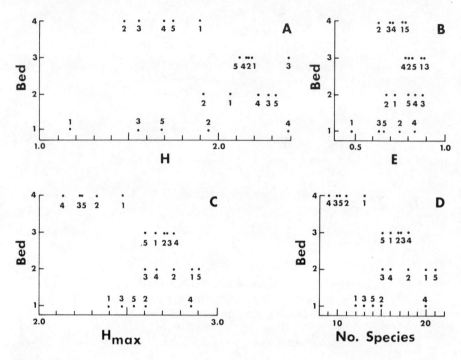

Figure 3
Summary of four parameters of the communities. Note that each parameter decreases from the third to the fourth bed. (A) Diversity indexes, H. (B) Equitability indexes, E. (C) Maximum diversity, H_{max}. (D) Number of species.

became more uniform, and on the average the abundance of species, diversity, and equitability increased as the environment stabilized and the uppermost part of the calcareous facies was deposited; (3) the ostracode assemblage decreased significantly in diversity and equitability as environmental conditions became unfavorable for most species in the late stages of shallowing that accompanied deposition of the upper Beil.

Three major groups of species were identified using R-mode cluster analysis (Fig. 6). Cluster A contained those species that were found commonly in the three lower beds. Of particular interest in this group are the three smaller clusters, A_1, A_2, and A_3. Subgroup A_1 contains those species found primarily in beds 2 and 3. Three of the species (*Pseudoparaparchites kansensis, Milleratia* ?, and *Fabalicypris glennensis*) were present only rarely and made up less than 1 percent of the assemblages in which they were found. Individuals of the species *Bairdia girtyi* and *Jonesina distenta* comprised only 1 to 5 percent of the ostracode fauna but were found in nearly every sample from beds 2 and 3. The two species were present only rarely in beds 1 and 4.

The species of subgroup A_2 included the taxa most characteristic of the lower beds. *Cryptobairdia seminalis* was by far the most common element of the first

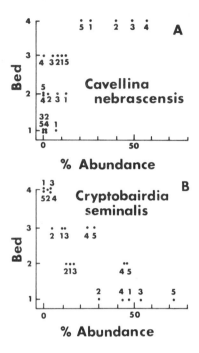

Figure 4
Relative abundances of two dominant species. (A) *Cavellina nebrascensis*. (B) *Cryptobairdia seminalis*.

bed, ranging from 35 to 72 percent of the ostracode fauna. As mentioned above, this species decreased in abundance in the higher strata until it was absent from most samples from bed 4. *Silenites silenus* followed a similar pattern, although in most samples it made up only about 10 percent of the fauna. *Bairdia pompilioides, Coronakirkbya moorei,* and *Kelletina robusta* each made up 6 percent or less of the fauna from the lower three beds and also decreased in abundance in stratigraphically higher intervals.

Subgroup A_3 had two species, *Fabalicypris acetalata* and *Orthobairdia texana*, which usually made up between 2 and 10 percent of the faunas of the lower three beds. However, they were absent from most samples from bed 4. Other species in cluster A were quite rare and weakly linked to the rest of the cluster.

Cluster B consisted of extremely rare species, usually with only one individual in the samples in which the species were present. Because of their rarity and the uncertainty of identification of some of the species, this group is regarded as an artifact of clustering for the most part and will not be discussed further here.

Cluster C comprised those species that generally attained their greatest abundance in bed 4. Fewer species were present in this cluster than in cluster A, and many of the species made up 10 percent or more of the assemblages. *Cavellina nebrascensis*

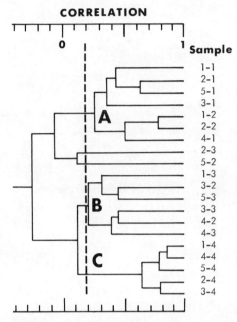

Figure 5
Dendrogram from Q-mode cluster analysis of a matrix of correlation coefficients; cophenetic correlation coefficient = 0.887. First integer in sample number is locality; second integer is stratigraphic interval, with 1 being the lowest bed and 4 the highest bed sampled.

was the most abundant species, but it did not dominate the assemblage as did *Cryptobairdia seminalis* in the lower beds. Both the distribution of individuals among the species and the smaller number of species present in bed 4 than in the lower beds resemble the pattern of ostracode diversity in modern marginal marine environments (e.g., Swain, 1955). This similarity suggests that ostracodes of the upper Beil responded to fluctuating environmental conditions with a decreased number of species but an increased number of individuals per species. Interestingly, at locality 2, bed U, which was above bed 4, showed a continued trend toward dominance of a few very abundant species. In bed U, *Amphissites centronotus, Bairdia beedei, Moorites minutus,* and *Cavellina nebrascensis* made up almost 90 percent of the assemblage, with no other species making up more than 5 percent of the sample.

Several authors have reported similar groups of ostracodes in other rock units in Kansas. Although some of the species differ, Lane (1964) and Bifano et al. (1974) observed the joint occurrence in Lower Permian shales of many of the same genera found in the upper Beil. Data presented by Jacques (1964) indicated that a similar

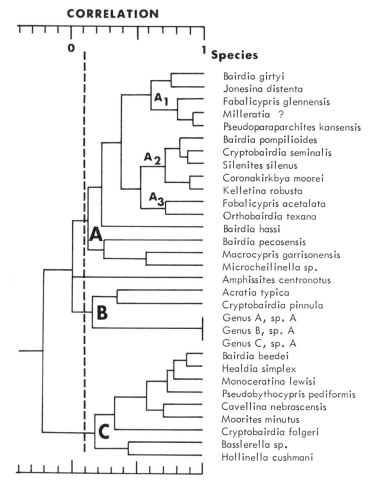

Figure 6
Dendrogram from R-mode cluster analysis of a matrix of correlation coefficients; cophenetic correlation coefficient = 0.787.

assemblage was present in shale of Missourian age in southeastern Kansas. These authors suggested that variations in the genera or species present in an assemblage were influenced by differences of salinity, conditions of the substrate, or depth. Unfortunately, it has not yet been possible to determine which factor or factors affected a given species.

Results of the principal coordinates analysis are summarized in Table 5. Seven components were extracted from the species-by-species distance matrix; the first two components accounted for 86 percent of the total intercollection variability. Scores for each species on the first two axes are given in Table 6. Six species that commonly comprised at least 10 percent of the assemblages in which they were

Table 5
Summary of principal coordinates analysis.

Principal Coordinate	Eigenvalue	Percentage of Variance Explained	Cumulative Percentage
I	115.335	55.73	55.73
II	63.089	30.48	86.21
III	10.444	5.05	91.26
IV	4.553	2.20	93.46
V	3.726	1.80	95.26
VI	2.458	1.19	96.45
VII	1.752	0.85	97.30

found loaded most heavily on the first coordinate axis, ranging from 2.3 to 4.4 (Fig. 7). Thus, the first component, which accounted for 56 percent of the intercollection variability, is a general abundance factor. The more abundant species loaded more heavily on the first component. Conversely, the least abundant species had the lowest loadings. The 11 species that projected between -1.65 and -1.50 never made up more than 1 percent of the assemblage of any sample.

The second component explained 30 percent of the total variability and represented the effect of stratigraphic interval on the collections. Species with the highest loadings (2.3 and 5.6) were found most abundantly in the lower beds; those with the lowest loadings (-1.0 to -3.3) were found most commonly in the uppermost beds. As Ebeling et al. (1970) found in their work, species that occurred at low, irregular frequencies or in only a few samples projected close to the origin; they also noted that such species contribute little to the explanation of variability among collections. In our study, 11 species projected between -0.01 and -0.44 and were considered uninterpretable. Some of these rare species may have been at the limits of their ecological tolerance and may have been much more abundant in other geographic areas or other stratigraphic intervals. However, such factors were not determinable in this study.

The results of the principal coordinates analysis showed a segregation of ostracode species by abundance and by stratigraphic interval in which the species were most commonly found. For example, *Cryptobairdia seminalis,* which as previously noted occurred in abundance in the calcareous facies, projected in the distant northeastern corner of the scatter diagram in group A. The other species abundant in the lower beds, *Silenites silenus* and *Orthobairdia texana,* also projected in the northeastern quadrant. Six species abundant in the upper beds, including *Bairdia beedei, Cavellina nebrascensis, Healdia simplex, Monoceratina lewisi, Moorites minutus,* and *Pseudobythocypris pediformis,* projected in the southeastern quadrant. These results suggest that two assemblages representing the end points in this spectrum may be defined on the basis of the species mentioned above. The first assemblage, including *Cryptobairdia seminalis, Silenites silenus,* and *Orthobairdia*

Table 6
Scores for species on first two principal coordinates.

Species	Principal Coordinate I	Principal Coordinate II
Acratia typica	-1.580	-0.328
Amphissites centronotus	-1.572	-0.333
Bairdia beedei	2.373	-1.329
Bairdia girtyi	-0.751	0.488
Bairdia hassi	1.391	-0.153
Bairdia pompilioides	-0.648	1.295
Bairdia pecosensis	-1.638	1.295
Basslerella sp.	-0.572	-0.108
Cavellina nebrascensis	3.840	-3.257
Coronakirkbya moorei	-1.177	0.778
Cryptobairdia folgeri	-0.423	-0.904
Cryptobairdia pinnula	-1.654	-0.424
Cryptobairdia seminalis	4.010	5.637
Fabalicypris acetalata	-0.199	0.833
Fabalicypris glennensis	-1.597	-0.245
Healdia simplex	3.041	-1.792
Hollinella cushmani	1.558	0.397
Kelletina robusta	-1.338	0.345
Macrocypris garrisonensis	-1.355	-0.288
Microcheilinella sp.	-1.090	-0.043
Milleratia?	-1.455	-0.018
Monoceratina lewisi	0.500	-0.773
Moorites minutus	4.381	-1.000
Jonesina distenta	0.465	0.755
Orthobairdia sp.	0.601	1.165
Pseudoparaparchites kansensis	-1.539	-0.139
Pseudobythocypris pediformis	3.396	-1.087
Silenites silenus	0.721	2.306
Genus A, sp. A	-1.650	-0.444
Genus B, sp. A	-1.650	-0.444
Genus C, sp. A	-1.650	-0.444

texana, is representative of ostracodes that lived in somewhat deeper water and in an area of greater carbonate deposition. The assemblage consisting of *Bairdia beedei, Cavellina nebrascensis, Healdia simplex, Monoceratina lewisi, Moorites minutus,* and *Pseudobythocypris pediformis* (with *Amphissites centronotus* if the assemblages from bed U are considered) was adapted to shallower water with intervals of increased influx of terrigenous sediments and probably greater turbidity. Additional studies of these and other species in adjacent units and in units where

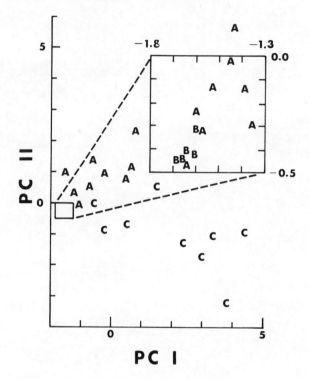

Figure 7
Two-dimensional principal coordinates ordination. Letters refer to letter designations given to clusters in Fig. 6.

environmental conditions are more readily interpretable will provide a test of the extent to which these species indeed represent distinct environments.

CONCLUSIONS

1. Diversity and equitability of ostracodes showed little variation within beds in an outcrop. Thus, the use of a single large sample from a bed in an outcrop was judged to be adequate for estimating diversity and equitability of that bed.

2. Diversity and equitability of the ostracode assemblages were variable at first and then became less variable while increasing slightly during the time of deposition of most of the interval studied. During late stages of regression, diversity and equitability were significantly reduced, presumably because the environment was more highly variable.

3. Changes of the environment during regression resulted in the replacement of species dominant in the lower part of the Beil. The assemblage dominated by *Cryptobairdia seminalis*, *Silenites silenus*, and *Orthobairdia texana* of the lower Beil, probably representative of an offshore marine environment, was gradually replaced by an assemblage in the upper Beil containing *Cavellina nebrascensis*, *Bairdia beedei*,

Pseudobythocypris pediformis, Moorites minutus, Healdia simplex, Monoceratina lewisi, and *Amphissites centronotus*. The most readily interpretable changes in the environment of deposition of the Beil during this time were a shoaling of water, periodic influx of greater amounts of terrigenous sediment, and possibly intermittent increases in mechanical energy conditions.

4. Q-mode cluster analysis showed that the highest similarities in faunal content among samples were within the lowermost and uppermost beds.

5. R-mode cluster analysis showed that three clusters of species were present, two of which were numerically important and were representative of depositional facies in the Beil.

6. The first two principal coordinates, roughly equivalent to relative abundance and stratigraphic position, explained 86 percent of the variability among samples.

ACKNOWLEDGMENTS

We are indebted to Michael J. Brady, Wakefield Dort, Jr., A. J. Rowell, and Robert J. Stanton, Jr., who read the manuscript and offered helpful suggestions for its improvement. K. C. Lohmann and David J. McBride assisted in the field. Conversations with Richard B. Koepnick and Richard H. Maerz were very helpful in providing added insight into the depositional history of the Beil.

This study was supported in part by a National Science Foundation grant to the University of Kansas for research in the fields of systematic and evolutionary biology (GB–8785, Robert S. Hoffman, principal investigator). The support included a traineeship for Michael D. Brondos for six weeks in the summer of 1972.

Computations were done with the Honeywell 635 computer at the University of Kansas Computation Center using in part the NT-SYS programs written by F. James Rohlf and his associates.

Unbroken specimens from this study have been deposited in the University of Kansas Museum of Invertebrate Paleontology and given numbers 1,047,298 to 1,054,397.

REFERENCES

Baird, G. C. 1971. Paleoecology of the Beil Limestone (Upper Pennsylvanian) in the northern midcontinent region. Unpublished M.S. thesis, University of Nebraska, 152p.

Benson, R. H. 1959. Ecology of Recent ostracodes of the Todos Santos Bay region, Baja California, Mexico. Univ. Kansas Paleont. Contrib. 23, Arthropoda, Art. 1, 80p.

———. 1961. Ecology of ostracode assemblages. *In* R. C. Moore (ed.), Treatise on invertebrate paleontology, Part Q, Arthropoda 3. Geological Society of America and University of Kansas Press, p. 956–963.

Bifano, F. V., A. L. Guber, and R. J. Cuffey. 1974. Ostracode paleoecology in shales of the Wreford megacyclothem (Lower Permian; Kansas and Oklahoma). Geol. Soc. Amer. Abst., 6:492.

Brillouin, L. 1962. Science and information theory (2nd ed.). Academic Press, Inc., New York, 347p.
Brondos, M. D. 1974. Diversity of assemblages of Late Paleozoic Ostracoda. Unpublished M.S. thesis, University of Kansas, 46p.
Brown, W. G. 1958. Stratigraphy of the Beil Limestone, Virgilian, of eastern Kansas. Unpublished M.S. thesis, University of Kansas, 188p.
Collinson, C. W. 1963. Collection and preparation of conodonts through mass production techniques. Ill. State Geol. Surv. Circ., 343:1-16.
———. 1965. Conodonts. In B. Kummel and D. Raup (eds.), Handbook of paleontological techniques. W. H. Freeman and Company, San Francisco, p. 94-102.
Cordell, R. J. 1952. Ostracodes from the Upper Pennsylvanian of Missouri. Part 1. The family Bairdiidae. Jour. Paleont., 26:74-112.
Ebeling, A. W., R. M. Ibara, R. J. Lavenberg, and F. J. Rohlf. 1970. Ecological groups of deep-sea animals off southern California. Bull. Los Angeles County Museum Nat. Hist. Science, 65:1-43.
Elofson, O. 1941. Zur Kenntnis der marinen Ostracoden Schwedens, mit besonderer Berucksichtigung des Skagerraks. Uppsala Univ. Zool. Bidrag, 19:215-534.
Farmer, J. D., and A. J. Rowell. 1973. Variation in the Bryozoan *Fistulipora decora* (Moore and Dudley) from the Beil Limestone of Kansas. In R. S. Boardman, A. H. Cheetham, and W. A. Oliver (eds.), Animal colonies. Dowden, Hutchison & Ross, Inc., Stroudsburg, Pa., p. 377-394.
Gibson, T. G., and M. A. Buzas. 1973. Species diversity: patterns in modern and Miocene Foraminifera of the eastern margin of North America. Geol. Soc. Amer. Bull., 84:217-238.
Gower, J. C. 1966. Some distance properties of latent root and vector methods used in multivariate analysis. Biometrika, 53:325-338.
Hattin, D. E. 1957. Depositional environment of the Wreford megacyclothem (Lower Permian) of Kansas. Kansas Geol. Surv. Bull., 124, 150p.
Hazel, J. E. 1970. Binary coefficients and clustering in biostratigraphy. Geol. Soc. Amer. Bull., 81:3237-3252.
———. 1975. Ostracode diversity in the Cape Hatteras, North Carolina, area. Jour. Paleont., 49:731-744.
Heckel, P. H., and J. F. Baesemann. 1975. Environmental interpretation of conodont distribution in Upper Pennsylvanian (Missourian) megacyclothems in eastern Kansas. Amer. Assoc. Petrol. Geol. Bull., 59:486-509.
Hulings, N. C., and H. S. Puri. 1964. The ecology of shallow water ostracods of the west coast of Florida. In H. S. Puri (ed.), Ostracodes as ecological and paleoecological indicators. Pubbl. Staz. Zool. Napoli, 33 supp.:308-344.
Imbrie, J. 1955. Quantitative lithofacies and biofacies study of Florena Shale (Permian) of Kansas. Amer. Assoc. Petrol. Geol. Bull., 39:649-670.
Jacques, T. E. 1964. Microfossil distribution in the Hickory Creek Shale, Wilson, Montgomery Counties, Kansas. Unpublished M.S. thesis, University of Kansas, 48p.
Jewett, J. M. 1951. Geologic structures in Kansas. Kansas Geol. Surv. Bull., 90-6: 105-172.
Johnson, W. D., Jr., and W. L. Adkison. 1967. Geology of eastern Shawnee County, Kansas and vicinity. U.S. Geol. Surv. Bull., 1215-A, 100p.
Kaesler, R. L. 1966. Quantitative re-evaluation of ecology and distribution of Recent Foraminifera and Ostracoda of Todos Santos Bay, Baja California, Mexico. Univ. Kansas Paleont. Contrib., Paper 10, 50p.
Kellett, B. 1933. Ostracodes of the Upper Pennsylvanian and the Lower Permian

strata of Kansas: I. The Aparchitidae, Beyrichidae, Glyptopleuridae, Kloedenellidae, Kirkbyidae, and Youngiellidae. Jour. Paleont., 7:59–108.

———. 1934. Ostracodes of the Upper Pennsylvanian and the Lower Permian strata of Kansas: II. The genus Bairdia. Jour. Paleont., 8:120–138.

———. 1935. Ostracodes of the Upper Pennsylvanian and the Lower Permian strata of Kansas: III. Bairdiidae (concluded), Cytherellidae, Cypridinidae, Entomoconchidae, Cytheridae, and Cypridae. Jour. Paleont., 9:132–166.

———. 1936. Carboniferous ostracodes. Jour. Paleont., 10:769–784.

Koepnick, R. B., and R. L. Kaesler. 1971. Intraspecific variation of morphology of *Triticites cullomensis* (Fusulinacea), a statistical analysis. Jour. Paleont., 45:881–887.

———, and R. L. Kaesler. 1974. Character correlations and morphologic variations of *Triticites cullomensis* (Fusulinacea). Jour. Paleont., 48:36–40.

Krutak, P. R. 1973. Modern ostracod species diversity patterns in brackish ecosystems, coastal Mississippi, U.S.A. Geol. Soc. Amer. Abst., 5:701.

———. 1975. Environmental variation in living and total populations of Holocene Foraminifera and Ostracoda, coastal Mississippi, U.S.A. Amer. Assoc. Petrol. Geol. Bull., 59:140–160.

Lane, N. G. 1958. Environment of deposition of the Grenola Limestone (Lower Permian) in southern Kansas. Kansas Geol. Surv. Bull., 130-3:117–164.

———. 1964. Paleoecology of the Council Grove Group (Lower Permian) in Kansas, based upon microfossil assemblages. Kansas Geol. Surv. Bull., 170-5:1–23.

McCrone, A. W. 1963. Paleoecology and biostratigraphy of the Red Eagle Cyclothem (Lower Permian) in Kansas. Kansas Geol. Surv. Bull., 164, 114p.

Merriam, D. F. 1963. The geologic history of Kansas. Kansas Geol. Surv. Bull., 162, 317p.

Moore, R. C. 1936. Stratigraphic classification of the Pennsylvanian rocks of Kansas. Kansas Geol. Surv. Bull., 22:1–256.

———. 1950. Late Paleozoic cyclic sedimentation in central United States. 18th Intern. Geol. Congr. London 1948, 4:5–16.

———. 1966. Paleoecological aspects of Kansas Pennsylvanian and Permian cyclothems. *In* D. F. Merriam (ed.), Symposium on cyclic sedimentation. Kansas Geol. Surv. Bull., 169-1:287–380.

Perkins, R. D., T. G. Perry, and D. E. Hattin. 1962. Some bryozoans from the Beil Limestone Member of the Lecompton Limestone (Virgilian) of Kansas. Kansas Geol. Surv. Bull., 157-5:1–25.

Pielou, E. C. 1966. The measurement of diversity in different types of biological collection. Jour. Theoret. Biol., 13:131–144.

———. 1969. An introduction to mathematical ecology. John Wiley & Sons, Inc. (Interscience Division), New York, 286p.

Pokorný, V. 1971. The diversity of fossil ostracode communities as an indicator of palaeogeographic conditions. *In* H. J. Oertli (ed.), Paléoécologie des Ostracodes. Bull. Centre Rech. Pau, S.N.P.A. 5(supp):45–62.

Puri, H. S., G. Bonaduce, and A. M. Gervasio. 1969. Distribution of Ostracoda in the Mediterranean. *In* J. W. Neale (ed.), the taxonomy, morphology, and ecology of Recent Ostracoda. Oliver & Boyd Ltd., Edinburgh, p. 356–411.

Purrington, W. 1948. Stratigraphic distribution of microfossils in the lower part of the Shawnee Group. Unpublished M.S. thesis, University of Kansas, 63p.

Rohlf, F. J. 1972. An empirical comparison of three ordination techniques in numerical taxonomy. Systematic Zool., 21:271–280.

Schrott, R. O. 1966. Paleoecology and stratigraphy fo the Lecompton megacyclothem (Late Pennsylvanian) in the Northern Midcontinent Region. Unpublished Ph.D. dissertation, University of Nebraska, 329p.

Sohn, I. G. 1960a. Paleozoic species of *Bairdia* and related genera. U.S. Geol. Surv. Profess. Paper 330–A:1–105.

——. 1960b. *Aechminella, Amphissites, Kirkbyella,* and related genera. U.S. Geol. Surv. Profess. Paper 330–B:106–160.

Sokal, R. R., and F. J. Rohlf. 1969. Biometry. W. H. Freeman and Company, San Francisco, 776p.

Swain, F. M. 1955. Ostracoda of San Antonio Bay, Texas. Jour. Paleont., 29:561–646.

Valentine, J. W. 1973. Evolutionary paleoecology of the marine biosphere. Prentice-Hall, Inc., Englewood Cliffs, N.J., 512p.

Van Morkhoven, F. P. C. M. 1962. Post-Paleozoic Ostracoda, v. 1. American Elsevier Publishing Company, Inc., New York, 204p.

von Bitter, P. H. 1972. Environmental control of conodont distribution in the Shawnee Group (Upper Pennsylvanian) of Eastern Kansas. Univ. Kansas Paleont. Contrib. Paper 59, 105p.

Woodwell, G. M., and H. H. Smith (eds.). 1969. Diversity and stability in ecological systems. Brookhaven Symp. Biol. 22, 264p.

Zangerl, R., and E. S. Richardson, Jr. 1963. The paleoecological history of two Pennsylvanian black shales. Fieldiana Geol. Mem., 4, 352p.

Zeller, D. E. N. (ed.). 1968. The stratigraphic succession in Kansas. Kansas Geol. Surv. Bull., 189, 81p.

Paleosynecology of Nonmarine Mollusca from the Green River and Wasatch Formations (Eocene), Southwestern Wyoming and Northwestern Colorado

John H. Hanley   U.S. Geological Survey

ABSTRACT

The complexly intertongued lacustrine and fluviatile Green River and Wasatch formations were deposited in early and middle Eocene basins of Wyoming, Colorado, and Utah. Lake Gosiute (Wyoming) and Lake Uinta (Colorado-Utah) and adjacent lowland areas supported diverse molluscan faunas.

The paleoecology of five recurring molluscan associations is based on the ecology of structurally similar living counterparts. The distribution of mollusks relative to lithostratigraphy defines five principal facies.

Deposits of the lowland area adjacent to Lake Gosiute are subdivided into pond, fluviatile, and paludal facies. The pond facies is represented by the *Physa-Biomphalaria-Omalodiscus* association. It is composed of pisidiid bivalves and a diverse association of aquatic pulmonate gastropods preserved in thin, lenticular limestone and marlstone lithosomes. The lowland-fluviatile facies is distinguished by two molluscan associations. The low-diversity *Plesielliptio* association in drab mudstone and sandstone represents a fluviatile habitat. The *Oreoconus* association consists of seven taxa of terrestrial gastropods, and suggests moist and dry, humid, calcium-rich alluvial plain habitats. The paludal facies lacks mollusks and is composed of carbonaceous shale and sandstone and lignite.

The littoral lacustrine facies is characterized by coquinal limestone and siltstone, sandstone, and drab shale containing *Goniobasis tenera, Viviparus trochiformis,* and *Plesielliptio* n. sp. A. This *Goniobasis-Viviparus* association delineates the geographic distribution of the shoreline fluctuations of Lake Gosiute. The littoral facies intertongues with the lowland fluviatile, paludal, and sublittoral lacustrine facies. Biostratonomic evidence suggests that the habitat was affected by shallow lacustrine currents that abraded and concentrated shells. However, an increase in the abundance of *Valvata, Hydrobia, Physa, Biomphalaria,* or *Omalodiscus* in this association at some localities reflects a sheltered, littoral habitat.

The sublittoral-lacustrine facies is composed of drab shale and oil shale dominated by the Pisidiidae-*Goniobasis-Valvata* association. Mollusks in the oil shale indicate deposition in a shallow, oxygenated environment.

INTRODUCTION

Location and Geologic Setting

This report concerns molluscan paleosynecology and lithostratigraphic relations of the Green River and Wasatch formations in southwestern Wyoming. The Green River Formation is a large lens of lacustrine strata, which originated during early and middle Eocene in Lake Gosiute in southwestern Wyoming, and in Lake Uinta in adjacent portions of Colorado and Utah (McDonald, 1972, Fig. 7). The Green River is enclosed by generally fine-grained fluviatile sediment assigned to the Wasatch formation below and the Bridger and Washakie formations above. The complex lithostratigraphic relations between these formations in Wyoming (Fig. 1) reflect climatic and tectonically induced changes in size and paleogeography of Lake Gosiute during its approximately 4 million year history.

In terms of the structural differentiation of basins in southwestern Wyoming by Love (1961, Fig. 1), the primary study area extends from the western Washakie Basin to the southeastern Green River basin, and from the southern margin of the Rock Springs uplift to northern Colorado (Fig. 2). In the vicinity of Pine Mountain (Fig. 3), lithostratigraphic relations and the distribution of Mollusca were studied in detail within the main body and Niland Tongue of the Wasatch Formation, and the Luman Tongue and Tipton Shale Member of the Green River Formation, in an area of 250 square miles (650 km^2) (Hanley, 1974, Pl. I). The east-west trending Uinta Mountains formed a barrier to the expansion of Lake Gosiute south of this region. Paleogeographic reconstructions (Roehler, 1965) indicate that the Pine Mountain area was repeatedly the site of nearshore, shallow-water lacustrine deposition. Thus, this area is ideal for delineating and studying the distribution of adjacent lacustrine and fluviatile environments of deposition and for studying the associated molluscan faunas.

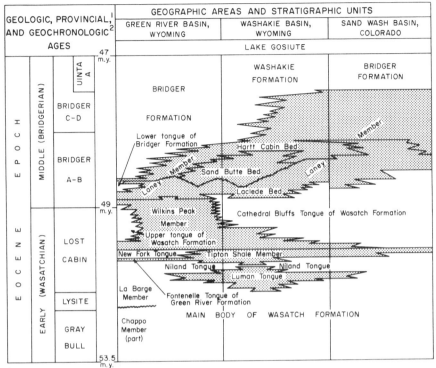

Figure 1
Stratigraphic relations of the Green River (stippled) and Wasatch formations, southwestern Wyoming. (From Roehler, 1972.)

Additional collections of Mollusca were made from the western margin of the Green River Basin south of Kemmerer, Wyoming (Figs. 2 and 3), and in the northwestern Piceance Creek Basin, Colorado (Fig. 3).

Previous Investigations

Early Tertiary nonmarine Mollusca of the Rocky Mountain region were first examined during territorial surveys and the explorations of the U.S. Geological Survey. Initial descriptions of the faunas were made by Meek (1860, 1872, 1876) and White (1876, 1878, 1880, 1883, 1886). Cockerell (1914, 1915) and Cockerell and Henderson (1912) described additional early Tertiary species. Cretaceous and Paleocene Mollusca in western Canada have been studied extensively by L. S. Russell for 50 years. Russell's work is largely summarized by Tozer (1956). Russell also studied Canadian Eocene mollusks (1952, 1955, 1957). LaRocque (1960) provided a syste-

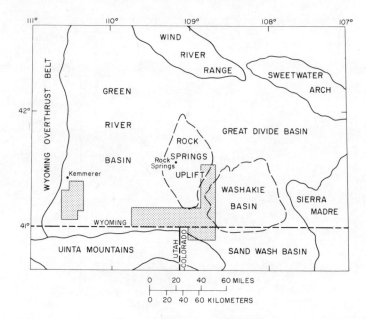

Figure 2
Geologic setting of principal Wyoming study areas (stippled).
(Base from Roehler, 1973.)

matic and paleoecologic analysis of mollusks from the Flagstaff Limestone (Paleocene-Eocene) in Utah. The Paleocene and Eocene Mollusca of Wyoming and Montana have been examined by McKenna et al. (1962), Moore (1960), and Yen (1946, 1948a, b). Mollusca from the Green River and Wasatch formations have not previously been studied in detail.

Purpose and Methods of Study

For over 100 years, early Tertiary nonmarine Mollusca of the Rocky Mountain region have been objects of systematic analysis, but their biostratigraphic and paleoenvironmental utility were virtually unexploited. In consideration of criteria for recognizing lacustrine strata, Picard and High (1972) suggested that mollusks are poor indicators of specific nonmarine environments. Recent work by Taylor (1960, 1966) and Hibbard and Taylor (1960), however, demonstrated the utility of late Cenozoic fossil nonmarine mollusks in paleoenvironmental interpretations.

The purpose of this report is to show the value of early Tertiary nonmarine mollusks in paleoenvironmental interpretations. To realize this goal, interpretations must be based on reliable stratigraphic and systematic foundations. Twenty-one detailed stratigraphic sections in the Green River and Wasatch formations south of the Rock Springs uplift (Hanley, 1974, Pl. 1) provide a detailed lithostratigraphic framework in which to consider the distribution of early and middle Eocene Mollusca. The mollusks have been subjected to detailed systematic analysis (Hanley,

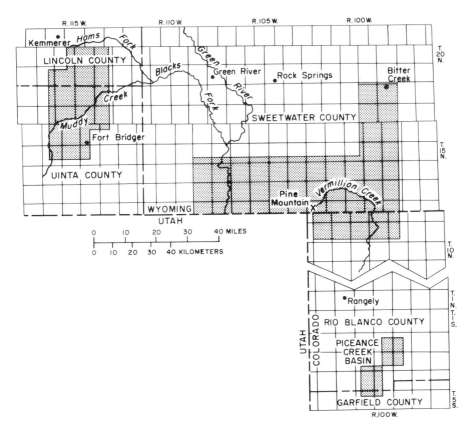

Figure 3
Locations of study areas (stippled) in Wyoming and Colorado.

1974, pp. 86-257). Morphologic bases for species discrimination in fossils are compared to concepts used for discriminating biological species in related, living taxa. Presumed ecophenotypic and genetic variations are differentiated, and species concepts are outlined for many taxa. These interpretations provide a reliable taxonomic base for paleosynecologic interpretations. Twenty-seven genera and forty species of mollusks are represented in the formations (Hanley, 1974, Table 1). Taphonomic analysis identifies taxa that are secondarily introduced into the fossil assemblage. Recurring molluscan associations are then defined.

Extant genera and some species of nonmarine mollusks are commonly eurytopic. Therefore, autecologic analysis of early Tertiary taxa that is based on the ecology of extant, related taxa often produces only generalizations. Refined paleoecologic interpretations are the product of synecologic analysis, involving the ecologic implications of molluscan biocoenoses (associations). Analysis of factors affecting the taxonomic composition, diversity, and abundance of Mollusca in the associations is critical in delineating lacustrine and adjacent continental environments. Paleoecology of these associations, based on the ecology of of structurally similar living counterparts, delineates fluviatile, alluvial-plain, pond, and littoral- and sublittoral-lacustrine habitats.

Distribution of molluscan associations in relation to the complex lithostratigraphy of the Green River and Wasatch formations defines five facies; these facies document the lacustrine-fluviatile depositional history of the formations.

TERMINOLOGY

Several terms used in this discussion are characterized by a variety of definitions. The intent of these terms in the present work is as follows:

Assemblage is applied to a collection of fossils prior to taxonomic, biostratonomic, and paleoecologic analyses.

Association is a subdivision of a community. In this report, the term is applied to distinct, recurring, taxonomic combinations of Mollusca. Mollusks are the most abundantly preserved representatives of the Eocene communities in the Rocky Mountain region. Their distribution varies directly with interpretable environmental parameters.

> *Community*—The sum total of interacting organisms, small and large, hard and soft bodied, which in combination are unique taxonomically, and distinct spatially, from surrounding communities. The community usually has certain distinct and coincident environmental boundaries (e.g., substrate, salinity belts, etc.) (Kauffman, 1974, p. 12.2).

TAPHONOMY OF MOLLUSCAN ASSEMBLAGES

Biostratonomy

Reconstruction of the preburial and postburial history of a fossil assemblage, involving biostratonomy and diagenesis, respectively, precedes paleoecologic interpretations. Johnson (1960) and Fagerstrom (1964) discussed numerous criteria by which the biostratonomic history of a fossil assemblage can be interpreted. Faunal composition, shell fragmentation, dissociation of hard parts, shell orientation, and texture and structure of the sedimentary aggregate are the criteria from which the origin of the molluscan assemblage can be inferred (Table 1). Any characteristic of a fossil assemblage can be formed by different processes, and should, therefore, be used in conjunction with numerous other criteria in delineating the origin of the assemblage.

The origin of the molluscan associations is described as "in place," "disturbed neighborhood," or "transported," based on Scott (1970). These terms reflect the degree of transport of the assemblage. The modifier "mixed" is applied to ecologically heterogeneous molluscan assemblages composed of aquatic and terrestrial mollusks. Disturbed-neighborhood molluscan assemblages dominate in the Green River and Wasatch formations. Mixed-disturbed-neighborhood assemblages and in-place assemblages rank second and third in abundance, respectively.

Table 1

Criteria for recognizing origin of fossil molluscan assemblages in Green River and Wasatch formations, southwestern Wyoming.

Characteristics of Assemblage	Origin of Assemblage[a]			
	In-Place Assemblage	Disturbed-Neighborhood Assemblage	Transported Assemblage	Mixed Assemblage (In place, disturbed neighborhood, or transported)
Faunal composition	Ecologically homogeneous	Ecologically homogeneous	Ecologically homogeneous	Ecologically heterogeneous
Fragmentation	Dominantly whole shells; small proportion of fragments	Many whole shells or many fragments, depending on extent of in situ littoral lacustrine abrasion	Many shell fragments	Exhibits characteristics of its particular manner of origin
Dissociation of skeletal parts				
Articulated vs. disarticulated	Equal or more articulated	Nearly all disarticulated	Disarticulated	Same as above
Left vs right valves	Approximately equal	Approximately equal	Not equal	Same as above
Orientation				
Convex up	Common	Common	Uncommon	Same as above
Convex down	Common	Common	Common	Same as above
Relative to bedding	Life position to parallel	Parallel to random	Parallel	Same as above
Parallel alignment	Uncommon	Uncommon	Common	Same as above
Texture and structure of sedimentary aggregate	Fossils typically associated with textures and structures consistent with presumed ecology	Fossils typically associated with textures and structures consistent with presumed ecology	Fossils associated with textures and structures inconsistent with presumed ecology	Same as above

[a] Origin from Scott (1970).

Diagenesis

Diagenesis is expressed in four ways in the assemblages studied during this investigation: (1) recrystallization, (2) dissolution, (3) silicification, and (4) distortion due to compaction of the sediment. In dense limestone and marlstone, the shells of mollusks are typically recrystallized. Generally, the recrystallization does not obscure growth lines or detail of the protoconch in gastropods. Shells in porous rocks, such as sandstone and siltstone, are typically removed through dissolution and are represented by internal and external molds. Silicification is common in some collections from the Laney Member of the Green River Formation. In some cases only the shells or matrix are silicified, but in other examples both fossils and matrix are affected. Deformation of mollusks from sediment compaction is common. Bivalves and gastropods are typically compressed perpendicular to the plane of commissure and to the axis of coiling, respectively. Heliciform gastropods are commonly compressed parallel to the axis of coiling.

DEFINITION AND ANALYSIS OF MOLLUSCAN ASSOCIATIONS

Five associations have been empirically defined after careful examination of the composition and biostratonomy of molluscan assemblages from the Green River and Wasatch formations (Table 2). Names of common taxa are applied to the associations (Pls. 1-3). Taxa represented only by poorly preserved specimens were not considered in the definition of the associations.

Quantitative techniques are often used to measure interspecific association. The Jaccard coefficient of association, for example, quantitatively defines recurring groups of organisms based on presence and absence data. Cluster analysis of the coefficients of association constructs recurrent groups of species (R-mode analysis) representing potential associations or communities, or recurrent groups of samples having similar species composition (Q-mode analysis) representing similar biofacies. These methods were used by Kaesler (1966), Scott (1970), and Warme (1969), among others, for paleoecologic analyses. Scott (1970) demonstrated that his empirically defined faunal associations closely paralleled those derived by Jaccard coefficient analysis. Most of the associations discussed in the present report are found at numerous stratigraphic levels within the Green River and Wasatch formations, and they are often repeated within measured sections. The frequency of recurrence and taxonomic constancy of these empirical associations suggest that they represent molluscan biocoenoses. Therefore, quantitative analysis of faunal assemblages was deemed unnecessary.

Recurrent groups of organisms can be analyzed relative to such structural characters as taxonomic composition, trophic relations, and rank, relative abundance, and distribution of taxa within associations. This report is primarily concerned with the composition, generalized rank abundance, and trophic relations within associations as evidence concerning paleoecology and paleoenvironments.

Table 2

Composition of molluscan associations from Green River and Wasatch formations, southwestern Wyoming and northwestern Colorado. Symbols: C, common; R, rare; —, absent; Pl, *Plesielliptio* association; G–Vi, *Goniobasis-Viviparus* association; Pi–G–V, Pisidiidae-*Goniobasis-Valvata* association; P–B–O, *Physa-Biomphalaria-Omalodiscus* association; Or, *Oreoconus* association

	Pl	G–Vi	Pi–G–V	P–B–O	Or
Bivalvia					
Unionidae					
Plesielliptio priscus	C	—	—	—	—
Plesielliptio n. sp. A	—	C	—	—	—
Unionidae: indet.	C	—	R	R	—
Pisidiidae					
Sphaerium sp.	—	—	R	R	—
Eupera-Pisidium sp. A	—	—	—	R	—
Pisidiidae: indet.	—	—	C	—	—
Gastropoda					
Pleuroceridae					
Goniobasis tenera	—	C	R	R	—
Goniobasis sp.	R	—	C	—	—
Hydrobiidae					
Hydrobia aff. *H. utahensis*	—	R	—	C	—
Hydrobia sp.	—	—	R	—	—
Valvatidae					
Valvata subumbilicata	—	R	—	R	—
Valvata cf. *V. filosa*	—	R	—	R	—
Valvata sp.	—	—	C	—	—
Viviparidae					
Viviparus trochiformis	—	C	—	—	—
Viviparus paludinaeformis	—	R	—	—	—
Viviparus sp.	—	—	R	—	—
Physidae					
Physa bridgerensis	—	—	—	C	—
Physa longiuscula?	—	R	—	C	—
Physa pleromatis	R	R	—	C	—
Physa sp. A	—	—	—	R	—
Planorbidae					
Biomphalaria aequalis	—	R	—	C	—
Biomphalaria storchi	—	R	—	C	—
Biomphalaria pseudoammonius	—	—	—	R	—
Drepanotrema? sp.	—	—	—	R	—
Gyraulus militaris	—	R	—	R	—
Omalodiscus cirrus	—	R	—	C	—
Promenetus sp. A	—	—	—	R	—

(*continued*)

Table 2 (continued)

	Pl	G-Vi	Pi-G-V	P-B-O	Or
Acroloxidae					
Acroloxus minutus	–	–	–	R	–
Lymnaeidae					
Lymnaea cf. *L. minuscula*	–	–	–	R	–
Lymnaea similis	–	–	–	R	–
Lymnaea sp. B	–	–	–	R	–
Pleurolimnaea tenuicosta	–	–	–	R	–
Bulimulidae					
Oreoconus n. sp. A	–	–	–	–	C
Urocoptidae					
Holospira n. sp. A	–	–	–	–	R
Endodontidae					
Discus ralstonensis	–	–	–	–	R
Oreohelicidae					
Oreohelix grangeri	–	–	–	–	R
Polygyridae					
Mesodon? sp.	–	–	–	–	R
Helicinidae					
Schasicheila n. sp. A	–	–	–	–	R
Grangerellidae					
Grangerella mcleodensis	–	–	–	–	R
Total number of taxa C =	2	3	3	7	1
R =	2	10	5	16	6

Explanation of Plate 1*

1-4 *Plesielliptio priscus* (Meek and Hayden): 1-2, Interior and exterior of right valve with straight ventral margin, hypotype USNM 209988, main body of Wasatch Formation, Sweetwater County, Wyoming, ×0.75. 3, Interior of right valve with prominent trigonal cardinal tooth, hypotype USNM 209989, main body of Wasatch Formation, Sweetwater County, Wyoming, ×1.5. 4, Dorsal margin of right valve with prominent double-looped juvenile umbonal sculpture and posteroventrally radiating costae, hypotype USNM 209993, main body of Wasatch Formation, Sweetwater County, Wyoming, ×5.

5 Pisidiidae: *Sphaerium* sp. Exterior, USNM 210027, Douglas Creek Member of Green River Formation, Rio Blanco County, Colorado, ×9.

6 *Hydrobia* aff. *H. utahensis* White. Apertural view, USNM 210049, Niland Tongue of Wasatch Formation, Sweetwater County, Wyoming, ×20.

7-8 *Valvata* cf. *V. filosa* Whiteaves. 7, apical view, USNM 210064, Niland Tongue of Wasatch Formation, Sweetwater County, Wyoming, ×15. 8, apertural view, USNM 210061, Niland Tongue of Wasatch Formation, Sweetwater County, Wyoming, ×15.

9-12 *Oreoconus* n. sp. A. 9, apertural view of immature specimen, USNM 210134, main body of Wasatch Formation, Sweetwater County, Wyoming, ×2. 10-12, pro-

file, abapertural, and apertural views of mature specimen, UMMZ 232201, Pass Peak Formation of Dorr (1969), Hoback Basin, Sublette County, Wyoming, × 1.75.

*In all plate explanations, USNM refers to the U.S. National Museum, Washington, D.C., and UMMZ refers to the University of Michigan Museum of Zoology, Ann Arbor, Michigan. A scale that was originally 2 cm in length is included on each plate.

PLATE 2

Explanation of Plate 2

1-4 *Plesielliptio* n. sp. A. 1, interior of right valve with lamellar cardinal tooth, USNM 210014, Luman Tongue of Green River Formation, Sweetwater County, Wyoming, ×1.25. 2, dorsal margin of left valve with double looped juvenile umbonal sculpture intersected obliquely by growth lines, USNM 210017, Luman Tongue of Green River Formation, Sweetwater County, Wyoming, ×6. 3, exterior of left valve with convex ventral margin and incomplete posterior shell tip, USNM 210000, Luman Tongue of Green River Formation, Sweetwater County, Wyoming, ×0.75. 4, interior of left valve with lamellar cardinal teeth, USNM 210000, Luman Tongue of Green River Formation, Sweetwater County, Wyoming, ×1.25.

5-11 *Goniobasis tenera* (Hall). 5, apertural view of mature specimen with incomplete juvenile whorls, hypotype USNM 210035, Fontenelle Tongue of Green River Formation, Lincoln County, Wyoming, ×2. 6, abapertural view, spire incomplete, hypotype USNM 210039, Fontenelle Tongue of Green River Formation, Lincoln County, Wyoming, ×2. 7, profile view, aperture and juvenile whorls incomplete, hypotype USNM 210032, Niland Tongue of Wasatch Formation, Sweetwater County, Wyoming, ×2. 8, abapertural view, spire incomplete, hypotype USNM 210036, Luman Tongue of Green River Formation, Sweetwater County, Wyoming, ×2. 9, abapertural view, juvenile whorls incomplete, hypotype USNM 210038, Luman Tongue of Green River Formation, Sweetwater County, Wyoming, ×2. 10, abapertural view, spire incomplete, hypotype USNM 210040, Luman Tongue of Green River Formation, Sweetwater County, Wyoming, ×2. 11, apertural view, aperture and juvenile whorls incomplete, hypotype USNM 210034, Luman Tongue of Green River Formation, Sweetwater County, Wyoming, ×2.

12-13 *Viviparus trochiformis* (Meek and Hayden). 12, apertural view of mature specimen, hypotype USNM 210071, Niland Tongue of Wasatch Formation, Sweetwater County, Wyoming, ×2. 13, abapertural view of immature specimen, hypotype USNM 210077, Luman Tongue of Green River Formation, Sweetwater County, Wyoming, ×3.

The composition of the molluscan associations and their distribution within the lithostratigraphic framework of the Green River and Wasatch formations (Fig. 4) are the principal bases for paleosynecologic and paleoenvironmental interpretations. The taxa that compose the associations collectively define habitats and environmental parameters more clearly than does any one taxon. Ecologic interpretations are facilitated by comparing the Eocene associations to structurally similar extant molluscan associations. The distribution of associations delineates the geographic relations between habitats, as well as environmental change through time in a given geographic area.

Relative abundance of species was not a factor in the analysis of associations for several reasons. In the field, no attempt was made to collect a random sample of fossils or a sample of particular weight or volume. The variable preservation of fossils and the localized outcrops were such that a randomized sampling method would have omitted well-preserved specimens. Similarly, variability in rock type, induration, and outcrop character prohibited consistent collection of samples of specific weight or volume. For these reasons, absolute abundance, density, and dispersion of fossils were not calculated. An alternative method of calculating relative abun-

248 *Structure and Classification of Paleocommunities*

PLATE 3

Explanation of Plate 3

1 *Omalodiscus cirrus* (White). Right side, USNM 210121, main body of Wasatch Formation, Sweetwater County, Wyoming, ×6.

2-3 *Physa pleromatis* White. Abapertural and apertural views, hypotype USNM 210094, main body of Wasatch Formation, Sweetwater County, Wyoming, ×2.

4-5 *Physa bridgerensis* Meek. Abapertural and apertural views, hypotype USNM 210082, main body of Wasatch Formation, Sweetwater County, Wyoming, ×1.75.

6-7 *Physa longiuscula?* (Meek and Hayden). 6, abapertural view, USNM 210088, main body of Wasatch Formation, Sweetwater County, Wyoming, ×10. 7, apertural view, USNM 210086, main body of Wasatch Formation, Sweetwater County, Wyoming, ×12.

8-10 *Biomphalaria aequalis* (White). 8, right side of mature specimen, hypotype USNM 210115, Laney Member of Green River Formation, Sweetwater County, Wyoming, ×7. 9, apertural view of immature specimen, hypotype USNM 210116, Laney Member of Green River Formation, Sweetwater County, Wyoming, ×12. 10, left side, aperture incomplete, hypotype USNM 210231, Laney Member of Green River Formation, Sweetwater County, Wyoming, ×10.

11-13 *Biomphalaria storchi* (Russell). Right side, apertural view, and left side of mature specimen, hypotype USNM 210120, main body of Wasatch Formation, Sweetwater County, Wyoming, ×3.

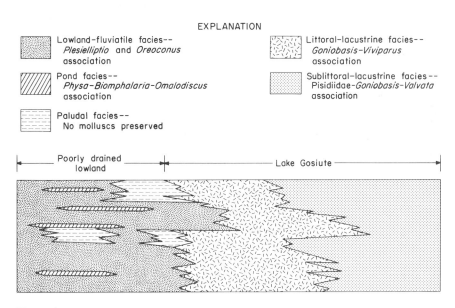

Figure 4
Schematic diagram showing distribution of the lowland-fluviatile, pond, and paludal facies of the Wasatch Formation and the littoral-lacustrine and sublittoral-lacustrine facies of the Green River Formation, as deposited during Early Eocene in Lake Gosiute and adjacent environments south of the Rock Springs uplift.

dance, proposed by Cottam et al. (1953), examined the number of specimens of each species relative to the total fauna rather than to a specific sample volume or weight. The method is based on the assumption that counting a minimum of 30 randomly selected specimens of a species is sufficient to determine relative abundance. In most of my collections, species are not represented by the minimum 30 specimens. The letters "C" and "R" in Table 2 refer to "common" and "rare" species, respectively. Although they are not derived quantitatively, they reflect my assessment of the generalized rank abundance of taxa within associations.

PALEOSYNECOLOGY OF MOLLUSCAN ASSOCIATIONS

Plesielliptio Association

The *Plesielliptio* association is dominated by the unionid bivalve *Plesielliptio priscus* (Meek and Hayden) (Pl. 1: 1-4). Shell thickness and form, and prominence of dentition in *P. priscus* (Pl. 1:1, 3) are characteristic of, but not restricted to, bivalves from flowing-water (lotic) habitats (McMichael and Hiscock, 1958). The straight to slightly sinuate ventral shell margin suggests a habitat in relatively swift water in small streams or rivers (Eagar, 1948). Lithostratigraphic relations support the inference that the *Plesielliptio* association occupied a river or stream habitat in the lowland adjacent to Lake Gosiute. The association is poorly defined by two collections from the lowland-fluviatile facies of the main body of the Wasatch Formation (Fig. 4).

Generally, unionids require fresh, clean, oxygenated, shallow, calcium-rich, permanent water habitats containing a current, pH greater than 7, stable substrate, food source, and at least seasonally warm temperatures. Although these generalizations are best applied to diverse associations of unionids, data suggest that they apply to *P. priscus*.

An interesting faunal analog to the *Plesielliptio* association is found in many streams in Virginia, which are characterized by a low-diversity, high-abundance fluviatile molluscan association dominated by *Elliptio complanatus* (Lightfoot). The variation in shell form in *P. priscus* is analogous to that in the living *E. complanatus* "complex" in the eastern United States.

Oreoconus Association

The *Oreoconus* association is dominated by terrestrial pulmonate gastropods, especially *Oreoconus* n. sp. A. In these hermaphroditic gastropods, the mantle cavity that normally contains the gills has been modified into a pulmonary sac. Oxygen may pass into the sac through a single opening, the pneumostome. The gills are replaced by a plexus of blood vessels in the roof of the pulmonary sac. Terrestrial pulmonates live in a variety of habitats, but in general they are present in moist, calcium- and organic-rich habitats. The most important factors affecting the distribution of terrestrial pulmonates are moisture and humidity. Diversity has been shown to increase with an increase in calcium and organic material in the habitat (Burch, 1955). Terrestrial snails also display remarkable capabilities to withstand desiccation. They withdraw into their shells, secrete a thin layer of mucus (the epiphragm) across the aperture to prevent moisture loss, and aestivate until equable conditions return. Some land snails may lie semidormant, sealed in their shells by mucus sheets, for up to five years during extended dry periods (Baker, 1958).

Although this terrestrial association is broadly defined, several generalizations can be based on synecology of the gastropods. *Oreohelix* and *Holospira* require notably calcium-rich habitats. *Holospira*, for example, typically lives today in small colonies restricted to limestone outcrops in hot, dry places. The genera *Mesodon* and *Schasicheila* typically prefer habitats within a deciduous forest characterized by rotting leaves or wood, in a moist, calcium-rich habitat.

The dominance of terrestrial gastropods in the lowland-fluviatile facies of the main body of the Wasatch Formation (Fig. 4), therefore, indicates the presence of moist, humid, calcium-rich habitats within a deciduous forest on the lowland alluvial plain adjacent to Lake Gosiute. A dry, calcium-rich habitat is suggested by *Holospira*. The trees and shrubs of the alluvial plain provided diverse habitats for terrestrial mollusks.

The *Oreoconus* association is broadly defined, because assemblages of well-preserved terrestrial gastropods are rare. The association is known only from the lowland-fluviatile facies of the main body of the Wasatch Formation (Fig. 4).

Physa-Biomphalaria-Omalodiscus Association

The association is composed of 9 genera and 16 species of aquatic pulmonate gastropods (see Pl. 3: 1-13 for common species), 2 bivalve species of the Pisidiidae (Pl. 1: 5), and 3 species of gill-breathing gastropods (see Pl. 1: 6 for the common species). Aquatic pulmonates and pisidiid bivalves coexist in a variety of recent habitats, which include shallow waters of lakes and locally impounded water in streams. A diverse assemblage of aquatic pulmonates and pisidiid bivalves, however, typically characterizes shallow, even temporary, lentic (quiet-water) habitats such as ponds. This association is most frequently represented in strata of the lowland-fluviatile facies of the main body and Niland Tongue of the Wasatch Formation.

Aquatic pulmonate gastropods have a pulmonary sac like their terrestrial counterparts, but some species (Planorbidae, Acroloxidae) have a secondary external gill termed a pseudobranch. These gastropods are able to respire subaqueously using their false gills. Other aquatic pulmonates (Physidae and Lymnaeidae) come to the surface and obtain atmospheric oxygen by bringing the pneumostome into contact with the air-water interface. This trait is advantageous in allowing them to tolerate reduced oxygen concentration in their aquatic habitat. Lymnaeid gastropods can also breathe oxygen directly from the water through their skins by a process termed cutaneous respiration.

Aquatic pulmonates often exhibit remarkable tolerance to short-term environmental stress, such as desiccation of their aquatic habitat. Many are able to reduce their metabolism and aestivate, as do their terrestrial counterparts. Pisidiid bivalves are also able to withstand desiccation of their habitat by burrowing into moist sediment, often beneath stones. Their ovoviviparity promotes survival of offspring, in that the young are released from the parent as fully developed juvenile bivalves.

High-dispersal potential of aquatic pulmonates and pisidiids promotes rapid colonization of local habitats. Aquatic pulmonates typically lay their eggs on shallow aquatic vegetation, thereby increasing the chance of aerial dispersal on the legs of water birds (Rees, 1965). Pisidiids are known to be locally dispersed by insects (Fernando, 1954; Rees, 1965). The abundance of aquatic pulmonates and pisidiids in the presumed pond habitat is enhanced by their hermaphroditic reproduction.

These factors reflect the excellent adaptation of aquatic pulmonate gastropods and pisidiid bivalves to life in local ephemeral aquatic habitats. The diverse association of aquatic pulmonates (16 species) in the *Physa-Biomphalaria-Omalodiscus* association suggests that the ponds may not have been seasonally ephemeral.

Hydrobia aff. *H. utahensis* (Pl. 1: 6) is the dominant (gill-breathing) mesogastropod in the association, which indicates its tolerance of a quiet-water habitat. *Valvata subumbilicata, V.* cf. *V. filosa,* and the eurytopic *Goniobasis tenera* are rare in the association. The external gill and hermaphroditic reproduction of *Valvata* are similar to characteristics of some aquatic pulmonates. The distribution of *Valvata* in pond, protected-littoral, and sublittoral habitats reflects its preference for quiet-water habitats.

Some molluscan assemblages from the pond habitat contain terrestrial pulmonate gastropods. *Oreoconus* n. sp. A is found in the main body of the Wasatch Formation in beds representing this habitat. *Vertigo arenula* (White) is present in a pond-habitat collection from the Laney Member of the Green River Formation. Such a faunal mixing substantiates the presence of moist, vegetated lowland adjacent to Lake Gosiute.

With the rare exception of *Goniobasis tenera,* no littoral-lacustrine mollusks are present in collections representing the pond habitat. The rare unionid bivalves are not definitely referable to species, although the best specimens resemble *Pleisielliptio priscus.* These factors suggest that the habitat was created by local expansion of streams to form ponds, rather than by lacustrine transgression and regression leaving ponded water. The possibility of an oxbow lake habitat cannot be excluded, but in no instance is there evidence associated with the limestone of a transition from a lotic (flowing-water) to a lentic molluscan fauna, which would indicate the change from a fluviatile habitat to that of an oxbow lake.

The four genera and seven species of gastropods that dominate this association are epifaunal browsers. Bambach and Walker (1971), Walker (1972), and Walker and Bambach (1974) have demonstrated that dominant taxa within paleocommunities may exhibit niche partitioning with regard to feeding level or feeding type. The dominance of several taxa with similar feeding habits may indicate that food was not a limiting factor in the habitat. In recent habitats, however, diverse associations of aquatic pulmonate gastropods are characterized by taxa that reach abundance peaks at different times of the year. Such seasonal variation in life cycles reduces competition among taxa, and it would be masked by the time-averaged nature of the fossil record.

Modern analogs to the presumed pond habitat were noted by Mozley (1954, p. 33) and Kenk (1949). Mozley cited the rich diversity of aquatic pulmonate gastropods in small pond-like lakes formed by expansion of streams. Kenk provided a detailed analysis of the composition and seasonal variation in a fauna of woodland pulmonate gastropods and pisidiid bivalves representing the same families and genera as the Eocene association. My observations of modern woodland pond and pool habitats indicate that they differ principally from the proposed Eocene habitat in their organic-rich, rather than carbonate-rich, substrate. Aquatic pulmonates thrive in submerged vegetation, as demonstrated in the woodland ponds of Michigan. Ponds adjacent to Lake Gosiute were undoubtedly rich in vegetation, although charophytes are the only fossil record of the vegetation.

Goniobasis-Viviparus Association

Goniobasis tenera and *Viviparus trochiformis* rank first and second in abundance, respectively, in this association. Competition between these dominant taxa may be reduced by niche partitioning in feeding mechanisms. *Goniobasis* is an epifaunal browser; *Viviparus* exhibits both browsing and ciliary feeding mechanisms.

Biostratonomy and lithostratigraphic relations of the *Goniobasis-Viviparus* association (Fig. 4) suggest that this association lived in a littoral-lacustrine habitat. Shells of *Plesielliptio* n. sp. A (Pl. 2: 1-4), *Goniobasis tenera* (Pl. 2: 5-11), and *Viviparus trochiformis* (Pl. 2: 12-13) are often concentrated in coquinal layers in the Luman Tongue of the Green River Formation. In addition, broken shells of these taxa typically exhibit rounded shell edges and only faint traces of shell ornamentation. These factors suggest that the mollusks were abraded by currents in a littoral-lacustrine habitat.

In some collections, nearly all shells of *Plesielliptio* n. sp. A are disarticulated and broken along the posteroventral margin. Many bivalves are represented only by the thick hinge portion of the shell, and the delicate juvenile whorls of *Goniobasis tenera* are typically not preserved. In these examples, it is not possible to differentiate preburial damage from breakage by weathering. The typically well preserved umbonal sculpture in *Plesielliptio* n. sp. A and shell ornamentation in *Goniobasis tenera* suggest that abrasion was not a significant factor in shell damage, even though the mollusks are concentrated in a coquinal layer. Disarticulation and breakage of *Plesielliptio* n. sp. A may have been promoted by subaerial exposure of the mollusks along the shoreline of Lake Gosiute. Under such conditions modern unionid bivalves typically crack along their ventral margin, and they are easily disarticulated by cracking of the brittle ligament.

Other shallow-water habitats occupied by this association were only slightly affected by currents. In one collection, for example, 60 percent of the specimens of *Plesielliptio* n. sp. A are articulated, and 21 percent of those disarticulated are oriented with their valves concave up. Many *Goniobasis tenera* exhibit juvenile whorls. These factors suggest that the habitat was little affected by currents, and that the planar orientation of *G. tenera* and *Viviparus trochiformis* in this collection was not significantly controlled by currents.

Shell form of *Plesielliptio* n. sp. A supports a lacustrine habitat. *Plesielliptio* n. sp. A is proportionally shorter and higher than *P. priscus,* and it exhibits a more curved ventral shell margin than does that species (Pl. 2: 3). Cardinal teeth in *Plesielliptio* n. sp. A (Pl. 2: 1,4) are lamellar and less strongly developed than the prominent trigonal teeth of *P. priscus.* Collectively, these morphologic features of *Plesielliptio* n. sp. A are typical of bivalves that inhabit lakes or slowly moving bodies of water (McMichael and Hiscock, 1958).

This association is distinctive in its lack of pisidiid bivalves and paucity of aquatic pulmonate gastropods, which are common in shallow-lacustrine habitats today. The mollusks probably could not tolerate the currents in the habitat, and currents possibly retarded the growth of submerged aquatic vegetation in which modern aquatic pulmonates thrive (Mozley, 1954). The abundance of aquatic pulmonates (e.g., *Biomphalaria, Gyraulus, Omalodiscus,* and *Physa*) and mesogastropods (e.g., *Hydrobia* and *Valvata*) in some collections may reflect a protected, shallow-lacustrine habitat. Mozley (1954) noted an increase in aquatic pulmonates from exposed to protected shallow-lacustrine habitats in Oneida Lake, New York.

This association is abundant in strata of the littoral-lacustrine facies of the Green River Formation south of the Rock Springs uplift. The association is also represented in strata of this facies in the Tipton Shale Member, Fontenelle Tongue, Douglas Creek Member, and Laney Member, all of the Green River Formation. It is present in lacustrine strata interbedded with the Niland Tongue of the Wasatch Formation.

Pisidiidae-*Goniobasis*-*Valvata* Association

Bivalves of the Pisidiidae and *Goniobasis* and *Valvata* characterize this association. It is distinguished from the *Goniobasis-Viviparus* association by the depauperate viviparid gastropods and unionid bivalves, and the more abundant *Valvata* and pisidiid bivalves.

Lithostratigraphic relations (Fig. 4) suggest that this association occupied a sublittoral habitat in Lake Gosiute adjacent to the littoral *Goniobasis-Viviparus* association. The association is dominant in gray shale and oil shale, which overlie and intertongue laterally with shallow-lacustrine sandstone and siltstone containing the littoral association. Interrelated changes in the lacustrine environment from littoral to sublittoral habitats include increased water depth; change from silt, sand, or gravel substrate to mud; decrease in current action; and probable increase in aquatic vegetation.

Living Pisidiidae and *Valvata* are found in lakes, ponds, rivers, and streams on a variety of substrates. They thrive in quiet-water habitats, typically associated with aquatic vegetation. The presumed sublittoral habitat of the fossil pisidiids (Pl. 1: 5) and *Valvata* is readily compatible with the ecology of related living species. The eurytopic nature of *Goniobasis* is substantiated by its abundance in this sublittoral-lacustrine habitat in addition to its littoral-lacustrine and fluviatile habitats. Unionid bivalves and *Viviparus* sp. are less common here than in the littoral-lacustrine habitat. The kerogen in the oil shale was derived from the undecayed remains of a rich algal growth, which formed a copropel ooze (Bradley, 1966). This unstable substrate may have provided inadequate support for unionid bivalves in this habitat.

This association characterizes strata representing the sublittoral-lacustrine facies in the Luman Tongue, Tipton Shale Member, and Laney Member of the Green River Formation.

CHARACTERISTICS, DISTRIBUTION, AND INTERPRETATION OF PRINCIPAL FACIES

Based on lithostratigraphic relations and the distribution of molluscan associations within the Luman Tongue and Tipton Shale Member of the Green River Formation, and the main body and Niland Tongue of the Wasatch Formation south of the Rock Springs uplift, I recognize five facies (Hanley, 1974, Pl. I). These facies correspond closely to Roehler's (1965) outline of environments of deposition in this region. Distribution of the lowland-fluviatile, pond, paludal, littoral-lacustrine, and sublittoral-lacustrine facies is shown schematically in Figure 4. Precise distribution of these facies within this stratigraphic interval is shown in Hanley (1974, Pl. I).

Lowland Drab Fluviatile Facies

Regional stratigraphic relations suggest that the gray-green mudstone and buff, fine-grained sandstone, the dominant rock types of the drab fluviatile facies, were deposited in lowland areas under poorly drained conditions. As Roehler (1965, p. 141) has observed, red colors of rocks at the edge of the basin uniformly give way basinward to drab-gray and green colors. I agree with his opinion that the color change reflects reduction of ferric iron to ferrous iron in poorly drained, low topographic areas toward the basin center. Oriel (1961, p. B152) also noted that because of poor drainage and reducing conditions sediments deposited close to Lake Gosiute lack the red color of sediments in well-drained alluvial soils of higher elevations.

The typical association of the drab fluviatile facies with limestone of the pond facies and with carbonaceous shale of the paludal facies indicates the poorly drained character of this lowland region. The drab fluviatile facies is characterized by two molluscan habitats: (1) rivers and streams inhabited by unionid bivalves such as *Plesielliptio priscus* of the low-diversity *Plesielliptio* association, and (2) the moist and dry, vegetated, calcium-rich alluvial plain inhabited by the more diverse terrestrial gastropods of the *Oreoconus* association.

Paludal Facies

Strata representing the paludal environment are proximally associated with marginal-lacustrine, floodplain, and pond deposits. Deposits of carbonaceous shale and sandstone, or lignite, often underlie or occur adjacent to lacustrine deposits. Carbonaceous strata are also frequently interbedded with strata representing lowland-fluviatile and pond facies. These relations indicate that swamps were common in poorly drained lowland areas, particularly adjacent to Lake Gosiute. This relationship is logical inasmuch as swamps are commonly distributed along the margins of lakes today. Transgression of Lake Gosiute is typically reflected by littoral-lacustrine deposits overlying paludal strata.

Mollusks have not been collected from strata representing the paludal facies. Either mollusks lived in this habitat and were not preserved or they found conditions unfavorable. I favor the former hypothesis for several reasons. Standing-water areas rich in aquatic vegetation are ideal habitats for aquatic pulmonate gastropods and pisidiid bivalves. In such environments, acidic conditions develop in association with decomposition and incomplete oxidation of organic debris. This can be particularly true in the substrate, where shells of mollusks can undergo postmortem dissolution.

Pond Facies

The pond facies is composed of limestone and marlstone interbedded with the paludal and lowland-fluviatile facies of the main body and Niland Tongue of the Wasatch Formation. Regional stratigraphic relations (Fig. 1) provide data regarding the paleoenvironmental setting of the limestone lithosomes that contain the very diverse *Physa-Biomphalaria-Omalodiscus* association. Lacustrine deposition of the Luman Tongue of the Green River Formation began in the Washakie Basin and gradually spread westward into the Green River Basin. The numerous limestone beds in the main body of the Wasatch Formation, therefore, formed adjacent to Lake Gosiute. Similarly, thin limestone beds containing the *Physa-Biomphalaria-Omalodiscus* association originated adjacent to lacustrine deposits interbedded with the Niland Tongue of the Wasatch Formation.

These local and regional stratigraphic relations suggest that the lenses represent a local, ephemeral, recurring environment in the lowland area adjacent to Lake Gosiute. Paleosynecology of the *Physa-Biomphalaria-Omalodiscus* association supports the interpretation of a pond environment. Because aquatic pulmonate gastropods and pisidiid bivalves are ideally adapted for existence in and dispersal into shallow, ephemeral ponds, this unstable environment is characterized by a very diverse molluscan association (Table 2).

Littoral-Lacustrine Facies

The littoral-lacustrine facies is characterized by coquinal limestone and siltstone and, locally, highly fossiliferous sandstone and shale. In the Luman Tongue of the Green River Formation, the facies complexly intertongues with the lowland drab fluviatile facies of the main body of the Wasatch Formation and the sublittoral-lacustrine facies of the Luman Tongue. The facies is readily recognized by the presence of the diverse *Goniobasis-Viviparus* molluscan association. The three dominant mollusks of this association are consistent indicators of littoral-lacustrine conditions in the Luman Tongue of the Green River Formation, and their distribution traces numerous complex shoreline fluctuations of Lake Gosiute during deposition of the Luman Tongue. Their presence in numerous layers, interbedded with fluviatile and paludal strata in the Niland Tongue of the Wasatch Formation clearly documents successive shorelines of several lacustrine transgressions. Similarly their presence in the widespread, time-transgressive "*Goniobasis* marker" at the base of the Tipton Shale Member of the Green River Formation reflects transgression of Lake Gosiute.

Sublittoral-Lacustrine Facies

This facies is characterized by drab shale and oil shale, which typically contains the low-diversity Pisidiidae-*Goniobasis*-*Valvata* association. The facies intertongues with the littoral-lacustrine facies and reflects maximum lacustrine transgression in the study area.

Several environments of deposition have been suggested for oil shale. Bradley (1948) proposed that oil shale formed in the deep, oxygen-poor hypolimnion of Lake Gosiute. Recently, Bradley (1973), Eugster and Surdam (1973), and Surdam and Wolfbauer (1973) suggested that oil shale formed in a shallow, oxygenated lacustrine environment. The presence of the Pisidiidae-*Goniobasis*-*Valvata* association in beds of low-grade oil shale supports the latter hypothesis.

CONCLUSIONS

1. Nonmarine mollusks are the most abundant macrofossils in the Green River and Wasatch formations in southwestern Wyoming.

2. Extant genera of nonmarine mollusks are typically eurytopic in their ecology. Paleoautecologic analysis of early Tertiary mollusks based on the ecology of extant related taxa typically yields only generalizations. Refined interpretations are derived from the paleosynecology of molluscan associations.

3. Analysis of factors affecting the taxonomic composition, diversity, and abundance of mollusks in associations is critical in the interpretation of lacustrine and adjacent continental environments. Paleosynecology of five molluscan associations, based on the ecology of structurally similar living counterparts, delineates fluviatile, alluvial-plain, pond, littoral-lacustrine, and sublittoral-lacustrine habitats.

4. The distribution of molluscan associations relative to the complex lithostratigraphy of the Green River and Wasatch formations defines five facies; these facies document the lacustrine-fluviatile depositional history of the formations.

5. The *Plesielliptio* and *Oreoconus* associations characterize fluviatile and alluvial-plain habitats, respectively, within strata of the lowland-fluviatile facies of the main body and Niland Tongue of the Wasatch Formation.

6. The diverse, locally distributed, *Physa-Biomphalaria-Omalodiscus* association reflects a ponded-water habitat in the poorly drained lowland adjacent to Lake Gosiute. The association is characteristic of the pond facies of the main body and Niland Tongue of the Wasatch Formation.

7. The *Goniobasis-Viviparus* association is a consistent indicator of a littoral-lacustrine habitat, and it documents numerous fluctuations in the shoreline of Lake Gosiute. The regional distribution of the association in strata of the littoral-lacustrine facies of the Luman Tongue and Tipton Shale Member of the Green River Formation, therefore, delineates the geographic extent of major lacustrine transgressions and regressions.

8. The presence of the Pisidiidae-*Goniobasis-Valvata* association in low-grade oil shale of the sublittoral-lacustrine facies of the Luman Tongue and Tipton Shale Member of the Green River Formation suggests formation of the oil shale in a shallow, sublittoral, oxygenated environment.

9. The integration of data derived from analysis of molluscan community structure and litostratigraphic relations provides a new approach to regional paleoenvironmental reconstruction in lower Tertiary continental and lacustrine strata.

ACKNOWLEDGMENTS

I am indebted to many individuals in the preparation of this paper. Paul O. McGrew, University of Wyoming, first showed me the importance of research in Tertiary nonmarine mollusks. Henry W. Roehler, U.S. Geological Survey, suggested the paleoecologic investigation reported in part here, and he led numerous field excursions to study the lithostratigraphic relationships of the Green River and Wasatch formations. Erle G. Kauffman, U.S. National Museum, served as my advisor and employer, allowing completion of my dissertation at the National Museum. I especially thank Donald W. Boyd, University of Wyoming, for his outstanding counseling as my primary dissertation director.

I have profited greatly from malacological discussions with Henry van der Schalie, Elmer G. Berry, John B. Burch, and John A. Dorr, Jr., the late Claude W. Hibbard, and George A. Te, all from the University of Michigan; Aurèle LaRocque and David H. Stansbery, The Ohio State University; J. P. E. Morrison, U.S. National Museum; Alan A. Solem, Chicago Field Museum of Natural History; J. G. J. Kuiper, Institut Néerlandais, Paris, France; Jiří Kříž, Geological Survey of Czechoslovakia; Heinz A. Kollman, Naturhistorisches Museum, Vienna, Austria; Annie V. J. Dhondt, Koninklijk Belgisch Instituut voor Natuurwetenschappen, Brussels, Belgium; and Norman F. Sohl, U.S. Geological Survey. Leo J. Hickey, U.S. National Museum, provided insight into early Tertiary paleoenvironments and paleobotany. John A. Dorr, Jr., and Donald B. Macurda graciously loaned fossil nonmarine Mollusca from the Museum of Paleontology, University of Michigan. Specimens were photographed by Vic Krantz, U.S. National Museum, and Jiří Kříž. Emmett Evanoff, Michael Elwood, and Steven Sloan served as field assistants during this investigation.

This research represents a portion of my Ph.D. dissertation in the Department of Geology, University of Wyoming. Fieldwork was supported in part by an NDEA Title IV Fellowship. A portion of this research was supported by a Smithsonian Institution Predoctoral Fellowship.

The manuscript was typed by Lora Johnson, Mary Lou R. Hanley, and Carol Barnhard. Aurèle LaRocque, Robert W. Scott, and Norman F. Sohl provided numerous excellent suggestions for its improvement.

REFERENCES

Baker, H. B. 1958. Land snail dispersal. Nautilus, 71:141-148.
Bambach, R. K., and K. R. Walker. 1971. Trophic relationships: an approach for analyzing the structure of fossil benthonic communities. Geol. Soc. Amer. Abst., 3:292-293.
Bradley, W. H. 1948. Limnology and the Eocene lakes of the Rocky Mountain region. Geol. Soc. Amer. Bull., 59:635-648.
——. 1966. Tropical lakes, copropel, and oil shale. Geol. Soc. Amer. Bull., 77:1333-1337.
——. 1973. Oil shale formed in a desert environment: Green River Formation, Wyoming. Geol. Soc. Amer. Bull., 84:1121-1124.
Burch, J. B. 1955. Some ecological factors of the soil affecting the distribution and abundance of land snails in eastern Virginia. Nautilus, 69:62-69.
Cockerell, T. D. A. 1914. Tertiary Mollusca from New Mexico and Wyoming. Amer. Museum Nat. Hist. Bull., 33:101-107.
——. 1915. New species of *Unio* from the Tertiary rocks of Wyoming. Amer. Museum Nat. Hist. Bull., 34:121-126.
——, and J. Henderson. 1912. Mollusca from the Tertiary strata of the west. Amer. Museum Nat. Hist. Bull., 31:229-234.
Cottam, G., J. T. Curtis, and B. W. Hale. 1953. Some sampling characteristics of a population of randomly dispersed invididuals. Ecology, 35:741-757.
Dorr, J. A., Jr. 1969. Mammalian and other fossils, early Eocene Pass Peak Formation, central western Wyoming. Univ. Mich. Museum Paleont. Contrib., 22:207-219.
Eagar, R. M. C. 1948. Variation in shape of shell with respect to ecological station. A review dealing with recent Unionidae and certain species of the Anthracosiidae in Upper Carboniferous times. Trans. Roy. Soc. Edin., 63:130-147.
Eugster, H. P., and R. C. Surdam. 1973. Depositional environment of the Green River Formation of Wyoming: a preliminary report. Geol. Soc. Amer. Bull., 84:1115-1120.
Fagerstrom, J. A. 1964. Fossil communities in paleoecology: their recognition and significance. Geol. Soc. Amer. Bull., 75:1197-1216.
Fernando, C. H. 1954. The possible dispersal of *Pisidium* by Corixidae (Hemiptera). Jour. Conch., 24:17-19.
Hanley, J. H. 1974. Systematics, paleoecology, and biostratigraphy of nonmarine Mollusca from the Green River and Wasatch formations (Eocene), southwestern Wyoming and northwestern Colorado. Unpublished Ph.D. dissertation, University of Wyoming, Laramie, Wyo., 285p.
Hibbard, C. W., and D. W. Taylor. 1960. Two late Pleistocene faunas from southwestern Kansas. Univ. Mich. Museum Paleont. Contrib., 16:1-223.
Johnson, R. G. 1960. Models and methods for analysis of the mode of formation of fossil assemblages. Geol. Soc. Amer. Bull., 71:1075-1086.
Kaesler, R. L. 1966. Quantitative re-evaluation of ecology and distribution of

Recent Foraminifera and Ostracoda of Todos Santos Bay, Baja California, Mexico. Kansas Univ. Paleont. Contrib. Paper 10, 50p.

Kauffman, E. G. 1974. Cretaceous assemblages, communities, and associations. Western Interior United States and Caribbean islands. Principles of benthic community analysis. Sedimenta IV, Comp. Sed. Lab., Univ. Miami, 12.1–12.27.

Kenk, R. 1949. The animal life of temporary and permanent ponds in southern Michigan. Univ. Mich. Museum Zool. Misc. Publ., 71:1–66.

LaRocque, A. 1960. Molluscan faunas of the Flagstaff Formation of central Utah. Geol. Soc. Amer. Mem., 78, 100p.

Love, J. D. 1961. Definition of Green River, Great Divide, and Washakie Basins, southwestern Wyoming. Amer. Assoc. Petrol. Geol. Bull., 45:1749–1755.

McDonald, R. E. 1972. Eocene and Paleocene rocks of the southern and central basins. In W. W. Mallory (ed.), Geologic Atlas of the Rocky Mountain Region. Rocky Mtn. Assoc. Geol., p. 243–256.

McKenna, M. C., P. Robinson, and D. W. Taylor. 1962. Notes on Eocene Mammalia and Mollusca from Tabernacle Butte, Wyoming. Amer. Museum Nat. Hist. Novitates, 2102:1–33.

——. D. E. Russell, R. M. West, C. C. Black, W. D. Turnbull, M. R., Dawson, and J. A. Lillegraven. 1973. K/Ar recalibration of Eocene North American land-mammal "ages" and European ages. Geol. Soc. Amer. Abs. with Programs, 5:733.

McMichael, D. F., and I. D. Hiscock. 1958. A monograph of the freshwater mussels (Mollusca: Pelecypoda) of the Australian Region. Australian Jour. Marine Freshwater Res., 9:372–508.

Meek, F. B. 1860. Descriptions of new fossil remains collected in Nebraska and Utah, by the exploring expeditions under the command of Capt. J. H. Simpson of the U.S. Topographical Engineers. Philadelphia Acad. Nat. Sci. Proc., 12:308–315.

——. 1872. Preliminary paleontological report, consisting of lists and descriptions of fossils, with remarks on the ages of the rocks in which they were found. U.S. Geol. Geogr. Surv. Territ. (Hayden Surv.) 6th Ann. Rept., p. 431–541.

——. 1876. A report on the invertebrate Cretaceous and Tertiary fossils of the upper Missouri country. U.S. Geol. Geogr. Surv. Territ. (Hayden Surv.) 9th Ann. Rept., p. 1–629.

Moore, R. G. 1960. A Paleocene fauna from the Hoback Formation, Wyoming. Unpublished Ph.D. dissertation, University of Michigan, Ann Arbor, Michigan, 160p.

Mozley, A. 1954. An introduction to molluscan ecology, distribution and population studies of fresh-water molluscs. H. K. Lewis and Co. Ltd., London, 71p.

Oriel, S. S. 1961. Tongues of the Wasatch and Green River formations, Fort Hill area, Wyoming. U.S. Geol. Surv. Profess. Paper, 424-B:B151–B152.

Picard, M. D., and L. R. High, Jr. 1972. Criteria for recognizing lacustrine rocks. In J. K. Rigby, and W. K. Hamblin (eds.). Recognition of Ancient Sedimentary Environments. Soc. Econ. Paleont. Mineral. Spec. Publ. 16:108–145.

Rees, W. J. 1965. The aerial dispersal of Mollusca. Proc. Malacol. Soc. London, 36:269–282.

Roehler, H. W. 1965. Early Tertiary depositional environments in the Rock Springs uplift area. Wyo. Geol. Assoc. Guidebook 19th Ann. Field Conf. p. 140–150.

——. 1972. A review of Eocene stratigraphy in the Washakie Basin, Wyoming. In Tertiary biostratigraphy of southern and western Wyoming. Adelphi Univ. Field Conf. Guidebook, p. 3–19.

———. 1973. Stratigraphic divisions and geologic history of the Laney Member of the Green River Formation in the Washakie Basin in southwestern Wyoming. U.S. Geol. Surv. Bull., 1372-E:E1-E28.

Russell, L. S. 1952. Molluscan fauna of the Kishenehn Formation, southeastern British Columbia. Natl. Museum Can. Bull., 126:120-141.

———. 1955. Additions to the molluscan fauna of the Kishenehn Formation, southeastern British Columbia and adjacent Montana. Nat. Museum Can. Bull., 136:102-119.

———. 1957. Mollusca from the Tertiary of Princeton, British Columbia. Ann. Rept. Natl. Museum Can. Bull., 147:84-95.

Scott, R. W. 1970. Paleoecology and paleontology of the Lower Cretaceous Kiowa Formation, Kansas. Univ. Kansas Paleont. Contrib. Art. 52 (Cretaceous 1), 94p.

Surdam, R. C., and C. A. Wolfbauer. 1973. Origin of oil shale in the Green River Formation, Wyoming. Wyo. Geol. Assoc. Guidebook, 25th Ann. Field Conf., p. 207-208.

Taylor, D. W. 1960. Late Cenozoic molluscan faunas from the High Plains. U.S. Geol. Sur. Profess. Paper 337, 94p.

———. 1966. Summary of North American Blancan nonmarine mollusks. Malacologia, 4:1-172.

Tozer, E. T. 1956. Uppermost Cretaceous and Paleocene non-marine molluscan faunas of western Alberta. Geol. Surv. Can. Mem., 280, 125p.

Walker, K. R. 1972. Trophic analysis: a method for studying the function of ancient communities. Jour. Paleont., 46:82-93.

———, and R. K. Bambach. 1974. Analysis of communities. Principles of benthic community analysis. Sedimenta IV, Comp. Sed. Lab., Univ. Miami, 2.1-2.20.

Warme, J. E. 1969. Live and dead molluscs in a coastal lagoon. Jour. Paleont., 43:141-150.

White, C. A. 1876. Descriptions of new species of invertebrate fossils from strata of the Carboniferous, Jurassic, Cretaceous, and Tertiary Periods. *In* J. W. Powell. Report on the geology of the eastern portion of the Uinta Mountains and a region of country adjacent thereto. U.S. Geol. Geogr. Surv. Territ., 218p.

———. 1878. Contributions to Invertebrate Paleontology No. 3: certain Tertiary Mollusca from Colorado, Utah, and Wyoming. U.S. Geol. Geogr. Surv. Territ. (F. V. Hayden) 12th Ann. Rept., 1:41-48.

———. 1880. Descriptions of new invertebrate fossils from the Mesozoic and Cenozoic rocks of Arkansas, Wyoming, Colorado, and Utah. U.S. Natl. Museum Proc., 3:157-162.

———. 1883. A review of the nonmarine fossil Mollusca of North America. U.S. Geol. Surv. 3rd Ann. Rept. (1881-1882). p. 403-550.

———. 1886. On the relation of the Laramie molluscan fauna to that of the succeeding freshwater Eocene and other groups. U.S. Geol. Surv. Bull., 34, 32p.

Wood, H. E., R. W. Chaney, J. Clark, E. H. Colbert, G. L. Jepsen, J. B. Reeside, Jr., and C. Stock. 1941. Nomenclature and correlation of the North American continental Tertiary. Geol. Soc. Amer. Bull., 52:1-48.

Yen, T. C. 1946. Eocene nonmarine gastropods from Hot Spring County, Wyoming. Jour. Paleon., 20:495-500.

———. 1948a. Eocene fresh-water Mollusca from Wyoming. Jour. Paleon., 22:634-640.

———. 1948b. Paleocene freshwater mollusks from southern Montana. U.S. Geol. Surv. Profess. Paper 214-C:35-50.

Depositional Systems and Marine Benthic Communities in the Floyd Shale, Upper Mississippian, Northwest Georgia

Thomas W. Broadhead   University of Iowa

ABSTRACT

Fossil marine communities have been classified by many workers in on-shore to offshore transitional sequences on marine shelves. This representation does not accommodate differences in community structure and composition parallel to the paleo-shoreline. Works of Craig (1955) and Ferguson (1962), used by Bretsky (1969) to generalize community structure in the Mississippian, may not be typical because each studied only one exposure, providing a view of a transgressive sequence at one point. Lack of work on a larger scale in Mississippian rocks resulted in a gap in Bretsky's (1969) chart in the "offshore" communities.

Five marine benthic communities identified from the Upper Mississippian Floyd Shale in northwest Georgia have been correlated with depositional systems and were used to interpret the depositional history of the Floyd Shale. Communities include (1) *Lingula*, (2) Bivalvia–Spiriferida–Productidina, (3) Fenestellidae, (4) *Pentremites*-Spiriferida–Fenestellidae, and (5) *Michelinia*-Rugosa. Community 1 occurs in siltstone and silty shale and is probably directly controlled by factors related to delta progradation, especially the distribution of delta front, prodelta, and interdistributary embayment facies. Communities 2 and 3 occur in shale and calcareous shale interpreted as semi- to well-protected bay facies developed along strike from principal delta lobes. Community 4 occurs in calcareous shale and calcilutite representing open bay or shelf deposits, which locally became well developed shoreward during destructional phases of delta lobes. Community 5 occurs in calcilutite to calcarenite units interpreted as carbonate banks.

INTRODUCTION

Communities of marine benthic invertebrates observed in the fossil record have received considerable attention during the last several years. Emphasis has been placed on description of communities and their variations in time and space, and integration of paleoecologic data previously derived from observations of communal taxa. The objective has been a meaningful reconstruction of many aspects of community dynamics.

Study of community paleoecology implies evaluation of both biotic relationships and physical factors controlling biotic distribution. In the Holocene marine environment, thorough evaluation of these factors by marine ecologists is time consuming and costly. Attempts by paleontologists and geologists to analyze communities of organisms are probably superficial in comparison with true ecosystem analysis because of incomplete data. Studies of invertebrate communities in the Paleozoic are commonly beset by deficiencies such as nonpreservation of much of the marine flora and soft-bodied fauna, or difficulties in recovery of many planktic organisms. Trophic levels represented in the preserved biota may not be obvious, and food-chain relationships commonly afford only speculation. In addition, evidence of physical environmental factors may be lacking. Precise magnitudes and short temporal variations of such physical parameters as temperature, salinity, and sedimentation rate are not detectable by present analytic techniques.

Most evaluations of the physical environment in community studies have been restricted to simple description of lithologies, but with little correlation of sediment patterns over large outcrop areas. A further refinement is the recognition of gross stratigraphic patterns, such as transgressive and regressive marine sequences accompanied by faunal changes and identified on the basis of changes in lithology. The association of groups of organisms with recent sedimentary environments has been recognized at least since the time of Parker's studies (1956, 1959) along the Gulf Coast, but specific environments correlated with fossil communities have not been emphasized until recently with the works of West (1972, Pennsylvanian of Oklahoma) and Bowen et al. (1974, Devonian of New York).

Various aspects of sedimentary environments and depositional systems are excellently documented in Holocene examples. Furthermore, many of these features have been recognized in rocks ranging in age from Precambrian to Pleistocene. Numerous studies of Upper Paleozoic rocks (e.g., Brown, 1969; Galloway and Brown, 1972; Brown et al., 1973) provide excellent guidelines and data for comparison and identification of such phenomena in other areas. Depositional systems of terrigenous clastic sediments provide a variety of physical environments, which vary greatly in ability to support a preservable invertebrate community. The Floyd Shale is an approximately 400-m sequence of Upper Mississippian terrigenous clastic and carbonate rocks that crop out in the Valley and Ridge area of northwestern Georgia and southeastern Tennessee, and in the Appalachian Plateau of northern Alabama. In Georgia, the Floyd Shale ranges in age from middle Meramecan through middle Chesteran and is the shoreward, clastic facies of the Tuscumbia Limestone and Monteagle Formation that crop out to the northwest on the Appalachian Plateau (Fig. 1). Lithologies in the terrigenous facies of the Floyd Shale range from quartz sandstone to clay shale. In the carbonate facies, lithologies range from calcereous shale and calcilutite to coarsely crystalline and skeletal calcarenite.

DEPOSITIONAL ENVIRONMENT

The greater thickness of the Floyd Shale and Hartselle Sandstone in the southern and eastern part of the outcrop belt (Cressler, 1970; McLemore, 1972) suggests that the sediment source was probably a land area to the east and south of the marine basin. Sediments were transported fluvially to the marine environment where deposition resulted in the irregular progradation of the shoreline. This, according to Fisher et al. (1969, p. 14), is the basic definition of a delta.

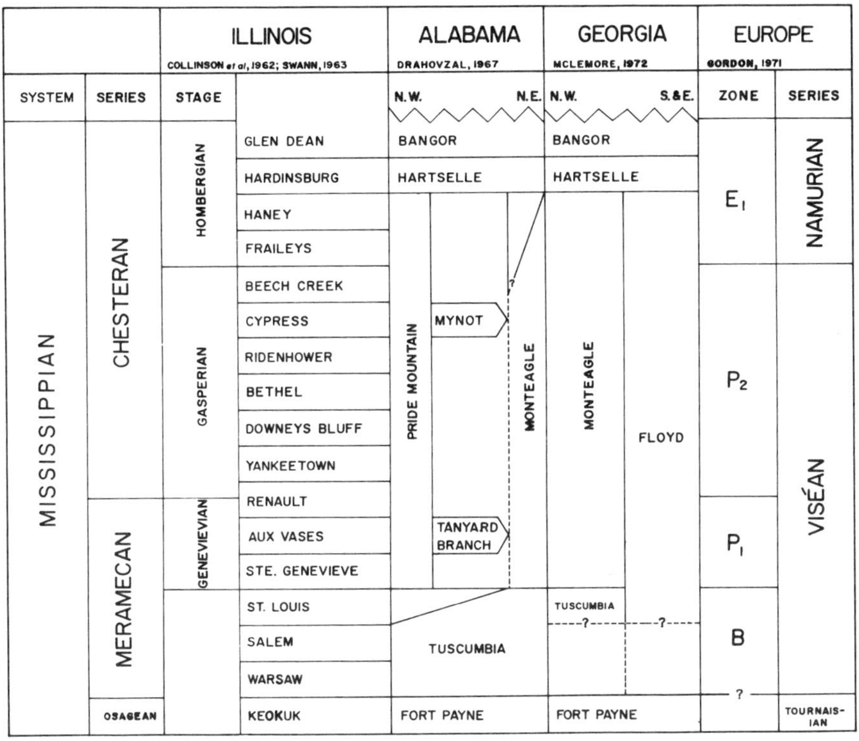

			ILLINOIS COLLINSON et al, 1962; SWANN, 1963	ALABAMA DRAHOVZAL, 1967		GEORGIA MCLEMORE, 1972		EUROPE GORDON, 1971	
SYSTEM	SERIES	STAGE		N.W.	N.E.	N.W.	S.&E.	ZONE	SERIES
MISSISSIPPIAN	CHESTERAN	HOMBERGIAN	GLEN DEAN	BANGOR		BANGOR		E_1	NAMURIAN
			HARDINSBURG	HARTSELLE		HARTSELLE			
			HANEY						
			FRAILEYS						
		GASPERIAN	BEECH CREEK	PRIDE MOUNTAIN	MYNOT	MONTEAGLE	MONTEAGLE	P_2	VISÉAN
			CYPRESS						
			RIDENHOWER						
			BETHEL						
			DOWNEYS BLUFF				FLOYD		
			YANKEETOWN						
	MERAMECAN	GENEVIEVIAN	RENAULT		TANYARD BRANCH			P_1	
			AUX VASES						
			STE. GENEVIEVE						
			ST. LOUIS			TUSCUMBIA		B	
			SALEM	TUSCUMBIA					
			WARSAW						
	OSAGEAN		KEOKUK	FORT PAYNE		FORT PAYNE		?	TOURNAISIAN

Figure 1
Regional correlation of the Floyd Shale in Georgia with rocks in Alabama and Illinois in the American and European time-stratigraphic framework.

The large volume of terrigenous clastic sediment that constitutes the Floyd Shale suggests that deltas responsible for Floyd deposition were of the "high-constructive" type, which develop under conditions of high sediment input relative to marine energy. Two basic types of high-constructive deltas (Fisher et al., 1969) are (1) high-constructive elongate, with sediment load high in mud, and sand facies prograding over relatively thick mud sequences, and (2) high-constructive lobate, which develop under similar conditions, but with relatively less mud load, and with sand facies prograding over thin mud sequences.

The basic component facies of a high-constructive delta during the constructional phase is summarized from Fisher et al. (1969, p. 15-19) as follows. At the base of the delta sequence and at the distal end of the facies tract is the prodelta facies, composed of clay and silt that settle from suspension. Overlying and geographically landward of the prodelta is the delta front, composed primarily of sand deposited by rivers flowing into the marine basin. The proximal or landward part of the delta front consists of a series of distributary channels, distributary mouth bars, and distal bars. The distal or seaward part of the delta front consists of marine reworked sheet sands that have been "spilled out" from the distributaries onto the proximal prodelta. With progradation, the vertical delta sequence coarsens upward in grain size from the clay and silt of the prodelta into the fine and medium sand of the delta front. Landward from the delta front, and overlying it in the vertical sequence, is the commonly extensive delta plain facies. The delta plain is mostly subaerially exposed and includes several component facies, including mud and organic-rich subaerial levees, marshes, and lakes and sand-filled distributary channels.

During the destruction of abandoned deltas the rate of sedimentation is drastically reduced, and the dominant processes affecting sediments are marine. Fisher et al. (1969, p. 19) reported that on abandonment "the prodelta area reverts to a normal shelf facies, marked by a much slower rate of deposition, and inhabited by a larger number of marine organisms." Delta-front sands that accumulated in shallow areas become extensively reworked by waves and other marine processes.

In Georgia, the Floyd Shale is overlain by the Hartselle Sandstone (Hartselle Sandstone Member of the Floyd Shale of Cressler, 1970). The gradational contact of the Floyd and Hartselle strongly suggests a genetic relationship between the two formations based on the model described above. In the western part of Floyd County at Judy Mountain, Cressler (1970, p. 48) reported the Hartselle to be about 300 f (91 m) of massively bedded very fine- to medium-grained sandstone and quartzite, siltstone, and quartz-pebble conglomerate, with siltstone common near the base. Cressler's description of the Hartselle, in addition to the presence of small channel sand bodies and thin "sheet sands" farther to the north, suggests that the Hartselle is the delta-front facies of Floyd deltas.

Ferm and Ehrlich (1967, p. 13) considered the Floyd Shale in Alabama to represent delta-front and prodelta facies. McLemore (1972, p. 73) agreed with Ferm and Ehrlich, but stated that in Georgia

> the great thickness and lack of shallow water sedimentary structures (ripple marks, mud cracks, etc.) seem to indicate that the Floyd was

deposited rather rapidly in deeper waters. The carbonaceous material and paucity of fossils appear to indicate that the general depositional environment was unfavorable to benthonic life and probably reducing.

McLemore's analysis seems to confirm Ferm and Ehrlich's interpretation. However, the absence of shallow-water sedimentary structures in most Floyd outcrops does not necessarily preclude deposition in shallow water. Bioturbation during periods of low sedimentation may disrupt ripple marks in the delta front and small cross beds that may form in the proximal prodelta facies (Fisher et al., 1969, p. 18). Ripple marks were observed at one locality, but mud cracks should not be expected in the delta front or prodelta because these facies are not subjected to subaerial exposure.

Abundant carbonaceous debris and low faunal abundance and diversity are characteristic of prodelta mudstones, as has been well documented from Pennsylvanian rocks of north-central Texas (Galloway and Brown, 1972; Brown et al., 1973). At many localities the Floyd is dark gray to black shale and siltstone that are unfossiliferous or contain only a few broken fragments of echinoderms and articulate brachiopods or locally abundant linguloid brachiopods. Reducing conditions commonly associated with prodelta facies are indicated in the Floyd by locally common siderite and pyrite nodules and limonite boxwork probably derived from the weathering of iron sulfides. An important aspect of several of the siderite–pyrite nodules is the abundance of goniatitic ammonoids and terrestrial plant fragments in excellent states of preservation. This association is similar to that in the Fayetteville Shale of Arkansas (Zangerl, 1971) in which siderite nodules apparently nucleated about decaying organic debris. Most fossils in nodules from the Floyd were highly motile or easily transported types; very few benthic fossils were observed.

The Floyd Shale is locally fossiliferous (124 of 392 localities studied), with many of the most diverse faunas occurring in the included carbonate rocks. Nevertheless, fossiliferous occurrences in the clastic facies commonly contain distinctive faunas that appear to correlate to a large degree with sedimentary environment. It is unlikely, based on the models mentioned above, that these zones are directly related to deltaic progradation. Rather, they may represent interdeltaic, lagoonal, shelf, or carbonate bank deposits similar to those in the Pennsylvanian of north-central Texas (Brown, 1969; Galloway and Brown, 1972; Brown et al., 1973). Toward the north and west, particularly in Catoosa County, Georgia, great thicknesses of the equivalent Tuscumbia and Monteagle limestones may represent shelf and bank carbonate rocks deposited in areas away from terrigenous influx. The local abundance of tabulate and rugose corals in this area suggests that water depth possibly did not exceed 50 m (Wells, 1957, p. 774).

The lack of extensive sandstone units in the Floyd Shale in Georgia makes interpretation of the genesis and depositional environment of the formation difficult, particularly if the study is restricted to the shale unit. Cressler's description of the overlying Hartselle Sandstone (1970, p. 48) suggested that the Floyd and Hartselle constitute a coarsening upward sequence. This strongly indicates that the Floyd, in part, represents a prodelta facies and that, at least in part, the Hartselle represents a delta-front facies. Rocks in or above the Hartselle that may be construed as delta plain or distributary facies probably once existed to the southeast along a facies tract typical of delta systems, but have not been recognized in Georgia, because of removal by erosion.

The general depositional history of the Floyd Shale and associated contemporaneous facies in Georgia may be summarized as follows (Fig. 2): delta progradation began initially during the Meramec, contemporaneous with deposition of the Tuscumbia Limestone on a shelf area to the northwest. During the late Meramec and early Chester, deltas, temporarily abandoned, were destroyed by marine processes and were replaced by offshore environments of the transgressive Monteagle Formation. Progradation in the area began again during the early Chester and continued through the middle Chester when Floyd deltas apparently reached their maximum geographic extent. Following abandonment, destructional phases occurred near the end of the middle Chester and continued into the late Chester, during which time delta-front Hartselle sands were reworked and the Bangor Limestone transgressed the foundering delta system.

Marine organisms living in this regime were greatly affected by changes in depositional environment. Areas of active sedimentation such as the delta-front and prodelta facies were unsuitable for many benthic organisms. During delta progradation, however, embayment and offshore shelf environments supported large and diverse faunas. Marine transgressions during delta destruction resulted in colonization of new areas by benthic organisms.

BENTHIC COMMUNITIES IN THE FLOYD SHALE

Fossil assemblages studied in the Floyd Shale comprise five major types identified on the basis of presence and relative abundance of component taxa. Many species and genera occur in more than one assemblage type, but their relative abundance and co-occurrence with other taxa facilitate identification of the five types. Overlapping distributions of genera and species in the assemblages of the Floyd Shale present a complex arrangement of biofacies. These biofacies have been combined primarily on the basis of assemblage characteristics into five types, which probably represent the preserved remains of marine benthic invertebrate communities. Diversity in the communities was evaluated at the generic level, and the lowest taxonomic level used to define these communities was the genus. Characterization of communities at the generic and higher taxonomic levels facilitates widespread geographic recognition; thus all five Floyd Shale communities may be regarded as "parallel communities" in the sense of Thorson (1957). The five communities are (1) *Lingula,* (2) Bivalvia-Spiriferida-Productidina, (3) Fenestellidae, (4) *Pentremites*-Spiriferida-Fenestellidae, and (5) *Michelinia*-Rugosa.

The inarticulate brachiopod, *Lingula,* in moderate to great abundance characterizes faunas of community 1. Associated rock types range from sandy siltstone to silty shale, which are commonly bioturbated and contain fragments of terrestrial

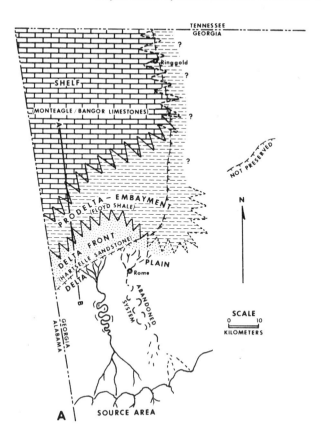

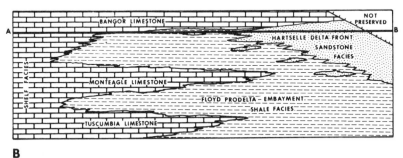

Figure 2
(A) Interpretation of middle Chesteran depositional environments in northwest Georgia showing the relationship of the Hartselle Sandstone and Floyd Shale during transgression of the Bangor Sea. (B) Diagrammatic (no scale) section of upper Mississippian rocks in Georgia. Line AB corresponds to position of facies shown in A. [Modified after Cressler (1970) and McLemore (1972).]

plants. This clastic-rich environment was probably developed as part of the prodelta mud or distal delta-front facies of Floyd deltas. Fluctuating salinity would have precluded stenohaline organisms because of the proximity to river outlets. Furthermore, the relatively high rate of sedimentation from settling clay and silt, although providing a source of food, would have resulted in burial of many sessile or nonburrowing animals. *Lingula* was well adapted to life in such an environment and may have lived in association with large numbers of worm-like organisms that inhabit similar Holocene environments. Orientation of most *Lingula* shells parallel to bedding suggests that currents or small turbidity flows from the delta front resulted in scouring.

The most common organisms in community 2 (Bivalvia-Spiriferida-Productidina) are the productid brachiopod *Inflatia* and several genera of bivalve mollusks, especially the nuculoid, *Phestia*, and the pterioid, *Aviculopecten*. Associated rock types range from silty shale to argillaceous limestone, with the more detrital-rich strata commonly containing fragments of terrestrial plants. This community is probably generally comparable to West's (1972) *Glabrocingulum* and transitional communities and is fairly common in the southeastern part of the outcrop area of the Floyd Shale. This community probably developed where abandoned delta lobes had foundered, and restriction of the embayments may have resulted from longshore transport of paralic sediments. Influx of freshwater into the embayment probably retarded colonization by more stenohaline organisms (e.g., corals, echinoderms, bryozoans), which are rare or absent at most localities. Also, a soft muddy substrate may have been unsuitable for larval attachment of many organisms, although shell debris may have commonly filled this requirement. Low sedimentation rate contributed to the growth of many sessile organisms. Although muds were reworked near the surface by burrowing bivalves, there is little indication of postmortem disturbance; few brachiopods are disarticulated, and many productid brachiopods are in inferred life position (Rudwick, 1970).

Community 3 (Fenestellidae) is dominated by fenestellid bryozoans and occurs in rocks ranging from slightly calcareous shale to siltstone. The community is named for the great abundance of fronds of the *"Fenestella"* type, most of which probably belonged to either *Fenestella* or *Archimedes*. This is the least common Floyd Shale community (six localities), and probably represents a sporadically developed environment on or marginal to the shelf, such as an embayment, distal prodelta, or mud bank facies. The assemblage in community 3 suggests that it was characterized by rapid colonization and growth of the fenestellids, which ultimately presented an effective nutrient filter that hampered the survival of lower-lying, suspension-feeding brachiopods. Accumulation of fronds on the bottom would have provided a good substrate for attachment of many larvae, but probably rendered the substrate unlivable for shallow-burrowing infauna.

Community 4 (*Pentremites*-Spiriferida-Fenestellidae) is the most generically diverse community in the Floyd Shale and includes a great variety of stenohaline organisms (e.g., echinoderms, corals, bryozoans), suggesting that it represents an environment with open marine circulation. The most common genera in this community are the blastoid, *Pentremites*, and the spiriferoid brachiopods, *Cleiothyridinia*,

Composita, Spirifer, and *Reticulariina,* which also occur in community 5, but with the exception of *Composita* and *Spirifer,* are not nearly as common in other communities. This community may be broadly similar to West's *Cleiothyridina* community, but differs largely in having fewer mollusks and more echinoderms and corals. Abundant articulated remains of pelmatozoan echinoderms, articulate brachiopods, and rare bivalves suggest little water agitation in this environment; it was probably largely developed below wave base.

Community 4 is the most widespread community in the Floyd Shale (48 localities) and characteristically occurs in rocks ranging from calcareous shale to calcilutite, which are commonly deeply weathered in outcrop, leaving a reddish-brown clay saprolite containing silicified fossils. The environment represented in community 4 was probably a vast carbonate- and mud-shelf facies over which Floyd delta lobes prograded and upon which carbonate banks developed from time to time. Circulation was probably excellent and sedimentation relatively slow, with an abundant food source in the plankton. At several localities, the faunas appear to be more nearly intermediate between community 4 and faunas of communities 2, 3, or 5, a circumstance that suggests local proximities of those associated environments to the open shelf.

Community 5 (*Michelinia*-Rugosa) commonly contains many of the faunal elements found in community 4, but diversity is more variable and generally lower. Where the rugose corals of community 4 are predominantly hapsiphyllids, those in community 5 characteristically include many dissepimented forms in addition to the hapsiphyllids. The tabulate, *Michelinia,* is relatively common in community 4 assemblages, but is more abundant in community 5. Other important communal organisms include blastoids (especially *Pentremites*), crinoids, and occasionally a wide variety of articulate brachiopods, including nearly all forms found in community 4. The most common associated rocks range from calcilutite to skeletal and oolitic calcarenite with weathering characteristics much like rocks containing community 4. Community 5 occurs in small carbonate buildups that are characteristic of carbonate-bank facies and commonly develop in clear, turbulent water. The wide range in diversity among faunas from various localities containing community 5 may represent differences in bathymetry relative to wave base, with the shallowest, most frequently agitated faunas characterized almost solely by tabulate and colonial rugose corals.

Because of the small thickness and areal extent of most Floyd Shale outcrops, transitions and changes from one community to another in one exposure are rare. A notable example, however, is the abrupt change from community 5 to 4 exposed in the quarry at Drag City near Ringgold, Georgia. In that quarry, about 25 m of dark gray, massively bedded oolitic and skeletal calcarenite containing abundant *Lithodrumus* sp., *Pentremites* sp., and unidentified inadunate crinoids is abruptly overlain by about 1 m of silty calcilutite to calcareous siltstone that contains numerous fragments of an arborescent lycophyte (aff. *Archaeosigillaria*). Above the plant-bearing beds, the rock grades upward into calcareous shale and calcilutite that contain abundant articulate brachiopods (*Composita, Cleiothyridina, Inflatia, Spirifer*), fenestellid bryozoans, hapsiphyllid corals, and a crinoid fauna dominated by

Agassizocrinus. The upper assemblage is characteristic of community 4 and developed in the area following the smothering of the community 5 fauna by the influx of the plant-bearing clastic sediment. Lack of Floyd exposures farther east in Catoosa and Whitfield counties procludes the possibility of identifying the paleoshore and possible source of the plants.

COMMUNITY DISTRIBUTION

Detailed analysis of community distribution within the Floyd Shale is particularly complicated by several factors: (1) complex geologic structure, (2) deep weathering, (3) time-transgressive lithic units (stratigraphic distribution) and (4) lack of detailed knowledge of the autecology of the organisms. The present geographic distribution of localities representing the five Floyd Shale communities is more simply presented (Fig. 3) and shows (1) the majority of localities containing communities 1 and 2 are in the south and east parts of the outcrop belt; (2) the widespread distribution of community 4, characterizing a broad, generalized environment; (3) the small distribution of community 3, suggesting a more restricted, sporadically developed environment; and (4) restricted occurrence of community 5 along the western part of the outcrop belt. Distribution of communities appears to be closely related to rock types that characterize particular sedimentary environments. This undoubtedly reflects the distribution of physical and biotic controls on the occurrence of communal organisms.

Distribution of many major taxa (Fig. 4) correlates with the bulk average inferred sediment type (Fig. 5). Communities 1 and 3 are characterized by low diversity, dominance by a single taxon (the genus *Lingula*, the family Fenestellidae) and trophic level (Fig. 5), and most commonly occur in rocks composed of a high percentage of terrigenous sediment. The association of more infaunal organisms with environments characterized by rapid sedimentation (e.g., the prodelta) or by probable changes in salinity (embayment) at least partly reflects some success in coping with unstable environmental factors. Grazing or deposit-feeding organisms in great abundance may suggest high levels of primary production or significant amounts of incoming nutrient-laden detritus. Several levels of epifaunal suspension feeders (e.g., community 4) are suggestive of stable environmental conditions and effective water circulation, but not necessarily of agitation.

Understanding of stratigraphic distribution of communities in the Floyd Shale has by necessity been based on biostratigraphic criteria where these are present in fossil assemblages. The three major transgressive phases associated with the Floyd Shale (Tuscumbia, Monteagle, and Bangor) are characterized by establishment of communities 4 and 5. Morphologic grades in *Pentremites* (i.e., progressive broadening and reduced convexity to increased concavity of ambulacra) and the occurrence of several species of crinoids and articulate brachiopods facilitate age identification of communities 4 and 5. Age determinations of community 2 are aided by the occurrence of some of the same brachiopod species found in community 4 and by the additional occurrence of ammonoid cephalopods (*Lyrogoniatites, Cravenoceras*) at many localities. Determination of the stratigraphic position of localities containing

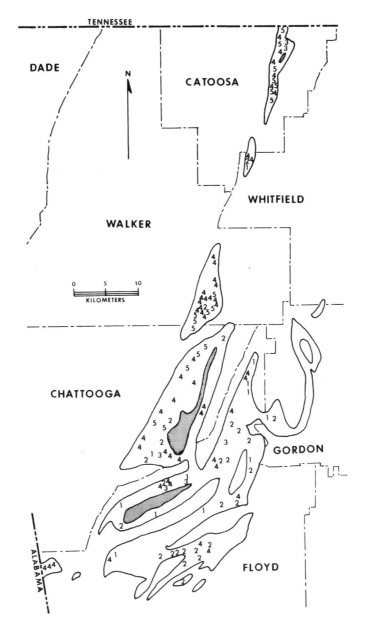

Figure 3
Distribution of localities containing benthic invertebrate communities in the Floyd Shale of Georgia. Numbers represent communities: 1, *Lingula*; 2, Bivalvia-Spiriferida-Productidina; 3, Fenestellidae; 4, *Pentremites*-Spiriferida-Fenestellidae; 5, *Michelinia*-Rugosa. Stippled areas indicate younger Paleozoic rocks.

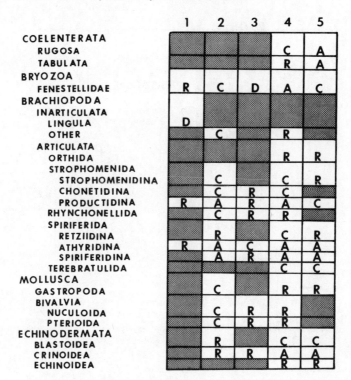

Figure 4
Distribution of important benthic invertebrate groups among Floyd Shale Communities: D, dominant, greater than 50 percent of specimens; A, abundant, 10 to 30 percent of specimens; C, common, 2 to 10 percent of specimens; R, rare, less than 2 percent (commonly one or two specimens); shaded areas, nonoccurrence.

communities 1 and 3 is complicated by the rarity or absence of any genus or species that could be used to correlate them with localities containing the other communities. Diagnostic articulate brachiopods are fairly common at a few localities containing community 3, but community 1 assemblages rarely contain fossils other than *Lingula* and are all but impossible to place in the stratigraphic framework. Thus, the occurrence of different communities in close geographic proximity is at least partly due to differences in stratigraphic position complicated by folding, faulting, and extreme weathering.

ENVIRONMENTAL INTERPRETATIONS

Generalized evolutionary and environmental patterns affecting Paleozoic communities of marine benthic invertebrates have been outlined and charted by Bretsky (1968, 1969). His arrangement depicted three major divisions of a generalized onshore to offshore transitional sequence on the marine shelf. Each division is char-

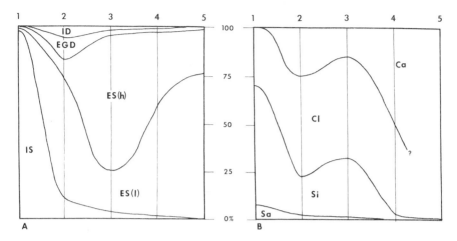

Figure 5
Gross average percentages of individuals in (A) trophic categories: IS, infaunal suspension-feeding; ID, infaunal deposit-feeding; EGD, epifaunal grazing and deposit-feeding; ES(h), epifaunal suspension-feeding (high-level); ES(1) epifaunal suspension-feeding (Low-level); and (B) inferred sediment type: Sa, sand; Si, silt; Cl, clay; Ca, carbonate. Data based on inferred trophic relationships and field description of lithotypes.

acterized by at least one "association" of communities identified from Paleozoic rocks either by Bretsky or by other workers.

For the Mississippian, Bretsky relied on the work of Craig (1955) and Ferguson (1962), who studied the Visean of Britain. Craig examined approximately 0.6 m of a calcareous shale from a single locality, and collected and analyzed about 5,000 fossils of which 40 percent were macrofossils. From this he identified two communities in the sense of Petersen (1913); (1) *Lingula squamiformis-Nuculopsis gibbosa-Sanguinolites costellatus* community from an inferred nearshore, rough-water environment, and (2) *Posidonia corrugata*-arenaceous foraminifer-*Waylandella cuneola* (ostracode) community from a slightly farther offshore environment. Bretsky has appropriately placed both of these communities in his Linguloid-Molluscan association.

Ferguson's study included one outcrop of a 2.8-m-thick shale unit representing a "marine transgression" (1962, p. 1090). Four topozones identified (from the base of the unit upward) are (1) *Lingula squamiformis-Streblopteria oranta*, (2) *L. squamiformis-Crurithyris urei*, (3) *Schizophoria resupinata-Eomarginifera longispina*, and (4) *E. longispina*-corals-bryozoans. Bretsky combined topozones 1 and 2 into the *Lingula-Nuculopsis* community of the Linguloid-Molluscan association, and 3 and 4 into the *Schizophoria-Eomarginifera* community of the Productid-Chonetid association.

Inherent problems exist from the use of these studies to summarize benthic community structure in Mississippian strata. Not the least of these problems is that each study was done on one exposure, which reduces the facility of interpretation of geographic distribution and variation within the communities. Second, the small thickness of rock studied by Craig and Ferguson may represent a time interval too brief and physical conditions too transitional to permit the establishment of stable communities. Craig's study included only the "nearshore" communities, but Ferguson's unit grades upward into a limestone with a "*Lithostrotion* reef" (1962, p. 1106). Nevertheless, there is no community that Bretsky felt should be placed in the "offshore" community associations, which leaves a gap between the top of the Atrypid-Bryozoan association of the Lower and Middle Paleozoic and the Fusulinid-Paleotextulariid association of the Upper Paleozoic (Pennsylvanian and Permian).

Recent work by West (1972) on the Pennsylvanian of Oklahoma and by Watkins (1973) on the Mississippian and Pennsylvanian of California have augmented knowledge of stratigraphic and geographic distribution of Carboniferous faunal associations and communities. West's (1972) identification of depositional environments containing his communities greatly enhances the environmental significance of community distribution. Watkins (1973) carefully analyzed the petrology of rocks containing his associations and communities, but did not postulate specific depositional environments.

Floyd Shale communities can be assigned to the generalized categories proposed by Bretsky. Community 1 would correspond to the Linguloid-Molluscan association, filling the nearshore environment. Community 2, although it contains a diverse molluscan fauna, could be placed in the Productid-Chonetid association if its inferred environmental position is disregarded. Community 4 corresponds to the offshore open shelf environment, filling the gap in Bretsky's chart. Communities 3 and 5 developed in specialized environments that probably lacked broad geographic continuity and need not be considered in the simplified onshore sequence.

An attempt has been made by Anderson (1971) to place more emphasis on tectonic factors related to depositional environment. His models included:

> (1) . . . low slope epeiric seas (in the order of one foot per mile) where waves impinge on the substrate and are dissipated some distance offshore producing a low energy subtidal zone onshore. (2) . . . epeiric seas where the zone of maximum wave and current agitation is at or near the shoreline. Such condition is associated with shores which are building out or prograding or with shores of higher slope (in the order of five to ten feet per mile). Progradation tends to fill low-energy onshore zones and steeper slopes do not leave room for them to develop as separate recognizable entities onshore from the zone of wave dissipation (p. 296).

Anderson correlated the number of recognizable communities with the type of model, there being more communities with the more stable model 1. In short, he emphasized water depth and kinetics relative to wave energy as a function of depth, maintaining the concept of onshore to offshore transitions. The addition of wave energy and depth as variables may still be insufficient to describe community distributions, even in a general overview.

The nonuniform influx of sediment into a marine basin effectively creates barriers between previously existing areas of similar environment and produces areas of variable geographic and temporal span with a wide range of physical environments. Parker's (1956) correlation of macrofauna and sedimentary environments in the Holocene Mississippi River delta area is an excellent example, which shows that no straight line can be drawn from the "onshore" to "offshore" areas that will include all environments. Thus, the juxtaposition of communities in a "normal" onshore to offshore sequence may be precluded, and any very small area studied may result in the sampling of nontypical assemblages. The "missing communities" in Ferguson's work might be explained in terms of depositional environment and lateral distribution of controlling facies, and may be expected to occur either farther offshore or in a lateral position, depending on local conditions of sedimentation.

ACKNOWLEDGMENTS

This paper is based on a part of the author's master's thesis at The University of Texas at Austin. I wish to thank James Sprinkle, L. F. Brown, Jr., and Keith Young of The University of Texas for advice and supervision of the thesis project. Brian F. Glenister and Philip H. Heckel of The University of Iowa provided criticism and suggestions in the preparations of this paper. Fieldwork was supported by a grant from the F. L. Whitney Fund of the Geology Foundation, The University of Texas at Austin, and maps were contributed by the Georgia Geological Survey.

REFERENCES

Anderson, E. J. 1971. Environmental models for Paleozoic communities. Lethaia, 4:287-302.
Bowen, Z. P., D. C. Rhoads, and A. L. McAlester. 1974. Marine benthic communities in the Upper Devonian of New York. Lethaia, 7:93-120.
Bretsky, P. W. 1968. Evolution of Paleozoic marine invertebrate communities. Science 159:1231-1233.
———. 1969. Evolution of Paleozoic benthic marine communities. Palaeogeog. Palaeoclimat., Palaeoecol., 6:45-59.
Brown, L. F. 1969. Geometry and distribution of fluvial and deltaic sandstones (Pennsylvanian and Permian), north-central Texas. Univ. Texas Bur. Econ. Geol. Circ. 69-4, 47p.
———, A. W. Cleaves, and A. W. Erxleben. 1973. Pennsylvanian depositional systems in north-central Texas: a guide for interpreting terrigenous clastic facies in a cratonic basin. Univ. Texas Bur. Econ. Geol. Guidebook 14, 122p.
Collinson, C., A. J. Scott, and C. B. Rexroad. 1962. Six charts showing biostratigraphic zones and correlations based on conodonts from the Devonian and

Mississippian rocks of the upper Mississippi Valey. Ill. State Geol. Surv. Circ. 328, 32p.

Craig, G. Y. 1955. The palaeoecology of the Top Hosie Shale (Lower Carboniferous) at a locality near Kilsyth. Quart. Jour. Geol. Soc. London, 110:103-119.

Cressler, C. W. 1970. Geology and ground-water resources of Floyd and Polk Counties, Georgia. Georgia Geol. Surv. I. C. 39, 95p.

Drahovzal, J. A. 1967. The biostratigraphy of Mississippian rocks in the Tennessee Valley. *In* W. E. Smith (ed.), A field guide to Mississippian sediments in northern Alabama and south-central Tennessee. 5th Annual Fieldtrip Guidebook, Alabama Geological Society, University of Alabama, p. 10-24.

Ferguson, L. 1962. The paleoecology of a Lower Carboniferous marine transgression. Jour. Paleont., 36:1090-1107.

Ferm, J. C., and R. Ehrlich. 1967. Petrology and stratigraphy of the Alabama coal fields. *In* A field guide to Carboniferous detrital rocks in northern Alabama. Geological Society of America, Coal Division. Guidebook, 1967 Field trip, Alabama Geological Society, University of Alabama, p. 11-15.

Fisher, W. L., L. F. Brown, A. J. Scott, and J. H. McGowen. 1969. Delta systems in the exploration for oil and gas. Univ. Texas Bur. Econ. Geol. Colloq. 75p.

Galloway, W. E., and L. F. Brown. 1972. Depositional systems and shelf-slope relationships in Upper Pennsylvanian rocks, north-central Texas. Univ. Texas Bur. Econ. Geol. I. C. 75, 62p.

Gordon, M. 1971. Carboniferous ammonoid zones of the south-central and western United States. 6th Congr. Intern. Strat. Geol. Carb. 2:817-826.

McLemore, W. H. 1972. Depositional environments of the Tuscumbia-Monteagle-Floyd interval in northwest Georgia and southeast Tennessee. Georgia Geol. Soc. Guidebook 11:69-73.

Parker, R. H. 1956. Macro-invertebrate assemblages as indicators of sedimentary environments in east Mississippi delta region. Amer. Assoc. Petrol. Geol. Bull., 40:295-376.

———. 1959. Macro-invertebrate assemblages of central Texas coastal bays and Laguna Madre. Amer. Assoc. Petrol. Geol. Bull., 43:2100-2166.

Petersen, C. G. J. 1913. Valuation of the sea. II. Animal communities of the sea-bottom and their importance for marine zoogeography. Rept. Danish Biol. Sta. 21: 44p.

Rudwick, M. J. S. 1970. Living and fossil brachiopods. Hutchinson Publishing Group Ltd., London, 199p.

Swann, D. H. 1963. Classification of Genevievian and Chesterian (Late Mississippian) rocks of Illinois. Ill. State Geol. Surv. Rept. Inv. 216, 91p.

Thorson, G. 1957. Bottom communities. *In* J. W. Hedgpeth (ed.), Treatise on marine ecology and paleoecology, v. 1, Marine ecology. Geol. Soc. Amer. Mem., 67:461-534.

Watkins, R. W. 1973. Carboniferous faunal associations and stratigraphy, Shasta County, Northern California. Amer. Assoc. Petrol. Geol. Bull., 57:1743-1764.

Wells, J. W. 1957. Corals. *In* J. W. Hedgpeth (ed.), Treatise on marine ecology and paleoecology, v. 2, Paleoecology. Geol. Soc. Amer. Mem., 67:773-782.

West, R. R. 1972. Relationship between community analysis and depositional environments: an example from the North American Carboniferous. 24th Intern. Geol. Congr. Proc. Sec. 7:130-146.

Zangerl, R. 1971. On the geologic significance of perfectly preserved fossils. North Amer. Paleont. Conv. 1969 Proc., I:1207-1222.

Taxonomic Index

Annelida
  Amaeana, 134
  Annelids, 43, 122, 128, 129
  Aricidea, 131, 140
  Boccardia, 137
  Ceratocephala, 137
  Chaetopteriid polychaetes, 118
  Chaetozone, 134, 140
  Cornulites, 60, 184, 191
  Cossura, 131, 137
  Diopatra, 140
  Glycera, 137, 140
  Goniada, 131, 133, 134, 139
  Haploscoloplos, 130, 133, 140
  Hesperone, 137
  Hirudinea, 128
  Lumbrineris, 137, 140
  Magelona, 134
  Marphysa, 137
  Nephtys, 131, 134, 137, 139
  Nereis, 70, 130
  Nothria, 133
  Oligochaeta, 128
  Paraonis, 137
  Pectinaria, 131, 137, 140
  Pholoe, 130, 137
  Poecilochaetus, 137
  Polychaetes, 76, 90, 91, 118, 119, 121, 128, 129
  Prionospio, 130, 133, 137, 139, 140
  Scoloplos, 140
  Serpula, 49, 50, 51, 52
  Spiophanes, 130, 134, 137, 139
  Sternaspis, 131, 137
  Sthenelanella, 131
  Terebellides, 137
  Thalenessa, 133, 140
  Tharyx, 131, 140

Arthropoda
  Acratia, 227, 229
  Aechminella, 234
  Ampelisca, 67, 70, 130, 133, 137
  Ampeliscid crustaceans, 118

Amphideutopus, 131
Amphipods, 70
Amphissites, 226, 227, 229, 231, 234
Argissa, 134
Arthropods, 122, 128, 129
Bairdia, 214, 224, 226, 227, 228, 229, 230, 234
Barnacles, 53, 182, 189
Basslerella, 227, 229
brine shrimp, 197
Byblis, 137
Callianassa, 137
Candona, 200, 203, 204, 205
Cavellina, 214, 222, 225, 226, 227, 228, 229, 230
Coronakirkbya, 225, 227, 229
Crustaceans, 43, 76, 90, 91, 119, 121, 128
Cryptobairdia, 214, 222, 224, 225, 226, 227, 228, 229, 230
cumaceans, 70
Cyclopcypris, 203, 204
Cypridopsis, 200, 203, 204
Cyprinotus, 200, 203, 204
Cytherissa, 203
Darwinula, 203, 204
Diastylopsis, 133, 139
Eucypris, 203, 204
Fabalicypris, 224, 225, 227, 229
Flexicalymene, 184, 192
Gnathia, 137
grubbing trilobites, 60
Haliophasma, 136
harpacticoid copepods, 70
Healdia, 214, 227, 228, 229, 231
Heterophoxus, 136
Hollinella, 227, 229
Ilyocypris, 200, 203, 204
Jonesina, 226, 227, 229
Kelletina, 225, 227, 229
Kirkbyella, 234
Limnocythere, 200, 203, 204, 205
Listriella, 136, 137
Lithoglyptid barnacles, 52
Macrocypris, 227, 229
Microcheilinella, 227, 229
Milleratia, 224, 227, 229
Monoculodes, 140

Moorites, 214, 226, 227, 228, 229, 231
Orthobairdia, 214, 225, 227, 228, 229, 230
Ostracode, 193, 213
Ostracodes, 130, 134, 183, 190
Paraphoxus, 133, 139
Pelocypris, 203, 204
Physocypria, 203, 204
Potamocypris, 203, 204
Pseudobythocypris, 214, 227, 228, 229, 231
Pseudoparaparchites, 227, 229
Pycnogonada, 128
Silenites, 214, 224, 225, 227, 228, 229, 230
Synchelidium, 134, 139
Tanaid, 131
Waylandella, 275

Brachiopoda
Acanthocrania, 183, 190
Acrospiriferid, 58
Athyridina, 274
Atrypid, 276
Brachiopods, 43, 55, 60, 97, 127, 129, 217, 218, 267, 270
Camarotoechia, 184, 191
"*Camarotoechia*," 60, 177, 184, 191
Chonetidina, 274, 275, 276
Cleiothyridina, 270, 271
Composita, 271
Crurithyris, 177, 183, 190, 217, 275
Dalejina, 60
Eocoelia, 58, 60, 61
Eomarginifera, 183, 190, 275
Glottidia, 130, 131, 135, 137, 141, 171, 174, 182, 189
Inflatia, 270, 271
Kingena, 48, 49, 50
Leptocoeliid, 58
Leptostrophia, 184, 191
Leptostrophid, 60
Lingula, 51, 58, 60, 61, 171, 176, 177, 178, 181, 182, 183, 184, 189, 190, 191, 263, 268, 270, 272, 273, 274, 275
Lingulid, 171
Lingulida, 128

Lingoloid, 58, 267, 275, 276
Orbiculoidea, 183, 184, 190, 191, 217
Orthida, 274
Orthorhynchula, 184, 191
Petrocrania, 183, 190
Productid brachiopods, 224, 270
Productidina, 263, 268, 270, 273, 274, 275, 276
Reticulariina, 271
Retziidina, 274
Rhipidomella, 183, 190
Rhynchonellid, 58
Rhynchonellida, 274
Schizophoria, 275
Spirifer, 271
Spiriferida, 263, 268, 270, 273, 274
Spiriferidina, 274
Strophochonetes, 60
Strophomenida, 274
Strophomenidina, 274
Terebratulida, 127
Zygospira, 184, 191

Bryozoa
  Archimedes, 270
  Bryozoans, 43, 52, 53, 55, 60, 127, 129, 184, 217, 218, 270, 275, 276
  Cheilostomata, 127
  Fenestella, 270
  Fenestellidae, 263, 268, 270, 271, 272, 273, 274
  Fistulipora, 232

Chordata
  Amphibians, 72
  Birds, 72, 74, 252
  Mammals, 72, 74
  Reptiles, 72
  Rodents, 72
  Sharks teeth, 217
  Ungulates, 72

Coelenterata
  Acropora, 12
  Actinaria, 127
  Agaricia, 12
  Anthozoans, 43, 127
  Ceriantharia, 127
  Coelenterates, 12, 129
  Colonial rugose corals, 271
  Conularia, 184, 192
  Coral, 73, 76, 218, 270, 275
  Diploria, 12
  Gorgonacea, 127
  Hapsiphyllid corals, 271
  Hydrozoans, 43
  Lithodrumus, 271
  Lithostrotion, 246
  Michelinia, 263, 268, 271, 273
  Millepora, 12
  Montastrea, 12
  Pennatulacea, 129
  Porites, 12
  Rugosa, 263, 268, 271, 273, 274
  Stromatoporoids, 43
  Tabulata, 274
  Tabulate rugose corals, 271

Echinodermata
  Agassizocrinus, 272
  Amphiodia, 130, 133, 136
  Amphipholis, 130, 134
  Asteroids, 43, 128
  Blastoids, 271, 274
  Brittlestars, 121
  Crinoids, 128, 183, 190, 217, 218, 271, 272, 274
  Dendraster, 92, 135, 140
  Echinoderms, 76, 90, 128, 129, 267, 270
  Echinoids, 38, 39, 40, 43, 128, 274
  Eocrinoids, 55
  Eremicaster, 81
  Holothurians, 43, 128
  Inadunate crinoids, 271
  Mellita, 77
  Ophiodermella, 141
  Ophiura, 70
  Ophiuroids, 43, 128
  Pelmatozoans, 43
  Pentremites, 263, 268, 270, 271, 272, 273

Echiurida
  Echiurids, 128, 129
  Listriolobus, 136

Taxonomic Index

Miscellaneous Taxa
  Conodonts, 217
  Graptoloids, 43

Mollusca, 77, 79, 90, 91, 93, 94, 98,
  109, 121, 128, 129, 143, 147,
  218, 276
  Amphineura, 43, 128, 129, 147
    Chaetodermatida, 131, 138, 147
  Bivalvia, 43, 75, 91, 97, 128, 129,
    131, 138, 147, 164, 165, 167,
    217, 263, 268, 270, 273
    Abra, 70, 98, 151
    Actinopteria, 183, 190
    Aequipecten, 77
    Aligena, 142
    Ambonychia, 184, 191
    Amygdalum, 182, 189
    Anadara, 67, 80
    Anodontia, 151, 182, 189
    Anomalocardia, 77, 151
    Apolymetis, 164, 165, 166, 167
    Arca, 67
    Arcid, 52, 56, 57, 60
    Arcopsis, 151
    Arcticids, 59
    Asthenothaerus, 135, 138, 142
    Aviculopecten, 183, 190, 217, 270
    Axinopsis, 131, 137
    Botula, 50, 52
    Brachidontes, 80, 151
    "*Breviarca*," 51, 52, 183, 189
    Byssally attached bivalves, 59, 60
    Cardiids, 59
    Cardita, 53, 137, 151
    Chione, 80, 92, 134, 141, 151,
      161, 164, 165, 167
    Chlamys, 141
    Clinocardium, 166, 167
    Codakia, 151
    Compasomyax, 131, 136, 137
    Corbula, 52, 176, 182, 183, 189,
      190
    Corbulamella, 56, 60
    Corbulid, 52, 56, 57, 59, 60, 182,
      183, 189, 190
    Crassinella, 52, 176, 182, 183,
      189, 190
    Crassostrea, 52, 53, 77, 182, 189

    Cryptomya, 92, 164, 165, 166,
      167
    Cucullaea, 53, 60
    Cumingia, 164, 165, 166, 167
    Cyclopecten, 81
    Cymatonota, 184, 192
    Cymbophora, 60
    Cyprimeria, 53
    Cyrtopleura, 80
    Deposit-feeding bivalves, 60
    Diplodonta, 92, 164, 165, 166
    Divaricella, 151
    Donax, 79, 80, 91, 164, 167
    Dosinia, 80
    Dosinopsis, 59
    Dunbarella, 183, 190
    Edmondia, 180, 184, 190
    Elliptio, 250
    Ensis, 141
    Ervilia, 151
    Eryoina, 151
    Eupera, 243
    Flaventia, 53, 182, 189
    Gervillia, 53
    Glomus, 81
    Heterodonax, 166, 167
    Homomya, 53
    Ilmatogyra, 47, 48, 50, 51
    Infaunal bivalve, 59
    Inoceramus, 52
    Ischyrodonta, 184, 191
    Laevicardium, 151, 164, 165, 166,
      167
    Leptopecten, 164, 165, 166, 167
    Lima, 49, 50, 53
    Limopsis, 56, 60
    Linearia, 59
    Lopha, 48, 50, 51, 52, 53, 182,
      189
    Lucina, 52, 151
    Lucinisca, 135, 141, 165, 166
    Lyrodesma, 184, 191
    Macoma, 131, 135, 137, 141, 151,
      161, 164, 165, 166, 167
    Mactrids, 59
    Malletia, 60
    Modiolopsis, 184, 191
    Modiolus, 53, 132, 135, 141, 182,
      189

Modulus, 151
Mulinia, 174, 182, 189
Mussels, 79
Myonera, 81
Mysella, 182, 189
Mytilid, 53, 54, 56, 69
Mytilus, 95, 165, 166, 167
Naiadites, 184, 191
Neithea, 48, 49, 50, 52
Nucula, 51, 52, 60, 67, 132, 137, 138
Nuculana, 51, 52, 59, 60, 77, 130, 135, 137, 138, 141, 183, 190, 311
Nuculites, 184, 191
Nuculoida, 274
Nuculopsis, 183, 190, 275
Ostrea, 53, 60, 164, 165, 166, 182, 189
Oxytoma, 56, 60
Oyster-boring bivalves, 48, 50
Oysters, 47, 48, 50, 52, 60, 79
Palaeoneilo, 60, 184, 191
Parvilucina, 151
Pecten, 164, 165, 166, 167
Periploma, 131, 138
Phacoides, 151
Phelopteria, 53, 56, 60
Phestia, 272
Pholadomya, 51, 183, 189
Phylloda, 80
Pinna, 183, 185, 190
Pisidiid, 236, 243, 249, 251, 252, 254, 258
Pisidium, 243, 259
Pitar, 142
Plesielliptio, 235, 236, 243, 244, 247, 249, 250, 252, 253, 254, 255, 257
Plicatula, 48, 49, 50, 52
Poromya, 135, 141
Posidonia, 275
Praenucula, 184, 191
Pronoella, 57
Protobranchs, 118
Protocardia, 52, 53, 56, 59, 60, 182, 189
Protothaca, 161, 164, 165, 166, 167

Pteriids, 49, 53, 54, 57, 59, 60
Pterinea, 184, 191
Petrioida, 274
Pteronitella, 60, 184, 191
Rangia, 77, 80
Rochefortia, 131, 132, 135, 138, 140
Sanguinolaria, 92, 164, 166, 167
Sanguinolites, 275
Saxicavella, 136, 137
Saxidomus, 164, 165, 166, 167
Scabrotrigonia, 53, 54, 57, 59, 176, 182, 183, 189, 190
Septimyalina, 183, 190
Siliqua, 135, 141
Solamen, 132
Solecurtus, 80
Solemya, 81
Solen, 132, 141
Sphaerium, 243, 244
Spisula, 79, 80, 92
Spondylus, 49
Streblopteria, 177, 183, 190, 275
Syncyclonema, 52
Syndosmya, 70
Tagelus, 151, 161, 164, 165, 166, 167
Tancredia, 56, 57, 59
Tancrediopsis, 178, 184, 191
Tellina, 70, 91, 131, 132, 133, 134, 138, 139, 140, 151, 165, 166, 167, 183, 190
Tellinids, 118
Tellinimera, 60
Tenuipteria, 60
Texigryphaea, 48, 49, 50, 52, 53, 60
Thracia, 138
Thyasira, 138
Tivela, 164, 167
Trachycardium, 53, 54, 56, 59, 176, 182, 183, 189
Transennella, 151
Tresus, 164, 165, 166, 167
Trigonia, 57, 59
Unio, 259
Unionids, 243, 255
Venus, 70
Vesicomya, 81

Volsellina, 183, 185, 190
Yoldia, 51, 118, 124, 183, 190
Zirfaea, 167
Cephalopoda, 43, 55, 128, 129
   Cravenoceras, 272
   Goniatitic ammonoids, 209
   Lyrogoniatites, 272
Gastropoda, 43, 55, 75, 95, 97, 128, 129, 131, 135, 147, 184, 191, 217, 274
   Acroloxus, 244
   Acteocina, 141
   Acteon, 131, 135
   Aglaja, 135, 141
   Alaba, 151
   Anachis, 151
   Anceya, 76
   Archaeogastropods, 43
   Astraea, 76, 151
   Atys, 151
   Balcis, 132, 135, 141, 151
   Biomphalaria, 235, 236, 243, 249, 251, 252, 254, 256, 257
   Bittium, 137, 138
   Boreotrophon, 81
   Browsing gastropods, 60
   Bucania, 184, 191
   Bucaniopsis, 183, 190
   Buccinum, 76
   Bulla, 151
   Busycon, 76
   Calyptraea, 151
   Cantharus, 76
   Cerithiopsis, 141
   Cerithium, 77, 151
   Chytra, 76
   Collisella, 151
   Columbella, 151
   Crassispira, 151
   Crepidula, 141, 151, 165, 166
   Crepipatella, 141
   Cylichna, 131, 135, 138
   Diodora, 151
   Discus, 244
   Distorsio, 80
   Donaldina, 183, 190
   Drepanochilus, 51, 59, 60, 176, 182, 183, 189
   Drepanotrema, 243
   Epitonium, 132
   Euspira, 182, 189
   Exilia, 81
   Fasciclaria, 151
   Glabrocingulum, 270
   Goniobasis, 236, 243, 247, 249, 252, 253, 254, 256, 257, 258
   Grangerella, 244
   Gyraulus, 243, 254
   Holospira, 244, 251
   Hydrobia, 236, 243, 244, 252
   Lacuna, 141
   Liospira, 184, 191
   Lischkeia, 81
   Lithoconus, 80
   Loxoplocus, 184, 191
   Lunatia, 92
   Lymnaea, 244
   Mangelia, 132, 135, 138, 140, 151
   Mesodon, 244, 251
   Mitrella, 141, 182, 189
   Nassarius, 92, 132, 135, 141, 151, 182, 189
   Natacopsis, 184, 191
   Natica, 151
   Neritina, 77, 80, 151
   Odostomia, 81, 132, 134, 142
   Oliva, 80, 151
   Olivella, 92, 132, 134, 139, 140, 142, 151
   Omalodiscus, 235, 236, 243, 249, 251, 252, 254, 256, 257
   Opalia, 76
   Oreoconus, 235, 243, 244, 245, 249, 250, 251, 252, 255, 257
   Oreohelix, 244, 251
   Otostoma, 182, 189
   Paramelania, 76
   Physa, 235, 236, 243, 249, 251, 252, 254, 256, 257
   Pirsila, 182, 189
   Plectonotus, 184, 191
   Plesioturrilites, 46
   Pleurolimnaea, 244
   Polinices, 76, 92, 132, 135, 142
   Promenetus, 243
   Prunum, 151
   Pulmonates, 251, 252, 253
   Pusia, 151

Taxonomic Index   285

Retusa, 141
Rissoina, 151
Schasicheila, 244, 251
Smaragdia, 151
Solariella, 81
Soleniscus, 183, 184, 190
Spekia, 76
Steiraxis, 81
Strombus, 80
Tegula, 151
Terebra, 135, 151
Tiphobia, 76
Tricolia, 141, 151
Tropidodiscus, 184, 191
Turbonilla, 131, 134, 141
Turritella, 49, 50, 51, 52, 53, 56, 57, 59, 67, 98, 182, 183, 189
Turritellids, 59
Valvata, 236, 243, 247, 249, 253, 254, 255, 256, 258
Volvulella, 131, 135, 142
Scaphopoda
  Cadulus, 131, 134, 137, 151
  Dentalium, 127, 151
  Scaphopods, 43, 60, 95, 97, 128, 129, 245

Nematoda, 128, 129, 131
  Nematodes, 70, 119, 139

Nemertinea
  Nemerteans, 119, 127, 129, 130, 133, 137, 139

Phoronida, 128, 129, 137

Plants, 267, 270, 271
  Alligator Weed, 73
  Archaeosigillaria, 271
  Aspen, 72, 73
  Birch, 72, 73
  Cactus, 73, 74
  Cattails, 73
  Chaparral, 73
  Eel Grass, 71
  Forbs, 73
  Hardwoods, 72, 73, 74, 84

Juniper, 74
Lichens, 74
Mosses, 74
Oak-hickory, 67
Oaks, 73, 74
Palms, 73
Papyrus, 73
Pecan, 74
Pines, 73
Reeds, 73
Salt grass, 73
Sedges, 73
Short grass, 73
Tamarack, 72, 73
Tropical rain forest, 72
Willow-sedge, 67
Yucca, 73

Platyhelminthes, 128, 129
  Turbellaria, 128

Porifera, 43, 53
  Cliona, 50
  Clionid, 52

Protozoa
  Cribratina [*Haplostiche*], 48
  Foraminiferida, 48, 109, 143, 148, 206, 275
  Fusulinids, 217, 218, 276
  Paleotextulariid, 276
  Triticites, 233

Rychocoela, 127

Sipunculida, 128, 129

Trace fossils
  Arenicolites, 57, 59
  Burrows, 184, 191
  Chondrites, 49, 50
  Coprolites, 217
  Horizontal burrow, 183, 190
  Ophiomorpha, 56, 57, 59
  Planolites, 48, 49, 50
  Skolithos, 57, 59
  Thalassinoides, 49, 50

Subject Index

Abundance, species
  rank, 89, 99, 119, 174, 180, 181, 252, 274
  relative, 78, 83, 84, 89, 95, 100, 101, 144, 148, 161, 214, 225
Actual trophic structure, 42
Addition by inference, 15, 23
Amensal association, 14, 21
Animal communities, 12, 69
Aquatic communities, 69, 75-84
Assemblage
  general, 11, 12, 14, 15, 18, 24, 70, 240, 268
  live-dead, 10, 91, 99, 100, 144, 157-159, 164-167
  preserved, preservable, shelled (fossil) (*see* Preservation of fauna)
  time-averaged, 100, 101, 158, 161, 172
Association
  classification of, 20, 21
  definition and/or use of, 7, 12-15, 19-21, 25, 31-33, 240, 242
Association-unit theory, 30
Autecology
  of Arthropoda, 189-192
  of Bivalvia, 48-53, 77, 79-81, 182, 189-191, 250-254, 270-276
  of Blastoidea, 270-276
  of Brachiopoda, 48, 174-185, 189-192, 270-276
  of Bryozoa, 270-276
  of Coelenterata, 12, 271-276
  of Crinoidea, 190, 271-276
  of Echinoidea, 47, 77
  of Foraminiferida, 48
  of Gastropoda, 52, 76, 80, 81, 182, 189-191, 250-254
  of Ostracoda, 203-205, 215, 216

Baja California, 215
Beil Limestone Member, 214-219

287

Subject Index

Biocoenosis, 3, 36, 108-110, 115, 239, 242
Biofacies, 32, 33, 112, 113, 115, 159, 268
Biogeographic units
  endemic center, 14, 15, 17
  province, 14-16
  realm, 14-16
  region, 14-16
  subprovince, 14, 15, 17
Biomass, 34, 45
Biome, 32, 72, 73
Biostratonomy, 240
Biota, global, 15, 16
Biotope, 10
Biovolume, 34
Bohemia, 216
Bridger Formation, 237
Brillouin diversity index, 175, 179, 199, 220
Browsers (*see* Collector)
Burica Peninsula, 81, 82

California, 276
  Gulf of, 79
  Mugu Lagoon, 92, 146, 155
  Southern, 90, 93, 109, 119
  Tijuana Slough, 92
Cambrian communities, 32
Carboniferous communities
  Lower, 175, 177, 181, 184, 190, 264, 268-277
  Upper (*see* Pennsylvanian communities)
Carnivore (*see* Predators)
Character species, 7, 31, 33
Chemical energy, 68, 79, 84
Classification, concept of, 31
Climatic events, 205-207
Cluster analysis, 33, 111, 112, 148-153, 159, 199, 204, 220, 223-225, 242
Collector, 41, 43, 118, 174, 252, 253
Community
  biologically accommodated, 37
  biotic, 36, 109, 159
  biotic (= taxonomic) composition of, 4, 89, 110
  classification, 8-13, 31, 72

  concepts of, 8-13, 31, 37, 70, 109, 144
  definition of, 3, 4, 7-9, 15, 18, 68, 70, 88, 109, 110, 145, 173, 240
  distribution, 83, 113, 273-277
  evolution, 3, 88, 99, 122, 172, 187, 214
  freshwater, 75, 84
  holistic concept, 3, 5, 9, 13, 24
  level-bottom, 7, 11, 35, 69
  life-habit, 12, 36
  names, 70, 113
  organism community, 36, 109
  parallel (=isocommunities), 25, 32, 268
  Petersen-Thorson, 11, 12, 20, 35, 173, 268
  physical factors, 8
  physically defined, 8, 37, 114
  preserved, preservable, shelled (fossil) (*see* Preservation of fauna)
  quantitatively defined, 11
  stability, 40
  structure, 4, 93, 116, 194, 242
  terrestrial, 69, 72
  unit boundaries, 4, 84
  unit theory, 30, 31
Competitive association, 14, 20
Confidence limits, 89, 90
Continental fragmentation, 54, 83
Correlation coefficient, 148-153
Cretaceous communities, 32, 46, 48-53, 56, 57, 116, 174, 175, 183, 189
Cyclic sequences, 217

Density, 154, 174, 181
Dependent associations, 14, 20
Deposit feeder (*see* Detritus feeders)
Depth zonation, 79-81, 84, 96, 114, 215
Detritus feeders, 41-43, 83, 116, 118, 175, 180-185, 275
Devonian communities, 32, 58, 116
Diagenesis, 5, 242
Dispersion index, 89
Diversity
  controlling factors, 40, 46, 83, 84, 158, 174, 185, 194, 195, 214

general, 69, 101, 154, 194, 214, 221-231
generic, 268
information theory, 37, 48, 179, 194, 199, 201, 214, 220-224
species, 37, 48, 51, 78, 83, 89, 174, 175, 179
Dominance diversity (*see* Diversity, information theory)
Dominance patterns, 7, 37, 119, 181
Dominant species, 7, 32, 34, 173
Draper Formation, 197

Ecologic analysis, 21
  classification, 15, 25
  data collecting, 15, 22
  inference, 15, 23, 24
  interpretation, 15, 23, 25
  logical stages of, 21, 22
Ecological units, 5-7, 13-21
  classification of, 2, 13-21
  definition of, 2, 6, 13-21
Ecology, definition, 3
Ecosystem, 14, 15, 17, 84
Edwards Formation, 74
Edwards Plateau, 74
Energy, 69, 79, 84
  flow system, 4, 101, 116
Environments
  alluvial plain, 251
  bay, 51, 95-97, 146
  carbonate shelf, 46, 216
  continental slope and deep-sea, 79-83
  deltaic, 46, 96, 113, 265-269
  desert, 72, 73
  estuarine, 46, 69, 76-78, 84, 95
  fluviatile, 249-251, 255
  forest, 72, 73
  grassland, 69, 72, 73
  hard-ground, 46
  heterogeneous, 40, 41, 158, 173, 195
  homogeneous, 40, 41
  lacustrine, 197, 203, 205, 249, 253, 256, 257
  lagoon, 146
  marshes, 93, 94, 146
  nearshore, 55, 59, 60, 78, 173, 216

  paludal, 249, 255
  pond, 249, 251, 256
  predictability of, 40, 46, 69, 78, 84
  rain forest, 69, 72, 73
  reef, 8, 10, 12, 34, 69, 146, 148-153
  shelf, 34, 78, 80, 91, 96, 98, 113, 148-153, 269
  shoreface, 46, 51, 55, 96
  stability (*see* Stability, environmental)
  strait, 148-153
  tidal creeks, flats, channels, 93, 94, 100, 148-153, 155
  tundra, 72-74

Eocene
  facies, 237, 249, 255-257
  mollusk associations, 242-250
Epifaunal communities, 12, 44
Equitability, 37, 89, 101, 174, 175, 179, 194, 200, 201, 214, 220-230
Evolution rate of species, 195, 205
Extinction of species, 207, 208

Feeding habits (*see* Trophic, feeding habits)
Fidelity, a measure, 31, 89
Fidelity of fossil record (*see* Preservation of fauna)
Florida, Charlotte Harbor, 174
Floyd Shale, 264, 265
Food-chain-dependent associations, 14, 20
Foraker Limestone, 176
Forest City Basin, 217
Formation (ecologic), 32, 34, 36
Fossil community, 108, 173
Fossil deposit, 15, 22, 24
Functional morphology, 174, 178, 187, 215

Geomorphology, 69, 78, 83
Georgia, 91, 265
Glen Rose Formation, 74
Grayson Formation, 46
Green River
  Basin, 237-239
  Formation, 236, 237, 249
Growth form, 35

Subject Index

Habitat, 32, 36, 70, 154, 160
  community, 9, 36, 70
Hadley Harbor, 71
Hartselle Sandstone, 265
Herbivores, 43
Holocene communities, 32, 70, 71,
  96–98, 174, 175, 182, 189
Homogeneity index, 37, 89

Identifier-organism associations, 14,
  20
Identifier species, 7, 9, 11
Immigration of species, 205, 207, 208
Individualistic dissent, theory of, 30
Infaunal communities, 12, 44
Interacting associations, 14, 20
Interaction of species, 4
Interdependent associations, 14, 20
Italy, 96

Jurassic communities, 55, 57, 59, 116

K-selection, 41
Kansas, 176, 177, 214–219
Kettleman Hills, 114
Kinetic energy, 68, 79, 84, 160
Kiowa Formation, 51, 176
Kruskal–Wallis test, 220, 221

Lake Baikal, 75
Lake Bonneville, 197, 198
  Group, 197
Lake Gosiute, 236, 249
Lake Tanganyika, 75
Lecompton Formation, 214
Life form, 35
Life-habit associations, 14, 20, 99
Lingula communities, 60, 171–192,
  268–270, 273, 275
Lithotope, 10
Little Cottonwood Formation, 197
Live-dead ratios, 91–98, 147, 155,
  157

Main Street Limestone, 46
Mann–Whitney–Wilcoxon U test, 220,
  223
Mississippi Delta, 113, 216, 277

Mississippian communities (*see* Carboniferous, Lower)
Mode of life
  boring, 50
  epifaunal, 44, 174, 179, 185, 275
  infaunal, 44, 174, 179, 185, 275
  vagrant, 44
Monteagle Formation, 265
Morphologic associations, 14, 21
Mutualism, 14, 20

Neutral associations, 14, 20
Niche, 42–44, 93, 185, 186
  diversification, 69, 181, 252
Nondependent associations, 14, 20
Nonobligate associations, 14, 21

Opportunistic species, 158
Ordovician communities, 10, 32, 58,
  175, 178, 181, 184, 191
Organism community, 36, 109

Paleocommunity, 5, 15, 18, 24, 87–
  101, 172
Paleosynecology, 250
Paleotemperatures, 202, 206
Parasitism, 14, 20
Pawpaw Formation, 46
Pennsylvanian assemblages, 213
Pennsylvania communities, 32, 276
Periodicity, 89
Permian communities, 175–177, 183,
  190
Physicochemical factors (*see* Community, physical factors)
Physiognomy, 32, 34
Plant communities, 12, 68, 69, 83,
  184
Pleistocene associations, 200–205
Pliocene communities, 81, 82, 114
Populations, 15, 78
Predator–prey associations, 14, 20
Predators, 42, 43, 118
Preservation of fauna, 42, 90, 99, 158,
  172, 173
Preservational associations, 14, 21
Principal coordinates analysis, 220,
  221, 227, 228

Subject Index

Proportionate similarity index (I), 161, 163
Protocooperation associations, 14, 21
Puerto Rico, 96

Q-mode (*see* Cluster analysis)

R-mode (*see* Cluster analysis)
r-selection, 41
Random associations, 14, 21
Recent (*see* Holocene communities)
Recurring species (*see* Taxa set)
Reedsville Formation, 178
Regression, marine, 215, 219
Ria de Arosa, Spain, 95-97, 100

Salt Lake Basin, Great, 194, 197
Sampling, 22, 111, 144, 146, 199, 214, 219, 220
San Francisco Bay, 115
Scotland, 95, 177
Sediment type (*see* Substrate, types)
Sere, 6, 15, 17, 38
Silurian communities, 32, 58, 175, 177, 178, 181, 184, 191
Solar energy, 68, 79, 84
Spatial variability, 41, 195
Stability
  community, 40
  environmental, 40, 45, 55, 78, 84, 159, 194, 195, 215, 219, 224
Statistical associations, 14, 21
Substrate
  niche, 42, 44
  types, 32, 38, 41, 44, 52, 59, 78, 79, 114, 115, 175, 180, 181, 185, 186, 216, 275
Succession, 6, 17, 32, 37, 41, 89, 195, 205
Suspension feeders, 42, 118, 175, 180-185, 275

Swallowers (*see* Detritus feeders)
Sweden, 215

Taphonomy, 10, 15, 23, 24, 30, 88, 99, 144, 172, 196, 240-242
Taxa set, 15, 23, 32, 33
Taxonomic associations, 14, 21
Taxonomic composition (*see* Community, biotic composition)
Ternary diagrams, 43, 44, 50, 52, 54, 59, 120
Terrestrial communities, 69, 72, 75, 83, 84
Texas Gulf Coast, 77, 215
Trace fossils, 48, 59, 91
Trophic
  category, 39
  classification, 38, 44, 117-119, 275
  controlling factors of, 39-41, 45, 84, 116, 121, 272
  data of, 45
  feeding habits, 32, 38, 39, 42, 117, 174
  structure, actual, 42, 119-122
  structure, general, 39, 45, 54, 84, 89, 101, 116
  structure, preserved, 42, 101, 117, 121, 175, 180-185
  structure through time, 54, 180
  theory, 38, 173, 185
Tucumcari Formation, 51
Tuscumbia Limestone, 265

Vagrant organisms, 44
Visean shale, 177, 265, 275
Vitality, 89

Wales, 114, 177, 178
Walnut Formation, 74
Wasatch Formation, 236, 237
Washakie Formation, 237

Yucatan, 95, 145, 146